Jordan Canonical Form and Dynamic Systems on Time Scales

Jordan Canonical Form and Dynamic Systems on Time Scales

By

Svetlin G. Georgiev

Cambridge
Scholars
Publishing

Jordan Canonical Form and Dynamic Systems on Time Scales

By Svetlin G. Georgiev

This book first published 2023

Cambridge Scholars Publishing

Lady Stephenson Library, Newcastle upon Tyne, NE6 2PA, UK

British Library Cataloguing in Publication Data
A catalogue record for this book is available from the British Library

ISBN (10): 1-5275-9475-0
ISBN (13): 978-1-5275-9475-3

Contents

Preface

The Jordan canonical form is one of the most important and useful concepts in linear algebra. This book is a development of the Jordan canonical form. The book is intended for senior undergraduate students and beginning graduate students of engineering and science courses. There are seven chapters in this book. The chapters in the book are pedagogically organized. Each chapter concludes with a section of practical problems.

In Chapter 1 we present necessary background material for vector spaces. Some of their properties are deducted. We introduce vector subspaces, sums of spaces and direct sums of spaces. Chapter 2 deals with finite-dimensional vector spaces. In Chapter 3 we introduce linear maps and we deduct some algebraic properties of the linear space $\mathscr{L}(V,W)$ for arbitrary vector spaces V and W. We define matrices and represent a linear map by a matrix. Chapter 4 is devoted to invertible linear maps and isomorphic vector spaces. They are defined products and quotients of vector spaces. Invariant subspaces, eigenvalues and eigenvectors are introduced in Chapter 5. They are investigated restrictions, quotient operators and upper-triangular matrices. In Chapter 6 we define kernels of powers of an operator, nilpotent operators, decomposition of operators, block diagonal matrices and we prove the Cayley-Hamilton theorem. We define characteristic and minimal polynomials for a linear operator. We present an algorithm to find the Jordan canonical form of a matrix. In Chapter 7 we show how to use the Jordan canonical form to solve first order linear dynamic systems on arbitrary times scales.

This book is addressed to a wide audience of specialists such as mathematicians, physicists, engineers and biologists. It can be used as a textbook at the graduate level and as a reference book for several disciplines.

Paris, November 2022 *Svetlin G. Georgiev*

Chapter 1

Vector Spaces

Throughout this book $\mathbb{F}^n$ stands for either $\mathbb{R}^n$ or $\mathbb{C}^n$. The elements of $\mathbb{F}^n$ are called scalars.

1.1 Definition of a Vector Space

Definition 1.1. *An addition on a set V is a function that assigns an element $u+v \in V$ for each pair of elements $u, v \in V$.*

Definition 1.2. *A scalar multiplication on a set V is a function that assigns an element $\lambda v \in V$ for each $\lambda \in \mathbb{F}$ and each $v \in V$.*

Now, we are ready to give a formal definition of a vector space.

Definition 1.3. *A vector space is a set V along an addition on V and a scalar multiplication on V such that the following properties hold:*

1. *(commutativity)*
$$u+v = v+u, \quad u, v \in V.$$
2. *(associativity)*
$$u+(v+w) = (u+v)+w, \quad u, v, w \in V.$$
3. *(additive identity)*
there exists an element $0 \in V$ such that
$$v+0 = v, \quad v \in V.$$

4. *(additive inverse)*

   for any $v \in V$, there exists $w \in V$ such that

$$v + w = 0.$$

5. *(multiplicative identity)*

$$1v = v, \quad v \in V.$$

6. *(distributive properties)*

$$\begin{aligned} a(u+v) &= au+av, \\ (a+b)u &= au+bu \end{aligned}$$

   for any $a, b \in \mathbb{F}$ and any $u, v \in V$.

The elements of a vector space are called vectors or points.

The scalar multiplication in a vector space depends on $\mathbb{F}$. Thus, when we need to be precise, we will say that V is a vector space over $\mathbb{F}$ instead of saying simply V is a vector space. For example, $\mathbb{R}^n$ is a vector space over $\mathbb{R}$, $\mathbb{C}^n$ is a vector space over $\mathbb{C}$.

Definition 1.4. *A vector space over $\mathbb{R}$ is called a real vector space.*

Definition 1.5. *A vector space over $\mathbb{C}$ is said to be a complex vector space.*

The simplest vector space contains only one point. In other words, $\{0\}$ is a vector space. With the usual operations of addition or scalar multiplication, $\mathbb{F}^n$ is a vector space. The example of $\mathbb{F}^n$ motivated our definition of a vector space.

Example 1.6. *F^∞ is defined to be the set of all sequences of elements of $\mathbb{F}$:*

$$\mathbb{F}^\infty = \{(x_1, x_2, \ldots) : x_j \in \mathbb{F}, \quad j \in \{1, 2, \ldots\}\}.$$

Addition and scalar multiplication on $\mathbb{F}^\infty$ are defined as follows:

$$(x_1, x_2, \ldots) + (y_1, y_2, \ldots) = (x_1 + y_1, x_2 + y_2, \ldots),$$

$$\lambda(x_1,x_2,\ldots) = (\lambda x_1,\lambda x_2,\ldots).$$

With these definitions we will show that $\mathbb{F}^\infty$ becomes a vector space over $\mathbb{F}$. Suppose that

$$x=(x_1,x_2,\ldots),\quad y=(y_1,y_2,\ldots),\quad z=(z_1,z_2,\ldots)\in\mathbb{F}^\infty$$

and $a,b\in\mathbb{F}$. Then, we have the following.

1.

$$\begin{aligned} x+y &= (x_1,x_2,\ldots)+(y_1,y_2,\ldots)\\ &= (x_1+y_1,x_2+y_2,\ldots)\\ &= (y_1+x_1,y_2+x_2,\ldots)\\ &= (y_1,y_2,\ldots)+(x_1,x_2,\ldots)\\ &= y+x. \end{aligned}$$

2.

$$\begin{aligned} x+(y+z) &= (x_1,x_2,\ldots)+((y_1,y_2,\ldots)+(z_1,z_2,\ldots))\\ &= (x_1,x_2,\ldots)+(y_1+z_1,y_2+z_2,\ldots)\\ &= (x_1+(y_1+z_1),x_2+(y_2+z_2),\ldots)\\ &= ((x_1+y_1)+z_1,(x_2+y_2)+z_2,\ldots)\\ &= (x_1+y_1,x_2+y_2,\ldots)+(z_1,z_2,\ldots)\\ &= ((x_1,x_2,\ldots)+(y_1,y_2,\ldots))+(z_1,z_2,\ldots)\\ &= (x+y)+z. \end{aligned}$$

3. *Let*

$$0=(0,0,\ldots).$$

Then

$$\begin{aligned} x+0 &= (x_1,x_2,\ldots)+(0,0,\ldots) \\ &= (x_1+0,x_2+0,\ldots) \\ &= (x_1,x_2,\ldots) \\ &= x. \end{aligned}$$

4. *Let*

$$w=(-x_1,-x_2,\ldots).$$

Then

$$\begin{aligned} x+w &= (x_1,x_2,\ldots)+(-x_1,-x_2,\ldots) \\ &= (x_1+(-x_1),x_2+(-x_2),\ldots) \\ &= (0,0,\ldots). \end{aligned}$$

5.

$$\begin{aligned} 1x &= 1(x_1,x_2,\ldots) \\ &= (1x_1,1x_2,\ldots) \\ &= (x_1,x_2,\ldots) \\ &= x. \end{aligned}$$

6.

$$a(x+y) = a((x_1,x_2,\ldots)+(y_1,y_2,\ldots))$$

$$\begin{aligned}
&= a(x_1+y_1, x_2+y_2, \ldots) \\
&= (a(x_1+y_1), a(x_2+y_2), \ldots) \\
&= (ax_1+ay_1, ax_2+ay_2, \ldots) \\
&= (ax_1, ax_2, \ldots) + (ay_1, ay_2, \ldots) \\
&= a(x_1, x_2, \ldots) + a(y_1, y_2, \ldots) \\
&= ax+ay.
\end{aligned}$$

Next,

$$\begin{aligned}
a(x+y) &= a((x_1, x_2, \ldots) + (y_1, y_2, \ldots)) \\
&= a(x_1+y_1, x_2+y_2, \ldots) \\
&= (a(x_1+y_1), a(x_2+y_2), \ldots) \\
&= (ax_1+ay_1, ax_2+ay_2, \ldots) \\
&= (ax_1, ax_2, \ldots) + (ay_1, ay_2, \ldots) \\
&= a(x_1, x_2, \ldots) + a(y_1, y_2, \ldots) \\
&= ax+ay.
\end{aligned}$$

Definition 1.7. *Let S be a set. With $\mathbb{F}^S$ we will denote the set of functions from S to $\mathbb{F}$. For $f, g \in \mathbb{F}^S$, the sum $f+g \in \mathbb{F}^S$ is the function defined by*

$$(f+g)(x) = f(x)+g(x), \quad x \in S.$$

For $\lambda \in \mathbb{F}$ and $f \in \mathbb{F}^S$, the product $\lambda f \in \mathbb{F}^S$ is the function

$$(\lambda f)(x) = \lambda f(x), \quad x \in S.$$

Example 1.8. $\mathbb{R}^{[0,1]}$ *is the set of all functions from* $[0,1]$ *to* $\mathbb{R}$.

Example 1.9. $\mathbb{F}^S$ *is a vector space over* $\mathbb{F}$. *Let* $f,g,h \in \mathbb{F}^S$ *and* $a,b \in \mathbb{F}$. *Then we have the following.*

1.

$$\begin{aligned}(f+g)(x) &= f(x)+g(x)\\ &= g(x)+f(x)\\ &= (g+f)(x), \quad x \in S.\end{aligned}$$

2.

$$\begin{aligned}(f+(g+h))(x) &= f(x)+(g+h)(x)\\ &= f(x)+(g(x)+h(x))\\ &= (f(x)+g(x))+h(x)\\ &= (f+g)(x)+h(x)\\ &= ((f+g)+h)(x), \quad x \in S.\end{aligned}$$

3. *Define* $0 : S \to \mathbb{F}$ *as follows:*

$$0(x) = 0, \quad x \in S.$$

Then

$$\begin{aligned}(f+0)(x) &= f(x)+0(x)\\ &= f(x)+0\\ &= f(x), \quad x \in S.\end{aligned}$$

4. *Define* $-f : S \to \mathbb{F}$ *as follows:*

$$(-f)(x) = -f(x), \quad x \in S.$$

Then

$$\begin{aligned}(f+(-f))(x) &= f(x)+(-f)(x)\\ &= f(x)-f(x)\\ &= 0\\ &= 0(x), \quad x \in S.\end{aligned}$$

5.

$$\begin{aligned}(1f)(x) &= 1f(x)\\ &= f(x), \quad x \in S.\end{aligned}$$

6.

$$\begin{aligned}(a(f+g))(x) &= a(f+g)(x)\\ &= a(f(x)+g(x))\\ &= af(x)+ag(x)\\ &= (af)(x)+(ag)(x)\\ &= (af+ag)(x), \quad x \in S.\end{aligned}$$

Next,

$$\begin{aligned}((a+b)f)(x) &= (a+b)f(x)\\ &= af(x)+bf(x)\end{aligned}$$

$$= (af)(x) + (bf)(x)$$

$$= (af + bf)(x), \quad x \in S.$$

In fact, the elements of $\mathbb{R}^{[0,1]}$ are real-valued functions on $[0,1]$, not lists. In general, a vector space is an abstract entity whose elements might be lists, functions or weird objects.

Our previous examples of vector spaces, $\mathbb{F}^n$ and $\mathbb{F}^\infty$, are special cases of the vector space $\mathbb{F}^S$ because a list of length n of numbers in $\mathbb{F}$ can be thought of as a function from $\{1,2,\ldots,n\}$ to $\mathbb{F}$ and a sequence of numbers in $\mathbb{F}$ can be thought of as a function from the set of positive integers to $\mathbb{F}$. In other words, we can think of $\mathbb{F}^n$ as $\mathbb{F}^{\{1,2,\ldots,n\}}$ and we can think of $\mathbb{F}^\infty$ as $\mathbb{F}^{\{1,2,\ldots\}}$.

1.2 Elementary Properties of Vector Spaces

In this section, we will deduce some properties of vector spaces.

Theorem 1.10. *A vector space has a unique additive identity.*

Proof. Suppose that 0_1 and 0_2 are two additive identities in the vector space V. Then

$$\begin{aligned} 0_1 &= 0_1 + 0_2 \\ &= 0_2 + 0_1 \\ &= 0_2. \end{aligned}$$

This completes the proof. □

Theorem 1.11. *Every element in a vector space has a unique additive inverse.*

Proof. Suppose that V is a vector space. Let $v \in V$ and $w_1, w_2 \in V$ be its additive inverse. Then

$$v + w_1 = 0,$$

$$v + w_2 = 0$$

and

$$\begin{aligned} w_1 &= w_1 + 0 \\ &= w_1 + (v + w_2) \\ &= (w_1 + v) + w_2 \\ &= (v + w_1) + w_2 \\ &= 0 + w_2 \\ &= w_2 + 0 \\ &= w_2. \end{aligned}$$

This completes the proof. □

Below, suppose that V is a vector space. Because additive inverses are unique, the following notation makes sense. Let $v, w \in V$.

1. $-v$ denotes the additive inverse of v.
2. $w - v$ is defined to be $w + (-v)$.

Theorem 1.12. *We have*

$$0v = 0, \quad v \in V,$$

where 0 *denotes a scalar (the number* $0 \in \mathbb{F}$*) on the left side of the equation and a vector(the additive identity of* V*) on right side of the equation.*

Proof. For $v \in V$, we have

$$\begin{aligned} 0v &= (0+0)v \\ &= 0v + 0v, \end{aligned}$$

whereupon

$$0v = 0.$$

This completes the proof. □

Theorem 1.13. *We have*

$$a0 = 0, \quad a \in \mathbb{F}, \tag{1.1}$$

where $0 \in V$ *is the additive identity of* V.

Proof. For $a \in \mathbb{F}$, we have

$$\begin{aligned} a0 &= a(0+0) \\ &= a0 + a0, \end{aligned}$$

whereupon we get (1.1). This completes the proof. □

Exercise 1.14. *Suppose that* $a \in \mathbb{F}$, $v \in V$, *and* $av = 0$. *Prove that* $a = 0$ *or* $v = 0$.

Theorem 1.15. *We have*

$$(-1)v = -v, \quad v \in V.$$

Proof. For $v \in V$, we have

$$\begin{aligned} v + (-1)v &= 1v + (-1)v \\ &= (1 + (-1))v \\ &= 0v \\ &= 0. \end{aligned}$$

Thus, $(-1)v$ is the additive inverse of v. This completes the proof. □

Exercise 1.16. *Prove that*

$$-(-v) = v, \quad v \in V.$$

Exercise 1.17. *Suppose that* $v, w \in V$. *Explain why there exists a unique* $x \in V$ *such that*

$$v + 3x = w.$$

1.3 Subspaces

Definition 1.18. *A subset U of V is called a subspace of V if U is also a vector space using the same addition and scalar multiplication as on V.*

Some mathematicians, use the term "linear subspace", which means the same as subspace.

Example 1.19. *The set*

$$\{(x_1, x_2, 0) : x_1, x_2 \in \mathbb{F}\}$$

is a subspace of $\mathbb{F}^3$.

Theorem 1.20. *A subset U of V is a subspace if and only if U satisfies the following three conditions:*

1. *$0 \in U$.*
2. *if $u, v \in U$, then $u + v \in U$.*
3. *if $a \in \mathbb{F}$ and $u \in U$, then $au \in U$.*

Proof.

1. Let U be a subspace of V. Then, by the definition for a subspace and by the definition for a vector space, it follows that U satisfies 1, 2, 3.
2. Let U satisfies 1, 2, 3. Take $u, v, w \in U$ and $a, b \in \mathbb{F}$ arbitrarily. Then $u, v, w \in V$ and hence,

$$\begin{aligned} u+v &= v+u, \\ u+(v+w) &= (u+v)+w, \\ a(u+v) &= au+av, \\ (a+b)u &= au+bu, \\ u+0 &= u. \end{aligned}$$

By the third condition, it follows that $(-1)u \in U$ and

$$u - u = u + (-1)u$$

$$= \quad 0.$$

Thus, U is a subspace of V. This completes the proof.

□

Remark 1.21. *The additive identity condition above could be replaced with the condition that U is nonempty. Then, taking $u \in U$, multiplying it by 0, and using the condition that U is closed under scalar multiplication would imply that $0 \in U$. However, if U is indeed a subspace of V, then the easiest way to show that U is nonempty is to show that $0 \in U$.*

Remark 1.22. *The three conditions in the result above usually enable us to determine quickly whether a given subset of V is a subspace of V.*

Example 1.23. *Consider the set*

$$U = \{(u_1, u_2, u_3, u_4) \in \mathbb{F}^4 : u_3 = 5u_4\}.$$

Let

$$u = (u_1, u_2, u_3, u_4), \quad v = (v_1, v_2, v_3, v_4) \in \mathbb{F}^4$$

and $a \in \mathbb{F}$ be arbitrarily chosen. Then

$$\begin{aligned} u_3 &= 5u_4, \\ v_3 &= 5v_4, \\ au_3 &= 5au_4 \end{aligned}$$

and

$$u_3 + v_3 = 5(u_4 + v_4).$$

Therefore $0 \in U$, $u + v \in U$ and $au \in U$. Hence and Theorem 1.20, it follows that U is a subspace of $\mathbb{F}^4$.

Note that $\{0\}$ is the smallest subspace of V and V is the largest subspace of V. Moreover, the empty set is not a subspace because it must contain an additive identity.

Example 1.24. *Let U be the set of real-valued continuous functions on $[0,1]$. Let also, $u, v \in U$ and $a \in \mathbb{F}$. Then, we have the following.*

1. *Since 0 is a continuous function on $[0,1]$, then $0 \in U$.*

2. *Since the sum of two continuous functions u and v on* $[0,1]$ *is a continuous function on* $[0,1]$, *we have that* $u+v \in U$.

3. *Since the multiplication of a continuous function* $u \in U$ *on* $[0,1]$ *by a scalar* $a \in \mathbb{F}$ *is a continuous function on* $[0,1]$, *we get* $au \in \mathbb{F}$.

Now, applying Theorem 1.20, we get that U is a subspace of $\mathbb{R}^{[0,1]}$.

Exercise 1.25. *Prove that the set of all differentiable real-valued functions on* $\mathbb{R}$ *is a subspace of* $\mathbb{R}^{\mathbb{R}}$.

1.4 Sum of Spaces

Definition 1.26. *Suppose that* U_1, …, U_m *are subsets of* V. *The sum of* U_1, …, U_m, *denoted by*

$$U_1 + \cdots + U_m,$$

is the set of all possible sums of elements of U_1, …, U_m. *More precisely,*

$$U_1 + \cdots + U_m = \{u_1 + \cdots + u_m : u_1 \in U_1, \ldots, u_m \in U_m\}.$$

Example 1.27. *Let*

$$U = \{(x,0,0) \in \mathbb{F}^3 : x \in \mathbb{F}\},$$

$$W = \{(0,y,0) \in \mathbb{F}^3 : y \in \mathbb{F}\}.$$

Take

$$(x,0,0) \in U, \quad (0,y,0) \in W.$$

Then

$$(x,0,0) + (0,y,0) = (x,y,0).$$

Therefore

$$U + W = \{(x,y,0) \in \mathbb{F}^3 : x,y \in \mathbb{F}\}.$$

Example 1.28. *Let*

$$U = \{(x,x,y,y) \in \mathbb{F}^4 : x,y \in \mathbb{F}\},$$

$$W = \{(x,x,x,y) \in \mathbb{F}^4 : x,y \in \mathbb{F}\}.$$

Take

$$(x_1,x_1,y_1,y_1) \in U, \quad (x_2,x_2,x_2,y_2) \in W.$$

Then

$$(x_1,x_1,y_1,y_1)+(x_2,x_2,x_2,y_2) = (x_1+x_2,x_1+x_2,y_1+x_2,y_1+y_2).$$

Therefore

$$U+W = \{(x,x,y,z) \in \mathbb{F}^4 : x,y,z \in \mathbb{F}\}.$$

Exercise 1.29. *Let*

$$U = \{(x,x,x,x,x) \in \mathbb{F}^5 : x \in \mathbb{F}\},$$

$$W = \{(x,y,y,y,x) \in \mathbb{F}^5 : x,y \in \mathbb{F}\}.$$

Find $U+W$.

Theorem 1.30. *Suppose* U_1, …, U_m *are subspaces of* V. *Then* $U_1+\cdots+U_m$ *is the smallest subspace of* V *containing* U_1, …, U_m.

Proof. Since $U_1, \ldots, U_m$ are subspaces of V, we have that $0 \in U_1, \ldots, 0 \in U_m$ and then

$$0+\cdots+0 = 0$$

$$\in U_1+\cdots+U_m.$$

Let $u, v \in U_1+\cdots+U_m$ be arbitrarily chosen. Then, there are $u_1, v_1 \in U_1$, …, $u_m, v_m \in U_m$ so that

$$u = u_1+\cdots+u_m,$$

$$v = v_1+\cdots+v_m.$$

Hence,

$$u+v = (u_1+v_1)+\cdots+(u_m+v_m). \tag{1.2}$$

Since U_1 is a subspace of V and $u_1, v_1 \in U_1$, we conclude that $u_1+v_1 \in U_1$, and so on. Because U_m is a subspace of V and $u_m, v_m \in U_m$, then $u_m+v_m \in U_m$. From here and from (1.2), we obtain that $u+v \in U_1+\cdots+U_m$ and

$U_1 + \cdots + U_m$ is closed under addition. Let $a \in F$ be arbitrarily chosen. Then

$$au = au_1 + \cdots + au_m. \tag{1.3}$$

Since U_1 is a subspace of V and $a \in \mathbb{F}$, $u_1 \in U_1$, we obtain $au_1 \in U_1$, and so forth. Because U_m is a subspace of V and $a \in \mathbb{F}$, $u_m \in U_m$, we find $au_m \in U_m$. Hence and (1.3), we conclude that $au \in U_1 + \cdots + U_m$ and $U_1 + \cdots + U_m$ is closed under scalar multiplication. Now, applying Theorem 1.20, we find that $U_1 + \cdots + U_m$ is a subspace of V. Let now, $w_1 \in U_1$ be arbitrarily chosen and fixed. Then, using that $0 \in U_2$, …, $0 \in U_m$, we find

$$\begin{aligned} w_1 &= w_1 + 0 + \cdots + 0 \\ &\in U_1 + \cdots + U_m. \end{aligned}$$

Because $w_1 \in U_1$ was arbitrarily chosen and we get that it is an element of $U_1 + \cdots + U_m$, we obtain the inclusion

$$U_1 \subseteq U_1 + \cdots + U_m.$$

Let $w_m \in U_m$ be arbitrarily chosen and fixed. Then, using that $0 \in U_1$, …, $0 \in U_{m-1}$, we obtain

$$\begin{aligned} w_m &= 0 + \cdots + 0 + w_m \\ &\in U_1 + \cdots + U_{m-1} + U_m. \end{aligned}$$

Because $w_m \in U_m$ was arbitrarily chosen and we get that it is an element of $U_1 + \cdots + U_m$, we find the inclusion

$$U_m \subseteq U_1 + \cdots + U_m.$$

Now, let W be a subspace of V so that

$$U_1, \ldots, U_m \subseteq W.$$

Because subspaces must contain all finite sums of their elements, we get

$$U_1 + \cdots + U_m \subseteq W.$$

Thus, $U_1 + \cdots + U_m$ is the smallest subspace of V containing U_1, …, U_m. This completes the proof. □

Remark 1.31. *Sums of subspaces in the theory of vector spaces are analogous to unions of subsets in the set theory. Given two subspaces of a vector space, the smallest subspace containing them is their sum. Analogously, given two subsets of a set, the smallest subset containing them is their union.*

1.5 Direct Sums

Definition 1.32. *Let U_1, …, U_m be subspaces of V. Then $U_1+\cdots+U_m$ is called a direct sum if each element of $U_1+\cdots+U_m$ can be represented in a unique way in the form*

$$u_1+\cdots+u_m,$$

where $u_1 \in U_1$, …, $u_m \in U_m$. If $U_1+\cdots+U_m$ is a direct sum, then $U_1\oplus\cdots\oplus U_m$ denotes $U_1+\cdots+U_m$, with $\oplus$ notation serving as an indication that this is a direct sum.

Example 1.33. *Let*

$$\begin{aligned} U &= \{(x,y,0)\in\mathbb{F}^3 : x,y\in\mathbb{F}\}, \\ W &= \{(0,0,z)\in\mathbb{F}^3 : z\in\mathbb{F}\}. \end{aligned}$$

Note that U and W are subspaces of $\mathbb{F}^3$. Next,

$$U+W=\{(x,y,z)\in\mathbb{F}^3 : x,y,z\in\mathbb{F}\}.$$

Assume that an element $u\in U+W$ has two representations

$$\begin{aligned} u &= (x_1,y_1,0)+(0,0,z_1)\in U+W, \\ u &= (x_2,y_2,0)+(0,0,z_2)\in U+W, \quad x_1,y_1,z_1,x_2,y_2,z_2\in\mathbb{F}. \end{aligned}$$

Then

$$\begin{aligned} (x_1,y_1,0)+(0,0,z_1) &= (x_1,y_1,z_1) \\ &= (x_2,y_2,0)+(0,0,z_2) \\ &= (x_2,y_2,z_2), \end{aligned}$$

whereupon

$$\begin{aligned} x_1 &= x_2, \\ y_1 &= y_2, \end{aligned}$$

$$z_1 = z_2,$$

and

$$(x_1, y_1, 0) = (x_2, y_2, 0),$$

$$(0, 0, z_1) = (0, 0, z_2).$$

Therefore $U \oplus W = \mathbb{F}^3$.

Example 1.34. *Let*

$$U_1 = \{(x, y, 0) \in \mathbb{F}^3 : x, y \in \mathbb{F}\},$$

$$U_2 = \{(0, 0, z) \in \mathbb{F}^3 : z \in \mathbb{F}\},$$

$$U_3 = \{(0, y, y) \in \mathbb{F}^3 : y \in \mathbb{F}\}.$$

We have that U_1, U_2 *and* U_3 *are subspaces of* $\mathbb{F}^3$. *Let*

$$u_1 = (x_1, y_1, 0) \in U_1,$$

$$u_2 = (0, 0, z_2) \in U_2,$$

$$u_3 = (0, y_3, y_3) \in U_3.$$

Then

$$u_1 + u_2 + u_3 = (x_1, y_1, 0) + (0, 0, z_2) + (0, y_3, y_3)$$

$$= (x_1, y_1 + y_3, z_2 + y_3)$$

and

$$U_1 + U_2 + U_3 = \mathbb{F}^3.$$

Note that

$$(0, 0, 0) \in U_1 + U_2 + U_3$$

and

$$(0, 0, 0) \in U_1,$$

$$\begin{aligned}(0,0,0) &\in U_2,\\ (0,0,0) &\in U_3,\\ (0,1,0) &\in U_1,\\ (0,0,1) &\in U_2,\\ (0,-1,-1) &\in U_3.\end{aligned}$$

Moreover,

$$(0,0,0) = (0,0,0)+(0,0,0)+(0,0,0)$$

and

$$(0,0,0) = (0,1,0)+(0,0,1)+(0,-1,-1).$$

Thus, $(0,0,0) \in U_1+U_2+U_3$ *has two different representations as a sum* $u_1+u_2+u_3$ *with* $u_1 \in U_1$, $u_2 \in U_2$ *and* $u_3 \in U_3$. *Therefore* $U_1+U_2+U_3$ *is not a direct sum.*

Exercise 1.35. *Let*

$$U = \{(x,y,y) \in \mathbb{F}^3 : x,y \in \mathbb{F}\}.$$

Find a subspace $W \in \mathbb{F}^3$ *so that*

$$U \oplus W = \mathbb{F}^3.$$

Theorem 1.36. *Suppose that* U_1, …, U_m *are subspaces of* V. *Then* $U_1 + \cdots + U_m$ *is a direct sum if and only if the only way to write* 0 *as a sum* $u_1 + \cdots + u_m$, *where* $u_1 \in U_1$, …, $u_m \in U_m$, *is by taking each* u_1, u_2, …, u_m *equal to* 0.

Proof. 1. Suppose that $U_1 + \cdots + U_m$ is a direct sum. Since U_1, …, U_m are subspaces of V, we have that

$$0 \in U_1, \quad \ldots, \quad 0 \in U_m, \quad 0 \in U_1 + \cdots + U_m.$$

By the definition of a direct sum, it follows that the only way to write 0 as a sum $u_1 + \cdots + u_m$, where $u_1 \in U_1$, …, $u_m \in U_m$, is by taking each u_1, u_2, …, u_m equal to 0.

2. Suppose that the only way to write 0 as a sum $u_1 + \cdots + u_m$, where $u_1 \in U_1$, …, $u_m \in U_m$, is by taking each u_1, u_2, …, u_m equal to 0. Let $v \in U_1 + \cdots + U_m$ be arbitrarily chosen and fixed. Let also,

$$u = u_1 + \cdots + u_m$$

and

$$u = v_1 + \cdots + v_m,$$

with $u_1, v_1 \in U_1$, …, $u_m, v_m \in U_m$. Subtracting these two equations, we find

$$0 = (u_1 - v_1) + \cdots + (u_m - v_m).$$

Since U_1, …, U_m are subspaces of V, we have

$$u_1 - v_1 \in U_1, \quad \ldots, \quad u_m - v_m \in U_m.$$

The equation above implies that

$$\begin{aligned} u_1 - v_1 &= 0, \\ &\vdots \\ u_m - v_m &= 0, \end{aligned}$$

whereupon

$$\begin{aligned} u_1 &= v_1, \\ &\vdots \\ u_m &= v_m. \end{aligned}$$

This completes the proof.

□

Theorem 1.37. *Suppose that U_1 and U_2 are subspaces of V. Then $U_1 + U_2$ is a direct sum if and only if $U_1 \cap U_2 = \{0\}$.*

Proof. 1. Suppose that $U_1 + U_2$ is a direct sum. Take $v \in U_1 \cap U_2$ arbitrarily. Then

$$0 = v + (-v),$$

where $v \in U_1$ and $-v \in U_2$. Because $U_1 + U_2$ is a direct sum, we have that 0 has a unique representation. Therefore $v = 0$. Since $v \in U_1 \cap U_2$ was arbitrarily chosen, we conclude that $U_1 \cap U_2 = \{0\}$.

2. Let $U_1 \cap U_2 = \{0\}$. We will prove that $U_1 + U_2$ is a direct sum. Suppose that

$$0 = u_1 + u_2,$$

where $u_1 \in U_1$, $u_2 \in U_2$. Then

$$u_2 = -u_1.$$

Since U_1 is a subspace of V, we have that $-u_1 \in U_1$. Therefore $u_2 \in U_1$ and $u_2 \in U_1 \cap U_2$. From here, $u_2 = 0$ and $u_1 = 0$. Thus, 0 has a unique representation as a sum $u_1 + u_2$ with $u_1 \in U_1$ and $u_2 \in U_2$. Hence and Theorem 1.36, we conclude that $U_1 + U_2$ is a direct sum. This completes the proof.

□

Remark 1.38. *The result above only deals with the case of two subspaces. When asking about a possible direct sum with more than two subspaces, it is not enough to test that each pair of the subspaces intersect only at $\{0\}$. To see this, consider Example 1.34. We have that*

$$U_1 \cap U_2 = \{0\}, \quad U_1 \cap U_3 = \{0\}, \quad U_2 \cap U_3 = \{0\}.$$

At the same time, we have that $U_1 + U_2 + U_3$ is not a direct sum.

Remark 1.39. *Sums of subspaces are analogous to unions of subsets. Similarly, direct sums of subspaces are analogous to disjoint unions of subsets. No two subspaces of a vector space can be disjointed, because both contain 0. So, disjointness is replaced, at least in the case of two subspaces, with the requirement that the intersection equals $\{0\}$.*

1.6 Advanced Practical Problems

Problem 1.40. *Prove that*

$$-(-(-v)) = -v,$$

$$-(-(-(-v))) \quad = \quad v, \quad v \in V.$$

Problem 1.41. *Suppose that $v, w \in V$. Explain why there exists a unique $x \in V$ such that*

$$2v - 5x = 3w.$$

Problem 1.42. *Prove that the set of all differentiable real-valued functions f on the interval $(0,3)$ such that*

$$f'(2) = b$$

is a subspace of $\mathbb{R}^{(0,3)}$ if and only if $b = 0$.

Problem 1.43. *Let ∞ and $-\infty$ be two distinct objects, neither of which is in $\mathbb{R}$. Define an addition and scalar multiplication on $\mathbb{R} \cup \{\infty\} \cup \{-\infty\}$ as you could guess from the notation. Specifically, the sum and product of two real numbers is as usual, and for $t \in \mathbb{R}$ define*

$$\begin{aligned}
t\infty &= \begin{cases} -\infty & \text{if} \quad t < 0 \\ 0 & \text{if} \quad t = 0 \\ \infty & \text{if} \quad t > 0, \end{cases} \\
t(-\infty) &= \begin{cases} \infty & \text{if} \quad t < 0 \\ 0 & \text{if} \quad t = 0 \\ -\infty & \text{if} \quad t > 0, \end{cases} \\
t + \infty &= \infty, \\
\infty + t &= \infty, \\
t + (-\infty) &= -\infty, \\
(-\infty) + t &= -\infty, \\
\infty + \infty &= \infty,
\end{aligned}$$

$$(-\infty)+(-\infty) = -\infty,$$

$$\infty+(-\infty) = 0.$$

Is $\mathbb{R}\cup\{\infty\}\cup\{-\infty\}$ *a vector space over* $\mathbb{R}$*? Explain.*

Problem 1.44. *Prove that the set*

$$\{(x_1,x_2,x_3,x_4,x_5)\in\mathbb{F}^5 : x_1+x_2=3x_5, \quad x_2=2x_4\}$$

is a subspace of $\mathbb{F}^5$.

Problem 1.45. *Prove that the set*

$$\{(x_1,x_2,x_3) : x_1=x_2+3\}$$

is not a subspace of $\mathbb{F}^3$.

Problem 1.46. *Let*

$$U = \{(x,y,-x)\in\mathbb{F}^3 : x,y\in\mathbb{F}\},$$

$$W = \{(x,2x,-x)\in\mathbb{F}^3 : x\in\mathbb{F}\}.$$

Find $U+W$.

Problem 1.47. *Let*

$$U_1 = \{(x,0,\ldots,0)\in\mathbb{F}^n : x\in\mathbb{F}\},$$

$$U_2 = \{(0,x,\ldots,0)\in\mathbb{F}^n : x\in\mathbb{F}\},$$

$$\vdots$$

$$U_n = \{(0,0,\ldots,0,x)\in\mathbb{F}^n : x\in\mathbb{F}\}.$$

Prove that

$$U_1\oplus\cdots\oplus U_n=\mathbb{F}^n.$$

Problem 1.48. *For each of the following subsets of* $\mathbb{F}^4$*, determine whether it is a subspace of* $\mathbb{F}^4$.

1.
$$\{(x_1,x_2,x_3,x_4) \in \mathbb{F}^4 : x_1 - 2x_2 + 4x_3 - x_5 = 0\}.$$

2.
$$\{(x_1,x_2,x_3,x_4) \in \mathbb{F}^4 : x_1 - 2x_2 + 4x_3 - x_5 = 2\}.$$

3.
$$\{(x_1,x_2,x_3,x_4) \in \mathbb{F}^4 : x_1x_2x_3x_4 = 0\}.$$

4.
$$\{(x_1,x_2,x_3,x_4) \in \mathbb{F}^4 : x_1 = 2x_2\}.$$

5.
$$\{(x_1,x_2,x_3,x_4) \in \mathbb{F}^4 : x_1 = 3x_3\}.$$

6.
$$\{(x_1,x_2,x_3,x_4) \in \mathbb{F}^4 : x_1 = 4x_4\}.$$

7.
$$\{(x_1,x_2,x_3,x_4) \in \mathbb{F}^4 : x_2 = 7x_3\}.$$

8.
$$\{(x_1,x_2,x_3,x_4) \in \mathbb{F}^4 : x_2 = 8x_4\}.$$

9.
$$\{(x_1,x_2,x_3,x_4) \in \mathbb{F}^4 : x_3 = -x_4\}.$$

10.
$$\{(x_1,x_2,x_3,x_4) \in \mathbb{F}^4 : x_1 = 2x_2x_3\}.$$

Problem 1.49. *Show that the set of all differentiable real-valued functions f on the interval $(-1,1)$ for which*

$$f'(0) = f\left(\frac{1}{2}\right)$$

is a subspace of $\mathbb{R}^{(-1,1)}$.

Problem 1.50. *Is $\mathbb{R}^2$ a subspace of $\mathbb{C}^2$?*

Problem 1.51. *Is*

$$\{(a,b,c) : \mathbb{R}^3 : a^3 = b^3\}$$

a subspace of $\mathbb{R}^3$?

Problem 1.52. *Is*

$$\{(a,b,c) : \mathbb{C}^3 : a^3 = b^3\}$$

a subspace of $\mathbb{C}^3$*?*

Problem 1.53. *Suppose that* U_1 *and* U_2 *are subspaces of* V*. Prove that* $U_1 \cap U_2$ *is a subspace of* V*.*

Problem 1.54. *Prove that the intersection of every collection of subspaces of* V *is a subspace of* V*.*

Problem 1.55. *Prove that the union of two subspaces of* V *is a subspace if and only if one of the subspaces is contained in the other.*

Problem 1.56. *Prove that the union of three subspaces of* V *is a subspace of* V *if and only if one of the subspaces contains the other two.*

Problem 1.57. *Suppose that* U *is a subspace of* V*. What is* $U+U$*?*

Problem 1.58. *Is the operation addition on the subspaces of* V *commutative?*

Problem 1.59. *Is the operation addition on the subspaces of* V *associative?*

Problem 1.60. *Does the operation addition on the subspaces of* V *have an additive identity?*

Problem 1.61. *Suppose*

$$U = \{(x,x,y,y) \in \mathbb{F}^4 : x,y \in \mathbb{F}\}.$$

Find a subspace W *of* $\mathbb{F}^4$ *such that*

$$U \oplus W = \mathbb{F}^4.$$

Problem 1.62. *Suppose*

$$U = \{(x,y,x+y,x-y,2x) \in \mathbb{F}^5 : x,y \in \mathbb{F}\}.$$

Find a subspace W *of* $\mathbb{F}^5$ *such that*

$$U \oplus W = \mathbb{F}^5.$$

Problem 1.63. *Suppose*

$$U = \{(x,y,x+y,x-y,2x) \in \mathbb{F}^5 : x,y \in \mathbb{F}\}.$$

Find subspaces W_1*,* W_2*,* W_3 *of* $\mathbb{F}^5$*, none of which equals* $\{0\}$ *such that*

$$U \oplus W_1 \oplus W_2 \oplus W_3 = \mathbb{F}^5.$$

Problem 1.64. *A function* $f : \mathbb{R} \to \mathbb{R}$ *is called even if*

$$f(x) = f(-x)$$

for any $x \in \mathbb{R}$. *A function* $f : \mathbb{R} \to \mathbb{R}$ *is called odd if*

$$f(x) = -f(-x)$$

for any $x \in \mathbb{R}$. *Let* U_e *denote the set of real-valued even functions on* $\mathbb{R}$ *and* U_0 *denote the set of real-valued odd functions on* $\mathbb{R}$. *Prove that*

$$U_e \oplus U_0 = \mathbb{R}^{\mathbb{R}}.$$

Chapter 2

Finite-Dimensional Vector Spaces

2.1 Linear Combinations and Span

Suppose that V is a vector space.

Definition 2.1. *A linear combination of a list of vectors $v_1, \ldots, v_m$ in V is a vector of the form*

$$a_1v_1 + \cdots + a_mv_m,$$

where $a_1, \ldots, a_m \in \mathbb{F}$.

Example 2.2. *Consider $\mathbb{F}^4$ and*

$$\begin{aligned} v_1 &= (-1,2,0,1), \\ v_2 &= (-3,1,1,4). \end{aligned}$$

Then

$$\begin{aligned} -2v_1 + 3v_2 &= -2(-1,2,0,1) + 3(-3,1,1,4) \\ &= (2,-4,0,-2) + (-9,3,3,12) \\ &= (-7,-1,3,10). \end{aligned}$$

Therefore the vector $(-7,-1,3,10)$ *is a linear combination of the vectors* v_1 *and* v_2.

Example 2.3. *Consider* $\mathbb{F}^3$ *and the vectors*

$$\begin{aligned} v_1 &= (2,1,-3), \\ v_2 &= (1,-2,4), \\ v_3 &= (17,-4,5). \end{aligned}$$

We will check if the vector v_3 *is a linear combination of the vectors* v_1 *and* v_2. *Assume that there are* $a_1, a_2 \in \mathbb{F}$ *so that*

$$v_3 = a_1 v_1 + a_2 v_2$$

or

$$\begin{aligned} (17,-4,5) &= a_1(2,1,-3) + a_2(1,-2,4) \\ &= (2a_1, a_1, -3a_1) + (a_2, -2a_2, 4a_2) \\ &= (2a_1 + a_2, a_1 - 2a_2, -3a_1 + 4a_2). \end{aligned}$$

By the last equations, we get the following system

$$\begin{aligned} 2a_1 + a_2 &= 17 \\ a_1 - 2a_2 &= -4 \\ -3a_1 + 4a_2 &= 5, \end{aligned}$$

whereupon

$$\begin{aligned} a_2 &= 17 - 2a_1 \\ a_1 - 2(17 - 2a_1) &= -4 \\ -3a_1 + 4(17 - 2a_1) &= 5, \end{aligned}$$

or

$$\begin{aligned} a_2 &= 17-2a_1 \\ a_1-34+4a_1 &= -4 \\ -3a_1+68-8a_1 &= 5, \end{aligned}$$

or

$$\begin{aligned} a_2 &= 17-2a_1 \\ 5a_1 &= 30 \\ -11a_1 &= -63, \end{aligned}$$

or

$$\begin{aligned} a_2 &= 17-2a_1 \\ a_1 &= 6 \\ a_1 &= \frac{63}{11}, \end{aligned}$$

which is impossible. Therefore v_3 is not a linear combination of the vectors v_1 and v_2.

Exercise 2.4. *Consider $\mathbb{F}^5$ and the vectors*

$$\begin{aligned} v_1 &= (-1,3,4,1,0), \\ v_2 &= (-3,2,1,0,-1), \\ v_3 &= (-4,1,-2,10,1), \\ v_4 &= (-3,1,1,0,1). \end{aligned}$$

1. *Check if v_4 is a linear combination of v_1, v_2 and v_3.*

2. *Check if v_1 is a linear combination of v_2 and v_3.*

3. *Check if v_2 is a linear combination of v_3 and v_4.*

Definition 2.5. *The set of all linear combinations of a list of vectors v_1, ..., v_m in V is called the span of v_1, ..., v_m, denoted*

$$span(v_1, \ldots, v_m).$$

In other words,

$$span(v_1, \ldots, v_m) = \{a_1 v_1 + \cdots + a_m v_m : a_1, \ldots, a_m \in \mathbb{F}\}.$$

Example 2.6. *Consider $\mathbb{F}^4$. By Example 2.2, it follows that*

$$(-7, -1, 3, 10) \in span((-1, 2, 0, 1), (-3, 1, 1, 4)).$$

Example 2.7. *Consider $\mathbb{F}^3$. By Example 2.3, it follows that*

$$(17, -4, 5) \notin span((2, 1, -3), (1, -2, 4)).$$

Exercise 2.8. *Consider $\mathbb{F}^4$. Check if*

$$(-1, -4, 9, 3) \in span((1, -2, 3, 0), (-1, 0, 1, 1)).$$

Theorem 2.9. *The span of a list of vectors in V is the smallest subspace of V containing all the vectors in the list.*

Proof. Let v_1, ..., v_m be a list of vectors in V. We will prove that $\text{span}(v_1, \ldots, v_m)$ is a subspace of V. Observe that

$$0 = 0v_1 + \cdots + 0v_m,$$

whereupon $0 \in \text{span}(v_1, \ldots, v_m)$. Let

$$u = a_1 v_1 + \cdots + a_m v_m \in \text{span}(v_1, \ldots, v_m),$$

$$w = b_1 v_1 + \cdots + b_m v_m \in \text{span}(v_1, \ldots, v_m),$$

where $a_1, \ldots, a_m, b_1, \ldots, b_m \in \mathbb{F}$. Then

$$u + w = a_1 v_1 + \cdots + a_m v_m$$

$$+a_1v_1+\cdots+b_mv_m$$

$$= (a_1+b_1)v_1+\cdots+(a_m+b_m)v_m.$$

Since $a_1+b_1 \in \mathbb{F}$, …, $a_m+b_m \in \mathbb{F}$, we conclude that

$$u+w \in \mathrm{span}(v_1,\ldots,v_m).$$

Take $a \in \mathbb{F}$ arbitrarily. Then

$$\begin{aligned} au &= a(a_1v_1+\cdots+a_mv_m) \\ &= aa_1v_1+\cdots+aa_mv_m. \end{aligned}$$

Because $aa_1 \in \mathbb{F}$, …, $aa_m \in \mathbb{F}$, we get that $au \in \mathrm{span}(v_1,\ldots,v_m)$. Hence and Theorem 1.20, we conclude that $\mathrm{span}(v_1,\ldots,v_m)$ is a subspace of V. Note that, for any $j \in \{1,\ldots,m\}$, we have

$$v_j = 0v_1+\cdots+0v_{j-1}+1v_j+0v_{j+1}+\cdots+0v_m$$

and $v_j \in \mathrm{span}(v_1,\ldots,v_m)$ for any $j \in \{1,\ldots,m\}$. Because the subspaces are closed under scalar multiplication and addition, every subspace of V containing v_1, …, v_m contains $\mathrm{span}(v_1,\ldots,v_m)$. So, $\mathrm{span}(v_1,\ldots,v_m)$ is the smallest subspace in V containing all the vectors v_1, …, v_m. This completes the proof. □

Definition 2.10. *If $span(v_1,\ldots,v_m) = V$, we say that the list v_1, …, v_m spans V.*

Theorem 2.11. *We have that the list*

$$(1,0,\ldots,0), \quad (0,1,0,\ldots,0), \quad \ldots, \quad (0,\ldots,0,1)$$

spans $\mathbb{F}^n$.

Proof. Let $(x_1,\ldots,x_n) \in \mathbb{F}^n$, $x_1 \in \mathbb{F}$, …, $x_n \in \mathbb{F}$, be arbitrarily chosen. Then

$$\begin{aligned} (x_1,\ldots,x_n) &= (x_1,0,\ldots,0)+(0,x_2,0,\ldots,0)+\cdots+(0,\ldots,0,x_n) \\ &= x_1(1,0,\ldots,0)+x_2(0,1,0,\ldots,0)+\cdots+x_n(0,\ldots,0,1). \end{aligned}$$

This completes the proof. □

Definition 2.12. *A vector space is called finite-dimensional if some list of vectors in it spans the space. Recall that by the definition every list has finite dimension.*

Example 2.13. *By Theorem 2.11, it follows that $\mathbb{F}^n$ is finite-dimensional.*

2.2 Polynomials

Definition 2.14. *A function $p : \mathbb{F} \to \mathbb{F}$ is called a polynomial with coefficients in $\mathbb{F}$ if there exist $a_0, \ldots, a_m \in \mathbb{F}$ such that*

$$p(z) = a_0 + a_1 z + a_2 z^2 + \cdots + a_m z^m$$

for any $z \in \mathbb{F}$. The set of all polynomials with coefficients in $\mathbb{F}$ will be denoted by $\mathscr{P}(\mathbb{F})$.

Let

$$\begin{aligned} p_1(z) &= a_0 + a_1 z + a_2 z^2 + \cdots + a_{m_1} z^{m_1} \in \mathscr{P}(\mathbb{F}), \\ p_2(z) &= b_0 + b_1 z + b_1 z^2 + \cdots + b_{m_2} z^{m_2} \in \mathscr{P}(\mathbb{F}), \end{aligned}$$

where $a_0, \ldots, a_{m_1}, b_0, \ldots, b_{m_2} \in \mathbb{F}$, $m_1 \geq m_2$. Define

$$\begin{aligned} p_1(z) + p_2(z) &= (a_0 + b_0) + (a_1 + b_1)z + (a_2 + b_2)z^2 + \cdots + (a_{m_2} + b_{m_2})z^{m_2} \\ &\quad + a_{m_2+1} z^{m_2+1} + \cdots + a_{m_1} z^{m_1}, \quad z \in \mathbb{F}, \end{aligned} \tag{2.1}$$

and

$$a p_1(z) = (a a_0) + (a a_1)z + (a a_2)z^2 + \cdots + (a a_{m_1})z^{m_1}, \quad z \in \mathbb{F}. \tag{2.2}$$

Example 2.15. *Let*

$$\begin{aligned} p_1(z) &= 1 + 2z + 3z^2 - z^3, \\ p_2(z) &= -2 + z + z^2, \quad z \in \mathbb{F}. \end{aligned}$$

Then

$$p_1(z) + p_2(z) = 1 + 2z + 3z^2 - z^3 + (-2) + z + z^2$$

$$\begin{aligned} &= (1+(-2))+(2+1)z+(3+1)z^2-z^3 \\ &= -1+3z+4z^2-z^3, \quad z \in \mathbb{F}, \end{aligned}$$

and

$$\begin{aligned} -2p_1(z) &= -2(1+2z+3z^2-z^3) \\ &= (-2\cdot 1)+(-2\cdot 2)z+(-2\cdot 3)z^2+(-2\cdot(-1))z^3 \\ &= -2-4z-6z^2+2z^3, \quad z \in \mathbb{F}, \end{aligned}$$

and

$$\begin{aligned} 3p_2(z) &= 3(-2+z+z^2) \\ &= (3\cdot(-2))+(3\cdot 1)z+(3\cdot 1)z^2 \\ &= -6+3z+3z^2, \quad z \in \mathbb{F}. \end{aligned}$$

Hence,

$$\begin{aligned} -2p_1(z)+3p_2(z) &= -2-4z-6z^2+2z^3-6+3z+3z^2 \\ &= (-2-6)+(-4+3)z+(-6+3)z^2+2z^3 \\ &= -8-z-3z^2+2z^3, \quad z \in \mathbb{F}. \end{aligned}$$

Exercise 2.16. *Let*

$$\begin{aligned} p_1(z) &= 3+5z-5z^2+8z^3, \\ p_2(z) &= 75-25z-12z^2+4z^3, \\ p_3(z) &= 12+20z-51z^2-5z^3+12z^4, \quad z \in \mathbb{F}. \end{aligned}$$

Find

1. $p_1(z)+p_2(z)$, $z \in \mathbb{F}$.
2. $p_1(z)+p_3(z)$, $z \in \mathbb{F}$.
3. $p_2(z)+p_3(z)$, $z \in \mathbb{F}$.
4. $p_1(z)+p_2(z)+p_3(z)$, $z \in \mathbb{F}$.
5. $2p_1(z)$, $z \in \mathbb{F}$.
6. $-3p_2(z)$, $z \in \mathbb{F}$.
7. $4p_3(z)$, $z \in \mathbb{F}$.
8. $2p_1(z)-3p_2(z)$, $z \in \mathbb{F}$.
9. $-3p_2(z)+4p_3(z)$, $z \in \mathbb{F}$.
10. $2p_1(z)-3p_2(z)+4p_3(z)$, $z \in \mathbb{F}$.

Theorem 2.17. *$\mathscr{P}(\mathbb{F})$ is a vector space over $\mathbb{F}$ with the usual operations of addition and scalar multiplication given by* (2.1) *and* (2.2), *respectively.*

Proof. Let

$$\begin{aligned} p_1(z) &= a_0+a_1z+a_2z^2+\cdots+a_{m_1}z^{m_1}, \\ p_2(z) &= b_0+b_1z+b_2z^2+\cdots+b_{m_2}z^{m_2}, \\ p_3(z) &= c_0+c_1z+c_2z^2+\cdots+c_{m_3}z^{m_3}, \quad z \in \mathbb{F}, \end{aligned}$$

where

$$a_0,\ldots,a_{m_1}, \quad b_0,\ldots,b_{m_2}, \quad c_0,\ldots,c_{m_3} \in \mathbb{F}, \quad m_1 \geq m_2 \geq m_3,$$

and $a,b \in \mathbb{F}$ be arbitrarily chosen. Then we have the following.

1.

$$\begin{aligned} p_1(z)+p_2(z) &= a_0+a_1z+a_2z^2+\cdots+a_{m_1}z^{m_1} \\ &\quad +b_0+b_1z+b_2z^2+\cdots+b_{m_2}z^{m_2} \\ &= a_0+a_1z+a_2z^2+\cdots+a_{m_2}z^{m_2} \end{aligned}$$

$$
\begin{aligned}
& +a_{m_2+1}z^{m_2+1}+\cdots+a_{m_1}z^{m_1} \\
& +b_0+b_1z+b_2z^2+\cdots+b_{m_2}z^{m_2} \\
= \quad & (a_0+b_0)+(a_1+b_1)z+(a_2+b_2)z^2+\cdots \\
& +(a_{m_2}+b_{m_2})z^{m_2}+a_{m_2+1}z^{m_2+1}+\cdots+a_{m_1}z^{m_1} \\
= \quad & (b_0+a_0)+(b_1+a_1)z+(b_2+a_2)z^2+\cdots \\
& +(b_{m_2}+a_{m_2})z^{m_2}+a_{m_2+1}z^{m_2+1}+\cdots+a_{m_1}z^{m_1} \\
= \quad & b_0+b_1z+b_2z^2+\cdots+b_{m_2}z^{m_2} \\
& +a_0+a_1z+a_2z^2+\cdots+a_{m_1}z^{m_1} \\
= \quad & p_2(z)+p_1(z), \quad z\in\mathbb{F}.
\end{aligned}
$$

2.

$$
\begin{aligned}
p_1(z)+(p_2(z)+p_3(z)) &= a_0+a_1z+a_2z^2+\cdots+a_{m_1}z^{m_1} \\
& +\left(b_0+b_1z+b_2z^2+\cdots+b_{m_2}z^{m_2}+c_0+c_1z+c_2z^2+\cdots+c_{m_3}z^{m_3}\right) \\
= \quad & a_0+a_1z+a_2z^2+\cdots+a_{m_3}z^{m_3}+a_{m_3+1}z^{m_3+1}+\cdots+a_{m_2}z^{m_2} \\
& +a_{m_2+1}z^{m_2+1}+\cdots+a_{m_1}z^{m_1} \\
& +\Big(b_0+b_1z+b_2z^2+\cdots+b_{m_3}z^{m_3}+b_{m_3+1}z^{m_3+1}+\cdots+b_{m_2}z^{m_2} \\
& +c_0+c_1z+c_2z^2+\cdots+c_{m_3}z^{m_3}\Big)
\end{aligned}
$$

$$
\begin{aligned}
&= a_0+a_1z+a_2z^2+\cdots+a_{m_3}z^{m_3}+a_{m_3+1}z^{m_3+1}+\cdots+a_{m_2}z^{m_2} \\
&\quad +a_{m_2+1}z^{m_2+1}+\cdots+a_{m_1}z^{m_1} \\
&\quad +\Big((b_0+c_0)+(b_1+c_1)z+(b_2+c_2)z^2+\cdots \\
&\quad +(b_{m_3}+c_{m_3})z^{m_3}+b_{m_3+1}z^{m_3+1}+\cdots+b_{m_2}z^{m_2}\Big) \\
&= (a_0+(b_0+c_0))+(a_1+(b_1+c_1))z+(a_2+(b_2+c_2))z^2 \\
&\quad +\cdots+(a_{m_3}+(b_{m_3}+c_{m_3}))z^{m_3}+(a_{m_3+1}+b_{m_3+1})z^{m_3+1} \\
&\quad +\cdots+(a_{m_2}+b_{m_2})z^{m_2}+a_{m_2+1}z^{m_2+1}+\cdots+a_{m_1}z^{m_1} \\
&= ((a_0+b_0)+c_0)+((a_1+b_1)+c_1)z+((a_2+b_2)+c_2)z^2 \\
&\quad +\cdots+((a_{m_3}+b_{m_3})+c_{m_3})z^{m_3}+(a_{m_3+1}+b_{m_3+1})z^{m_3+1} \\
&\quad +\cdots+(a_{m_2}+b_{m_2})z^{m_2}+a_{m_2+1}z^{m_2+1}+\cdots+a_{m_1}z^{m_1} \\
&= (a_0+b_0)+(a_1+b_1)z+(a_2+b_2)z^2+\cdots+(a_{m_3}+b_{m_3})z^{m_3} \\
&\quad +(a_{m_3+1}+b_{m_3+1})z^{m_3+1}+\cdots+(a_{m_2}+b_{m_2})z^{m_2} \\
&\quad +a_{m_2+1}z^{m_2+1}+\cdots+a_{m_1}z^{m_1} \\
&\quad +c_0+c_1z+c_2z^2+\cdots+c_{m_3}z^{m_3} \\
&= (p_1(z)+p_2(z))+p_3(z), \quad z\in\mathbb{F}.
\end{aligned}
$$

3. Set

$$0=0+0z+oz^2+\cdots+0z^{m_1}, \quad z\in\mathbb{F}.$$

Then

$$\begin{aligned} p_1(z)+0 &= a_0+a_1z+a_2z^2+\cdots+a_{m_1}z^{m_1}+0+0z+0z^2+\cdots+0z^{m_1} \\ &= (a_0+0)+(a_1+0)z+(a_2+0)z^2+\cdots+(a_{m_1}+0)z^{m_1} \\ &= a_0+a_1z+a_2z^2+\cdots+a_{m_1}z^{m_1} \\ &= p_1(z), \quad z\in\mathbb{F}. \end{aligned}$$

4. Define

$$q(z)=(-a_0)+(-a_1)z+(-a_2)z^2+\cdots+(-a_{m_1})z^{m_1}, \quad z\in\mathbb{F}.$$

Then

$$\begin{aligned} p_1(z)+q(z) &= a_0+a_1z+a_2z^2+\cdots+a_{m_1}z^{m_1} \\ &\quad +(-a_0)+(-a_1)z+(-a_2)z^2+\cdots+(-a_{m_1})z^{m_1} \\ &= (a_0+(-a_0))+(a_1+(-a_1))z+(a_2+(-a_2))z^2 \\ &\quad +\cdots+(a_{m_1}+(-a_{m_1}))z^{m_1} \\ &= 0+0z+0z^2+\cdots+0z^{m_1}, \quad z\in\mathbb{F}. \end{aligned}$$

5. Now, using the computations in 1, we get

$$\begin{aligned} a(p_1(z)+p_2(z)) &= a((a_0+b_0)+(a_1+b_1)z+(a_2+b_2)z^2 \\ &\quad +\cdots+(a_{m_2}+b_{m_2})z^{m_2}+a_{m_2+1}z^{m_2+1}+\cdots+a_{m_1}z^{m_1}) \\ &= (a(a_0+b_0))+(a(a_1+b_1))z+(a(a_2+b_2))z^2+\cdots \\ &\quad +(a(a_{m_2}+b_{m_2}))z^{m_2}+(aa_{m_2+1})z^{m_2+1}+\cdots+(aa_{m_1})z^{m_1} \end{aligned}$$

$$\begin{aligned} &= (aa_0+ab_0)+(aa_1+ab_1)z+(aa_2+ab_2)z^2+\cdots \\ &\quad +(aa_{m_2}+ab_{m_2})z^{m_2}+(aa_{m_2+1})z^{m_2+1}+\cdots+(aa_{m_1})z^{m_1} \\ &= (aa_0)+(aa_1)z+(aa_2)z^2+\cdots+(aa_{m_1})z^{m_1} \\ &\quad +(ab_0)+(ab_1)z+(ab_2)z^2+\cdots+(ab_{m_2})z^{m_2} \\ &= ap_1(z)+ap_2(z), \quad z\in\mathbb{F}. \end{aligned}$$

6.

$$\begin{aligned} (a+b)p_1(z) &= (a+b)(a_0+a_1z+a_2z^2+\cdots+a_{m_1}z^{m_1}) \\ &= ((a+b)a_0)+((a+b)a_1)z+((a+b)a_2)z^2 \\ &\quad +\cdots+((a+b)a_{m_1})z^{m_1} \\ &= (aa_0+ba_0)+(aa_1+ba_1)z+(aa_2+ba_2)z^2 \\ &\quad +\cdots+(aa_{m_1}+ba_{m_1})z^{m_1} \\ &= (aa_0)+(aa_1)z+(aa_2)z^2+\cdots+(aa_{m_1})z^{m_1} \\ &\quad +(ba_0)+(ba_1)z+(ba_2)z^2+\cdots+(ba_{m_1})z^{m_1} \\ &= ap_1(z)+bp_1(z), \quad z\in\mathbb{F}. \end{aligned}$$

Therefore, all axioms for a vector space are fulfilled. Thus, $\mathscr{P}(\mathbb{F})$ is a vector space. This completes the proof.

□

By Theorem 2.17, it follows that $\mathscr{P}(\mathbb{F})$ is a subspace of $\mathbb{F}^{\mathbb{F}}$. If a polynomial $p\in\mathscr{P}(\mathbb{F})$ is represented by two sets of coefficients, i.e.,

$$p(z)=a_0+a_1z+a_2^2+\cdots+a_{m_1}z^{m_1}, \quad z\in\mathbb{F},$$

and

$$p(z) = b_0 + b_1 z + b_2 z^2 + \cdots + b_{m_2} z^{m_2}, \quad z \in \mathbb{F},$$

where $a_0, \ldots, a_{m_1}, b_0, \ldots, b_{m_2} \in \mathbb{F}$, $m_1 \geq m_2$, then

$$\begin{aligned}
0 &= 0 + 0z + 0z^2 + \cdots + 0z^{m_1} \\
&= p_1(z) - p_2(z) \\
&= a_0 + a_1 z + a_2 z^2 + \cdots + a_{m_1} z^{m_1} \\
&\quad -b_0 - b_1 z - b_2 z^2 - \cdots - b_{m_2} z^{m_2} \\
&= a_0 + a_1 z + a_2 z^2 + \cdots + a_{m_2} z^{m_2} + a_{m_2+1} z^{m_2+1} + \cdots + a_{m_1} z^{m_1} \\
&\quad -b_0 - b_1 z - b_2 z^2 - \cdots - b_{m_2} z^{m_2} \\
&= (a_0 - b_0) + (a_1 - b_1) z + (a_2 - b_2) z^2 + \cdots \\
&\quad +(a_{m_2} - b_{m_2}) z^{m_2} + a_{m_2+1} z^{m_2+1} + \cdots + a_{m_1} z^{m_1}, \quad z \in \mathbb{F}.
\end{aligned}$$

Now, using that 0 has a unique representation, we get

$$\begin{aligned}
a_0 - b_0 &= 0 \\
a_1 - b_1 &= 0 \\
a_2 - b_2 &= 0 \\
&\vdots \\
a_{m_2} - b_{m_2} &= 0 \\
a_{m_2+1} &= 0
\end{aligned}$$

$$\vdots$$

$$a_{m_1} = 0.$$

or

$$\begin{aligned}
a_0 &= b_0 \\
a_1 &= b_1 \\
a_2 &= b_2 \\
&\vdots \\
a_{m_2} &= b_{m_2} \\
a_{m_2+1} &= 0 \\
&\vdots \\
a_{m_1} &= 0.
\end{aligned}$$

Hence,

$$p(z) = a_0 + a_1 z + a_2 z^2 + \cdots + a_{m_2} z^{m_2}, \quad z \in \mathbb{F}.$$

Consequently, the coefficients of any polynomial are uniquely determined by the polynomial.

Definition 2.18. *A polynomial $p \in \mathscr{P}(\mathbb{F})$ is said to have degree m if there exist scalars $a_0, \ldots, a_m \in \mathbb{F}$ with $a_m \neq 0$ such that*

$$p(z) = a_0 + a_1 z + a_2 z^2 + \cdots + a_m z^m, \quad z \in \mathbb{F}.$$

If p has a degree m, we will write $\deg p = m$.The polynomial that is identically 0 is said to have degree $-\infty$. For m a nonnegative integer, $\mathscr{P}_m(\mathbb{F})$ denotes the set of all polynomials with coefficients in $\mathbb{F}$ and degree at most m.

By the above definition, it follows that for each nonnegative integer m, using that $-\infty < m$, we have that $0 \in \mathscr{P}_m(\mathbb{F})$.

Example 2.19. *The polynomial*

$$p(z) = 1 + 3z + z^7, \quad z \in \mathbb{F},$$

has a degree 7.

Exercise 2.20. *Determine the degree of the following polynomials.*

1. $p(z) = -1 - 2z - 2z^3 + 5z^4$, $z \in \mathbb{F}$.
2. $p(z) = -27 + 27z - = 9z^2 + z^3$, $z \in \mathbb{F}$.
3. $p(z) = -1 + 9z^2 + 4z^3 - 12z^4$, $z \in \mathbb{F}$.
4. $p(z) = -20z^4 - 13z^3 + 20z^2 + 7z + 6$, $z \in \mathbb{F}$.
5. $p(z) = z^4 - z^2 + 3$, $z \in \mathbb{F}$.
6. $p(z) = z^4 + 7z^3 + 18z^2 + 20z + 8$, $z \in \mathbb{F}$.
7. $p(z) = z^5 + z^4 + z^3 + z^2 + z + 1$, $z \in \mathbb{F}$.
8. $p(z) = z^5 - z^3 + 2z^2 + z + 1$, $z \in \mathbb{F}$.
9. $p(z) = z^6 - 3z^5 - 4z^3 + z - 1$, $z \in \mathbb{F}$.
10. $p(z) = z^7 - z^6 - 3z^3 + z + 1$, $z \in \mathbb{F}$.

Note that

$$\mathscr{P}_m(\mathbb{F}) = \operatorname{span}(1, z, z^2, \ldots, z^m).$$

Therefore $\mathscr{P}_m(\mathbb{F})$ is a finite-dimensional vector space for each nonnegative integer m.

2.3 Infinite-Dimensional Vector Spaces

Definition 2.21. *A vector space is said to be infinite-dimensional if it is not finite-dimensional.*

Theorem 2.22. *The vector space $\mathscr{P}(\mathbb{F})$ is infinite-dimensional.*

Proof. Consider a list of elements of $\mathscr{P}(\mathbb{F})$. Let m denote the highest degree of the polynomials in this list. Then every polynomial in the span of this list has a degree at most m. Therefore z^{m+1} does not belong to this list. Hence, no list spans $\mathscr{P}(\mathbb{F})$ and $\mathscr{P}(\mathbb{F})$ is infinite-dimensional. This completes the proof. □

2.4 Linear Independence and Linear Dependence

Definition 2.23. *A list $v_1, \ldots, v_m$ of vectors in V is called linearly independent if the only choice of $a_1, \ldots, a_m \in \mathbb{F}$ such that*

$$a_1 v_1 + \cdots + a_m v_m = 0$$

is

$$a_1 = \ldots = a_m = 0.$$

Otherwise, the list $v_1, \ldots, v_m$ is said to be linearly dependent. The empty list is declared to be linearly independent.

Example 2.24. *Consider $\mathbb{F}^3$ and*

$$\begin{aligned} v_1 &= (1,4,6), \\ v_2 &= (1,-1,1), \\ v_3 &= (1,1,3). \end{aligned}$$

Assume that there are $a_1, a_2, a_3 \in \mathbb{F}$ so that

$$a_1 v_1 + a_2 v_2 + a_3 v_3 = 0.$$

Then

$$\begin{aligned} 0 &= (0,0,0) \\ &= a_1(1,4,6) + a_2(1,-1,1) + a_3(1,1,3) \\ &= (a_1, 4a_1, 6a_1) + (a_2, -a_2, a_2) + (a_3, a_3, 3a_3) \\ &= (a_1 + a_2 + a_3, 4a_1 - a_2 + a_3, 6a_1 + a_2 + 3a_3). \end{aligned}$$

Hence, we get the system

$$\begin{aligned} a_1 + a_2 + a_3 &= 0 \\ 4a_1 - a_2 + a_3 &= 0 \end{aligned}$$

$$6a_1 + a_2 + 3a_3 = 0,$$

whereupon

$$\begin{aligned} a_3 &= -a_1 - a_2 \\ 4a_1 - a_2 - a_1 - a_2 &= 0 \\ 6a_1 + a_2 - 3a_1 - 3a_2 &= 0, \end{aligned}$$

or

$$\begin{aligned} a_3 &= -a_1 - a_2 \\ 3a_1 - 2a_2 &= 0 \\ 3a_1 - 2a_2 &= 0, \end{aligned}$$

or

$$\begin{aligned} a_3 &= -a_1 - a_2 \\ 3a_1 - 2a_2 &= 0, \end{aligned}$$

or

$$\begin{aligned} a_3 &= -a_1 - a_2 \\ a_2 &= \frac{3}{2}a_1, \end{aligned}$$

or

$$\begin{aligned} a_3 &= -a_1 - \frac{3}{2}a_1 \\ a_2 &= \frac{3}{2}a_1, \end{aligned}$$

or

$$a_3 = -\frac{5}{2}a_1$$

$$a_2 = \frac{3}{2}a_1.$$

Take

$$\begin{aligned} a_1 &= 2, \\ a_2 &= 3, \\ a_3 &= -5. \end{aligned}$$

Then a_1, a_2 and a_3 satisfy the last system, and

$$\begin{aligned} (a_1, a_2, a_3) &= (2, 3, -5) \\ &\neq (0, 0, 0), \end{aligned}$$

and

$$\begin{aligned} 2v_1 + 3v_2 - 6v_3 &= 2(1, 4, 6) + 3(1, -1, 1) - 5(1, 1, 3) \\ &= (2, 8, 12) + (3, -3, 3) + (-5, -5, -15) \\ &= (5, 5, 15) + (-5, -5, -15) \\ &= (0, 0, 0). \end{aligned}$$

Thus, the considered list v_1, v_2, v_3 is linearly dependent.

Example 2.25. *Consider $\mathbb{F}^3$ and*

$$\begin{aligned} v_1 &= (1, -1, 2), \\ v_2 &= (-1, 1, -1), \\ v_3 &= (2, -1, 1). \end{aligned}$$

Assume that there are $a_1, a_2, a_3 \in \mathbb{F}$ so that

$$a_1 v_1 + a_2 v_2 + a_3 v_3 = 0.$$

Then

$$\begin{aligned} 0 &= (0,0,0) \\ &= a_1(1,-1,2)+a_2(-1,1,-1)+a_3(2,-1,1) \\ &= (a_1,-a_1,2a_1)+(-a_2,a_2,-a_2)+(2a_3,-a_3,a_3) \\ &= (a_1-a_2+2a_3,-a_1+a_2-a_3,2a_1-a_2+a_3). \end{aligned}$$

Thus, we get the system

$$\begin{aligned} a_1-a_2+2a_3 &= 0 \\ -a_1+a_2-a_3 &= 0 \\ 2a_1-a_2+a_3 &= 0, \end{aligned}$$

whereupon

$$\begin{aligned} a_1 &= a_2-2a_3 \\ -a_2+2a_3+a_2-a_3 &= 0 \\ 2a_2-4a_3-a_2+a_3 &= 0, \end{aligned}$$

or

$$\begin{aligned} a_1 &= a_2-2a_3 \\ a_3 &= 0 \\ a_2 &= 3a_3, \end{aligned}$$

or

$$a_1 = 0$$

$$a_2 = 0$$

$$a_3 = 0.$$

Therefore the considered list v_1, v_2, v_3 is linearly independent.

Exercise 2.26. *Consider $\mathbb{F}^3$. Check if the following lists of vectors are linearly independent or linearly dependent.*

1.

$$v_1 = (-1,0,1),$$

$$v_2 = (-3,0,2),$$

$$v_3 = (1,1,1).$$

2.

$$v_1 = (0,0,1),$$

$$v_2 = (1,0,1),$$

$$v_3 = (1,1,1).$$

3.

$$v_1 = (1,2,3),$$

$$v_2 = (4,5,6),$$

$$v_3 = (7,8,9).$$

4.

$$v_1 = (2,-3,1),$$

$$v_2 = (3,-1,5),$$

$$v_3 = (1,-4,3).$$

5.

$$\begin{aligned} v_1 &= (5,4,3), \\ v_2 &= (3,3,2), \\ v_3 &= (8,1,3). \end{aligned}$$

Example 2.27. *Consider $\mathscr{P}_2(\mathbb{F})$ and*

$$\begin{aligned} p_1(z) &= 1, \\ p_2(z) &= z-1, \\ p_3(z) &= (z+3)^2, \quad z \in \mathbb{F}. \end{aligned}$$

Assume that there are $a_1, a_2, a_3 \in \mathbb{F}$ such that

$$a_1 p_1(z) + a_2 p_2(z) + a_3 p_3(z) = 0, \quad z \in \mathbb{F}.$$

We have

$$\begin{aligned} 0 &= 0 + 0z + 0z^2 \\ &= a_1 + a_2(z-1) + a_3(z+3)^2 \\ &= a_1 - a_2 + a_2 z + a_3(z^2 + 6z + 9) \\ &= a_1 - a_2 + a_2 z + a_3 z^2 + 6a_3 z + 9a_3 \\ &= (a_1 - a_2 + 9a_3) + (a_2 + 6a_3)z + a_3 z^2, \quad z \in \mathbb{F}. \end{aligned}$$

Since the last equation holds for any $z \in \mathbb{F}$ and the coefficients of any polynomial are uniquely determined by the polynomial, we get the system

$$\begin{aligned} a_1 - a_2 + 9a_3 &= 0 \\ a_2 + 6a_3 &= 0 \end{aligned}$$

$$a_3 = 0,$$

whereupon

$$a_1 = 0$$

$$a_2 = 0$$

$$a_3 = 0.$$

Therefore the list p_1, p_2, p_3 of polynomials of $\mathscr{P}_2(\mathbb{F})$ is linearly independent.

Example 2.28. *Consider $\mathscr{P}_2(\mathbb{F})$ and*

$$p_1(z) = 4z^2 - 3z - 1,$$

$$p_2(z) = 4z - 3,$$

$$p_3(z) = 4z^2 + 9z - 10, \quad z \in \mathbb{F}.$$

Assume that there are $a_1, a_2, a_3 \in \mathbb{F}$ so that

$$a_1 p_1(z) + a_2 p_2(z) + a_3 p_3(z) = 0, \quad z \in \mathbb{F}.$$

We have

$$\begin{aligned} 0 &= 0 + 0z + 0z^2 \\ &= a_1(4z^2 - 3z - 1) + a_2(4z - 3) + a_3(4z^2 + 9z - 10) \\ &= 4a_1z^2 - 3a_1z - a_1 + 4a_2z - 3a_2 + 4a_3z^2 + 9a_3z - 10a_3 \\ &= -a_1 - 3a_2 - 10a_3 + (-3a_1 + 4a_2 + 9a_3)z + (4a_1 + 4a_3)z^2, \quad z \in \mathbb{F}. \end{aligned}$$

Since the last equation holds for any $z \in \mathbb{F}$ and the coefficients of any polynomial are uniquely determined by the polynomial, we obtain the following system

$$-a_1 - 3a_2 - 10a_3 = 0$$

$$\begin{aligned} -3a_1 + 4a_2 + 9a_3 &= 0 \\ 4a_1 + 4a_3 &= 0, \end{aligned}$$

whereupon

$$\begin{aligned} a_1 &= -a_3 \\ -a_1 - 3a_2 + 10a_1 &= 0 \\ -3a_1 + 4a_2 - 9a_1 &= 0, \end{aligned}$$

or

$$\begin{aligned} a_1 &= -a_3 \\ 3a_2 &= 9a_1 \\ 4a_2 &= 12a_1, \end{aligned}$$

or

$$\begin{aligned} a_1 &= -a_3 \\ a_2 &= 3a_1. \end{aligned}$$

Take

$$\begin{aligned} a_1 &= 1, \\ a_2 &= 3, \\ a_3 &= -1. \end{aligned}$$

Then

$$(a_1, a_2, a_3) = (1, 3, -1)$$

$$\neq \quad (0,0,0)$$

and

$$\begin{aligned} a_1p_1(z)+a_2p_2(z)+a_3p_3(z) &= 4z^2-3z-1+3(4z-3)-(4z^2+9z-10) \\ &= 4z^2-3z-1+12z-9-4z^2-9z+10 \\ &= 0, \quad z\in\mathbb{F}. \end{aligned}$$

Consequently the list p_1, p_2, p_3 of polynomials of $\mathscr{P}_2(\mathbb{F})$ is linearly independent.

Exercise 2.29. *Consider $\mathscr{P}_2(\mathbb{F})$. Check if the following lists of polynomials in $\mathscr{P}_2(\mathbb{F})$ are linearly independent or linearly dependent.*

1.

$$\begin{aligned} p_1(z) &= z+1, \\ p_2(z) &= -3, \\ p_3(z) &= (1-2z)^2, \quad z\in\mathbb{F}. \end{aligned}$$

2.

$$\begin{aligned} p_1(z) &= z^2+4z, \\ p_2(z) &= 2z^2-z+4, \\ p_3(z) &= 4z^2-4z+1, \quad z\in\mathbb{F}. \end{aligned}$$

3.

$$\begin{aligned} p_1(z) &= 4z^2-3z+2, \\ p_2(z) &= -3z^2+2z-3, \\ p_3(z) &= 7z^2-5z-2, \quad z\in\mathbb{F}. \end{aligned}$$

4.

$$\begin{aligned} p_1(z) &= z^2 - 7z + 11, \\ p_2(z) &= z^2 + 4, \\ p_3(z) &= 2z - 3, \quad z \in \mathbb{F}. \end{aligned}$$

5.

$$\begin{aligned} p_1(z) &= 3z - 4, \\ p_2(z) &= 3z^2 - 2z - 3, \\ p_3(z) &= z^2 + 3z + 3, \quad z \in \mathbb{F}. \end{aligned}$$

Theorem 2.30. *The list $v_1, \ldots, v_m$ of vectors in V is linearly independent if and only if each vector in V has only one representation as a linear combination of the list v_1, …, v_m.*

Proof. 1. Let the list $v_1, \ldots, v_m$ be linearly independent. Assume that for $v \in V$ we have the representations

$$v = a_1 v_1 + \cdots + a_m v_m$$

and

$$v = b_1 v_1 + \cdots + b_m v_m,$$

where $a_1, \ldots, a_m$, b_1, …, $b_m \in \mathbb{F}$, and

$$(a_1, \ldots, a_m) \neq (b_1, \ldots, b_m). \tag{2.3}$$

We have

$$\begin{aligned} 0 &= v - v \\ &= a_1 v_1 + \cdots + a_m v_m \\ &\quad - b_1 v_1 - \cdots - b_m v_m \end{aligned}$$

$$= (a_1 - b_1)v_1 + \cdots + (a_m - b_m)v_m.$$

Since the list $v_1, \ldots, v_m$ is linearly independent, the last equation holds if

$$\begin{aligned} a_1 - b_1 &= 0 \\ &\vdots \\ a_m - b_m &= 0, \end{aligned}$$

or

$$\begin{aligned} a_1 &= b_1 \\ &\vdots \\ a_m &= b_m. \end{aligned}$$

Thus,

$$(a_1, \ldots, a_m) = (b_1, \ldots, b_m),$$

which contradicts with (2.3). Therefore each vector in V has only one representation as a linear combination of the list $v_1, \ldots, v_m$.

2. Let each vector in V has only one representation as a linear combination of the list v_1, …, v_m. We have

$$0 = 0v_1 + \cdots + 0v_m.$$

Hence, the equation

$$0 = a_1v_1 + \cdots + a_mv_m,$$

$a_1, \ldots, a_m \in \mathbb{F}$, holds if

$$\begin{aligned} a_1 &= 0 \\ &\vdots \end{aligned}$$

$$a_m = 0.$$

By the definition, it follows that the list $v_1, \dots, v_m$ is linearly independent. This completes the proof.

□

Theorem 2.31. *Let the list $v_1, \dots, v_m$ of vectors in V be linearly dependent. Then there exists $j \in \{1, \dots, m\}$ such that $v_j \in span(v_1, \dots, v_{j-1})$ and if v_j is removed from the list $v_1, \dots, v_m$, the span of the remaining list equals $span(v_1, \dots, v_m)$.*

Proof. Since the list $v_1, \dots, v_m$ is linearly dependent, there are scalars $a_1, \dots, a_m \in \mathbb{F}$, not all 0, so that

$$0 = a_1 v_1 + \cdots + a_m v_m$$

and

$$(a_1, \dots, a_m) \neq (0, \dots, 0).$$

Let j be the largest element of the set $\{1, \dots, m\}$ so that $a_j \neq 0$. Then

$$a_{j+1} = 0$$

$$\vdots$$

$$a_m = 0.$$

1. Let $j \neq 1$. Then

$$a_j v_j = -a_1 v_1 - \cdots - a_{j-1} v_{j-1} - a_{j+1} v_{j+1} - \cdots - a_m v_m,$$

whereupon

$$v_j = -\frac{a_1}{a_j} v_1 - \cdots - \frac{a_{j-1}}{a_j} v_{j-1} - \frac{a_{j+1}}{a_j} v_{j+1} - \cdots - \frac{a_m}{a_j} v_m. \quad (2.4)$$

Thus, $v_j \in \text{span}(v_1, \dots, v_{j-1})$.

2. Let $j = 1$. Then

$$a_1 v_1 = 0,$$

whereupon

$$v_1 = 0$$

and $v_1 \in \text{span}() = \{0\}$.

Now, we will prove the remaining part of the assertion.

1. Let $j \neq 1$. Take $u \in \mathrm{span}(v_1, \ldots, v_{j-1}, v_{j+1}, \ldots, v_m)$ arbitrarily. Then, there are scalars $a_1, \ldots, a_{j-1}, a_{j+1}, \ldots, a_m \in \mathbb{F}$ so that

$$\begin{aligned} u &= a_1 v_1 + \cdots + a_{j-1} v_{j-1} + a_{j+1} v_{j+1} + \cdots + a_m v_m \\ &= a_1 v_1 + \cdots + a_{j-1} v_{j-1} + 0 v_j + a_{j+1} v_{j+1} + \cdots + a_m v_m. \end{aligned}$$

Hence, $u \in \mathrm{span}(v_1, \ldots, v_m)$. Since $u \in \mathrm{span}(v_1, \ldots, v_{j-1}, v_{j+1}, \ldots, v_m)$ was arbitrarily chosen and we get that it is an element of $\mathrm{span}(v_1, \ldots, v_m)$, we find the inclusion

$$\mathrm{span}(v_1, \ldots, v_{j-1}, v_{j+1}, \ldots, v_m) \subseteq \mathrm{span}(v_1, \ldots, v_m). \tag{2.5}$$

Let now, $v \in \mathrm{span}(v_1, \ldots, v_m)$. Then there are scalars $b_1, \ldots, b_m \in \mathbb{F}$ so that

$$v = b_1 v_1 + \cdots + b_m v_m.$$

Hence, using (2.4), we obtain

$$\begin{aligned} v &= b_1 v_1 + \cdots + b_{j-1} v_{j-1} + b_j v_j + b_{j+1} v_{j+1} + \cdots + b_m v_m \\ &= b_1 v_1 + \cdots + b_{j-1} v_{j-1} \\ &\quad + b_j \left(-\frac{a_1}{a_j} v_1 - \frac{a_{j-1}}{a_j} v_{j-1} - \frac{a_{j+1}}{a_j} v_{j+1} - \cdots - \frac{a_m}{a_j} v_m \right) \\ &\quad + b_{j+1} v_{j+1} + \cdots + b_m v_m \\ &= \left(b_1 - b_j \frac{a_1}{a_j} \right) v_1 + \cdots + \left(b_{j-1} - \frac{a_{j-1}}{a_j} \right) v_{j-1} \\ &\quad + b_{j+1} v_{j+1} + \cdots + b_m v_m, \end{aligned}$$

whereupon

$$v \in \mathrm{span}(v_1, \ldots, v_{j-1}, v_{j+1}, \ldots, v_m).$$

Since $v \in \mathrm{span}(v_1, \ldots, v_m)$ was arbitrarily chosen and we get that it is an element of $\mathrm{span}(v_1, \ldots, v_{j-1}, v_{j+1}, \ldots, v_m)$, we obtain the inclusion

$$\mathrm{span}(v_1, \ldots, v_m) \subseteq \mathrm{span}(v_1, \ldots, v_{j-1}, v_{j+1}, \ldots, v_m).$$

By the last inclusion and (2.5), we obtain the equality

$$\mathrm{span}(v_1,\ldots,v_m) = \mathrm{span}(v_1,\ldots,v_{j-1},v_{j+1},\ldots,v_m).$$

2. Let $j = 1$. Then

$$v_1 = 0.$$

Hence, for any $a_1,\ldots,a_m \in \mathbb{F}$, we have

$$a_1v_1 + \cdots + a_mv_m = a_2v_2 + \cdots + a_mv_m.$$

Therefore

$$\mathrm{span}(v_1,\ldots,v_m) = \mathrm{span}(v_2,\ldots,v_m).$$

This completes the proof.

□

Theorem 2.32. *In a finite-dimensional vector space, the length of any linearly independent list of vectors is less than or equal to the length of any spanning list of vectors.*

Proof. Let $u_1,\ldots,u_m$ be a linearly independent list of vectors in V and $B = \{w_1,\ldots,w_n\}$ spans V. Since B spans V, if we add to B a new vector from V we will get a linearly dependent list. In particular,

$$u_1, w_1, \ldots, w_n$$

is a linearly dependent list in V. By Theorem 2.31, it follows that we can remove some w's from this list, so that the new list of length n consisting of u_1 and the remaining w's spans V. The new list we will denote again by B. Now, we add u_2 to B and we get a list of length $(n+1)$ which is linearly dependent. By Theorem 2.31, it follows that we can remove some vector of this new list so that the obtained list spans V. Since u_1 and u_2 are linearly independent, this vector is one of the w's, not one of u's. We can remove that w's, so that the obtained new list B consisting of u_1, u_2 and remaining w's spans V, and so on. After m steps, we have added all u's and the process stop. By Theorem 2.31, it follows that there is a vector that we can remove it. Because the list u_1, …, u_m is linearly independent, this vector is one of the w's, not one of u's. Thus, there are at least as many w's as u's and then $m \leq n$. This completes the proof. □

2.5 Finite-Dimensional Vector Subspaces

Theorem 2.33. *Let V be a finite-dimensional vector space and U be its subspace. Then U is finite-dimensional.*

Proof. We will consider the following cases.

1. Let $U = \{0\}$. Then U is finite-dimensional.

2. Let $U \neq \{0\}$. Then, we choose a nonzero vector $v_1 \in U$. If

$$U = \text{span}(v_1),$$

then U is finite-dimensional and the proof is completed. Otherwise, we choose a vector $v_2 \notin \text{span}(v_1)$, $v_2 \in U$. If

$$U = \text{span}(v_1, v_2),$$

then U is finite-dimensional and we are done, and so on. If

$$U = \text{span}(v_1, \ldots, v_{j-1}),$$

then U is finite-dimensional and the proof is completed. Let

$$U \neq \text{span}(v_1, \ldots, v_{j-1}).$$

Then, we take $v_j \in U$ so that

$$v_j \notin \text{span}(v_1, \ldots, v_{j-1}).$$

After each step we have constructed a list of vectors such that no vector in this list which is in the span of the previous vectors. Thus, after each step we have constructed a linearly independent list. This linearly independent list cannot be longer than any spanning list of V. So, the process terminates and U is finite-dimensional. This completes the proof.

□

2.6 Bases

Definition 2.34. *A basis in V is a list of vectors in V that is linearly independent and spans V.*

Example 2.35. *Consider* $\mathbb{F}^2$ *and*

$$\begin{aligned} v_1 &= (1,2), \\ v_2 &= (3,5). \end{aligned}$$

Assume that there are scalars $a_1, a_2 \in \mathbb{F}$ *so that*

$$a_1 v_1 + a_2 v_2 = 0.$$

We have

$$\begin{aligned} 0 &= (0,0) \\ &= a_1(1,2) + a_2(3,5) \\ &= (a_1, 2a_1) + (3a_2, 5a_2) \\ &= (a_1 + 3a_2, 2a_1 + 5a_2). \end{aligned}$$

Hence, we get the system

$$\begin{aligned} a_1 + 3a_2 &= 0 \\ 2a_1 + 5a_2 &= 0, \end{aligned}$$

whereupon

$$\begin{aligned} a_1 &= -3a_2 \\ -6a_2 + a_2 &= 0, \end{aligned}$$

or

$$\begin{aligned} a_1 &= 0 \\ a_2 &= 0. \end{aligned}$$

Therefore the list v_1, v_2 *is linearly independent. Let* $u = (u_1, u_2) \in \mathbb{F}^2$ *be arbitrarily chosen. We will search* $b_1, b_2 \in \mathbb{F}$ *so that*

$$u = b_1 u_1 + b_2 u_2.$$

We have

$$\begin{aligned} u &= (u_1, u_2) \\ &= b_1(1,2) + b_2(3,5) \\ &= (b_1, 2b_1) + (3b_2, 5b_2) \\ &= (b_1 + 3b_2, 2b_1 + 5b_2). \end{aligned}$$

From here, we get the system

$$\begin{aligned} b_1 + 3b_2 &= u_1 \\ 2b_1 + 5b_2 &= u_2, \end{aligned}$$

or

$$\begin{aligned} b_1 &= u_1 - 3b_2 \\ 2(u_1 - 3b_2) + 5b_2 &= u_2, \end{aligned}$$

or

$$\begin{aligned} b_1 &= u_1 - 3b_2 \\ 2u_1 - 6b_2 + 5b_2 &= u_2, \end{aligned}$$

or

$$\begin{aligned} b_1 &= u_1 - 3b_2 \\ b_2 &= 2u_1 - u_2, \end{aligned}$$

or

$$\begin{aligned} b_1 &= u_1 - 3(2u_1 - u_2) \\ b_2 &= 2u_1 - u_2, \end{aligned}$$

or

$$b_1 = u_1 - 6u_1 + 3u_2$$

$$b_2 = 2u_1 - u_2,$$

or

$$b_1 = -5u_1 + 3u_2$$

$$b_2 = 2u_1 - u_2.$$

Consequently

$$u = (-5u_1 + 3u_2)v_1 + (2u_1 - u_2)v_2.$$

Since $u \in \mathbb{F}^2$ *was arbitrarily chosen, we conclude that any vector in* $\mathbb{F}^2$ *can be represented in a unique way as a linear combination of* v_1 *and* v_2*. Thus, the list* v_1, v_2 *is a basis of* $\mathbb{F}^2$*.*

Example 2.36. *Consider* $\mathbb{F}^3$ *and*

$$v_1 = (1, 2, -4),$$

$$v_2 = (7, -5, 6).$$

Let $a_1, a_2 \in \mathbb{F}$ *be such that*

$$0 = a_1v_1 + a_2v_2.$$

We have

$$\begin{aligned} 0 &= (0,0) \\ &= a_1(1,2,-4) + a_2(7,-5,6) \\ &= (a_1, 2a_1, -4a_1) + (7a_2, -5a_2, 6a_2) \\ &= (a_1 + 7a_2, 2a_1 - 5a_2, -4a_1 + 6a_2), \end{aligned}$$

whereupon we get the system

$$a_1 + 7a_2 = 0$$

$$\begin{aligned} 2a_1 - 5a_2 &= 0 \\ -4a_1 + 6a_2 &= 0, \end{aligned}$$

or

$$\begin{aligned} a_1 &= -7a_2 \\ -14a_2 - 5a_2 &= 0 \\ 28a_2 + 6a_2 &= 0, \end{aligned}$$

or

$$\begin{aligned} a_1 &= 0 \\ a_2 &= 0. \end{aligned}$$

Therefore the list v_1, v_2 *is linearly independent in* $\mathbb{F}^2$. *Let* $u = (u_1, u_2, u_3) \in \mathbb{F}^3$ *be arbitrarily chosen. We will search* $b_1, b_2 \in \mathbb{F}$ *so that*

$$u = b_1 v_1 + b_2 v_2.$$

We have

$$\begin{aligned} u &= (u_1, u_2, u_3) \\ &= b_1(1, 2, -4) + b_2(7, -5, 6) \\ &= (b_1, 2b_1, -4b_1) + (7b_2, -5b_2, 6b_2) \\ &= (b_1 + 7b_2, 2b_1 - 5b_2, -4b_1 + 6b_2). \end{aligned}$$

Hence,

$$b_1 + 7b_2 = u_1$$

$$\begin{aligned}2b_1-5b_2 &= u_2\\ -4b_1+6b_2 &= u_3,\end{aligned}$$

or

$$\begin{aligned}b_1 &= u_1-7b_2\\ 2(u_1-7b_2)-5b_2 &= u_2\\ -4(u_1-7b_2)+6b_2 &= u_3,\end{aligned}$$

or

$$\begin{aligned}b_1 &= u_1-7b_2\\ 2u_1-14b_2-5b_2 &= u_2\\ -4u_1+28b_2+6b_2 &= u_3,\end{aligned}$$

or

$$\begin{aligned}b_1 &= u_1-7b_2\\ 19b_2 &= 2u_1-u_2\\ -4u_1+34b_2 &= u_3,\end{aligned}$$

or

$$\begin{aligned}b_1 &= u_1-7b_2\\ b_2 &= \frac{2u_1-u_2}{19}\\ b_2 &= \frac{u_3+4u_1}{34}.\end{aligned}$$

By the last two equations we find

$$\frac{2u_1-u_2}{19}=\frac{u_3+4u_1}{34}$$

or

$$68u_1 - 34u_2 = 19u_3 + 76u_1,$$

or

$$8u_1 + 34u_2 + 19u_3 = 0. \tag{2.6}$$

Thus, only the vectors $u = (u_1, u_2, u_3) \in \mathbb{F}^3$ *for which* (2.6) *holds can be represented as a linear combination of* v_1 *and* v_2. *Therefore*

$$span(v_1, v_2) \neq \mathbb{F}^3$$

and the list v_1, v_2 *is not a basis in* $\mathbb{F}^3$.

Exercise 2.37. *Consider* $\mathbb{F}^2$ *and*

$$\begin{aligned} v_1 &= (1,2), \\ v_2 &= (3,5), \\ v_3 &= (4,13). \end{aligned}$$

Prove that the list v_1, v_2, v_3 *is not a basis in* $\mathbb{F}^3$.

Theorem 2.38. *A list* $v_1, \ldots, v_m$ *of vectors in* V *is a basis if and only if any element of* V *can be written uniquely as a linear combination of* $v_1, \ldots, v_m$.

Proof. 1. Let the list $v_1, \ldots, v_m$ be a basis of V. Assume that for $v \in V$ we have two representations

$$v = b_1 v_1 + \cdots + b_m v_m$$

and

$$v = c_1 v_1 + \cdots + c_m v_m,$$

where $b_1, \ldots, b_m$, c_1, …, $c_m \in \mathbb{F}$, and

$$(b_1, \ldots, b_m) \neq (c_1, \ldots, c_m). \tag{2.7}$$

Observe that

$$\begin{aligned} 0 &= v - v \\ &= b_1 v_1 + \cdots + b_m v_m \end{aligned}$$

$$-c_1v_1 - \cdots - c_mv_m$$

$$= (b_1 - c_1)v_1 + \cdots + (b_m - c_m)v_m.$$

Since the list $v_1, \ldots, v_m$ is a basis, we have that it is linearly independent. Thus, the last equation holds if

$$b_1 - c_1 = 0$$

$$\vdots$$

$$b_m - c_m = 0,$$

or

$$b_1 = c_1$$

$$\vdots$$

$$b_m = c_m.$$

Consequently

$$(b_1, \ldots, b_m) = (c_1, \ldots, c_m),$$

which contradicts with (2.7). Hence, we conclude that each vector in V has only one representation as a linear combination of the list $v_1, \ldots, v_m$.

2. Let each vector in V has only one representation as a linear combination of the list v_1, ..., v_m. Then the list $v_1, \ldots, v_m$ spans V. Because

$$0 = 0v_1 + \cdots + 0v_m,$$

the equation

$$0 = b_1v_1 + \cdots + b_mv_m,$$

$b_1, \ldots, b_m \in \mathbb{F}$, holds if

$$b_1 = 0$$

$$\vdots$$

$$b_m = 0.$$

By the definition, it follows that the list $v_1, \ldots, v_m$ is linearly independent. Now, using that the list $v_1, \ldots, v_m$ spans V, we conclude that this list is a basis of V. This completes the proof.

□

Theorem 2.39. *Every spanning list in a vector space can be reduced to a basis of the vector space.*

Proof. Let $B = \{v_1, \ldots, v_m\}$ be a spanning list in V. If $v_1 = 0$, then we remove it from B. If $v_1 \neq 0$, then we leave B unchanged. If $v_2 \in \text{span}(v_1)$, then we remove it from B. If $v_2 \notin \text{span}(v_1)$, then we leave B unchanged, and so on. If $v_j \in \text{span}(v_1, \ldots, v_{j-1})$, then we remove it from B. If $v_j \notin \text{span}(v_1, \ldots, v_{j-1})$, we leave B unchanged. We stop the process after m steps. The last list B spans V because the original list spanned B and we have discarded only vectors that were already in the span of the previous vectors. The process ensures that no vector in the last list B is in the span of the previous vectors. Now, applying Theorem 2.31, we get that the last list B is linearly independent. Consequently the last list B is a basis of V. This completes the proof. □

Corollary 2.40. *Every finite-dimensional vector space has a basis.*

Proof. Let V be a finite-dimensional vector space. By the definition, it follows that there is a list of vectors of V that spans V. Now, applying Theorem 2.31, we get that V has a basis. This completes the proof. □

Example 2.41. *Consider $\mathbb{F}^4$ and*

$$v_1 = (1,0,0,-1),$$

$$v_2 = (2,1,1,0),$$

$$v_3 = (1,1,1,1),$$

$$v_4 = (1,2,3,4),$$

$$v_5 = (0,1,2,3).$$

Firstly, we will show that

$$span(v_1,v_2,v_3,v_4,v_5) = \mathbb{F}^4.$$

Let

$$u = (u_1,u_2,u_3,u_4) \in \mathbb{F}^4$$

be arbitrarily chosen and fixed. Let also, $a_1,a_2,a_3,a_4,a_5 \in \mathbb{F}$ *be such that*

$$u = a_1v_1 + a_2v_2 + a_3v_3 + a_4v_4 + a_5v_5.$$

We have

$$\begin{aligned} u &= (u_1,u_2,u_3,u_4) \\ &= a_1(1,0,0,-1) + a_2(2,1,1,0) + a_3(1,1,1,1) \\ &\quad +a_4(1,2,3,4) + a_5(0,1,2,3) \\ &= (a_1,0,0,-a_1) + (2a_2,a_2,a_2,0) + (a_3,a_3,a_3,a_3) \\ &\quad +(a_4,2a_4,3a_4,4a_4) + (0,a_5,2a_5,3a_5) \\ &= (a_1+2a_2+a_3+a_4, a_2+a_3+2a_4+a_5, \\ &\quad a_2+a_3+3a_4+2a_5, -a_1+a_3+4a_4+3a_5). \end{aligned}$$

Thus, we get the system

$$\begin{aligned} a_1+2a_2+a_3+a_4 &= u_1 \\ a_2+a_3+2a_4+a_5 &= u_2 \\ a_2+a_3+3a_4+2a_5 &= u_3 \end{aligned}$$

$$-a_1 + a_3 + 4a_4 + 3a_5 = u_4,$$

whereupon

$$\begin{aligned} a_4 + a_5 &= u_3 - u_2 \\ a_1 + 2a_2 + a_3 + a_4 &= u_1 \\ a_2 + a_3 + 2a_4 + a_5 &= u_2 \\ -a_1 + a_3 + 4a_4 + 3a_5 &= u_4, \end{aligned}$$

or

$$\begin{aligned} a_4 &= u_3 - u_2 - a_5 \\ a_1 + 2a_2 + a_3 + u_3 - u_2 - a_5 &= u_1 \\ a_2 + a_3 + 2u_3 - 2u_2 - 2a_5 + a_5 &= u_2 \\ -a_1 + a_3 + 4u_3 - 4u_2 - 4a_5 + 3a_5 &= u_4, \end{aligned}$$

or

$$\begin{aligned} a_4 &= u_3 - u_2 - a_5 \\ a_1 + 2a_2 + a_3 - a_5 &= u_1 + u_2 - u_3 \\ a_2 + a_3 - a_5 &= 3u_2 - 2u_3 \\ -a_1 + a_3 - a_5 &= 4u_2 - 4u_3 + u_4, \end{aligned}$$

or

$$\begin{aligned} a_4 &= u_3 - u_2 - a_5 \\ a_1 &= -2a_2 - a_3 + a_5 + u_1 + u_2 - u_3 \end{aligned}$$

$$\begin{aligned} a_2 &= a_3 + a_5 + 3u_2 - 2u_3 \\ -2a_2 - a_3 + a_5 + u_1 + u_2 - u_3 + a_3 - a_5 &= 4u_2 - 4u_3 + u_4, \end{aligned}$$

or

$$\begin{aligned} a_4 &= u_3 - u_2 - a_5 \\ a_1 &= -2a_2 - a_3 + a_5 + u_1 + u_2 - u_3 \\ a_2 &= a_3 + a_5 + 3u_2 - 2u_3 \\ 2a_2 &= u_1 - 3u_2 + 3u_3 - u_4, \end{aligned}$$

or

$$\begin{aligned} a_4 &= u_3 - u_2 - a_5 \\ a_1 &= -u_1 + 3u_2 - 3u_3 + u_4 - a_3 + a_5 + u_1 + u_2 - u_3 \\ \frac{1}{2}u_1 - \frac{3}{2}u_2 + \frac{3}{2}u_3 - \frac{1}{2}u_4 &= a_3 + a_5 + 3u_2 - 2u_3 \\ a_2 &= \frac{1}{2}u_1 - \frac{3}{2}u_2 + \frac{3}{2}u_3 - \frac{1}{2}u_4, \end{aligned}$$

or

$$\begin{aligned} a_4 &= u_3 - u_2 - a_5 \\ a_2 &= \frac{1}{2}u_1 - \frac{3}{2}u_2 + \frac{3}{2}u_3 - \frac{1}{2}u_4 \\ a_1 &= 4u_2 - 4u_3 + u_4 - a_3 + a_5 \\ a_3 &= \frac{1}{2}u_1 - \frac{9}{2}u_2 + \frac{7}{2}u_3 - \frac{1}{2}u_4 - a_5, \end{aligned}$$

or

$$a_1 = 4u_2 - 4u_3 + u_4 - \frac{1}{2}u_1 + \frac{9}{2}u_2 - \frac{7}{2}u_3 + \frac{1}{2}u_4 + a_5 + a_5$$

$$\begin{aligned} a_2 &= \frac{1}{2}u_1 - \frac{3}{2}u_2 + \frac{3}{2}u_3 - \frac{1}{2}u_4 \\ a_3 &= \frac{1}{2}u_1 - \frac{9}{2}u_2 + \frac{7}{2}u_3 - \frac{1}{2}u_4 - a_5 \\ a_4 &= u_3 - u_2 - a_5. \end{aligned}$$

Therefore

$$\begin{aligned} a_1 &= -\frac{1}{2}u_1 + \frac{17}{2}u_2 - \frac{15}{2}u_3 + \frac{3}{2}u_4 + 2a_5 \\ a_2 &= \frac{1}{2}u_1 - \frac{3}{2}u_2 + \frac{3}{2}u_3 - \frac{1}{2}u_4 \\ a_3 &= \frac{1}{2}u_1 - \frac{9}{2}u_2 + \frac{7}{2}u_3 - \frac{1}{2}u_4 - a_5 \\ a_4 &= u_3 - u_2 - a_5, \end{aligned}$$

and then

$$span(v_1, v_2, v_3, v_4, v_5) = \mathbb{F}^4.$$

We have that $v_1 \neq 0$ *and* $v_2 \notin span(v_1)$. *Now, we will check if*

$$v_3 \in span(v_1, v_2). \tag{2.8}$$

Let $c_1, c_2 \in \mathbb{F}$ *be such that*

$$v_3 = c_1 v_1 + c_2 v_2.$$

Then

$$\begin{aligned} v_3 &= (1,1,1,1) \\ &= c_1(1,0,0,-1) + c_2(2,1,1,0) \\ &= (c_1,0,0,-c_1) + (2c_2,c_2,c_2,0) \\ &= (c_1+2c_2,c_2,c_2,-c_1), \end{aligned}$$

whereupon

$$\begin{aligned} c_1+2c_2 &= 1 \\ c_2 &= 1 \\ -c_1 &= 1, \end{aligned}$$

or

$$\begin{aligned} c_1 &= -1 \\ c_2 &= 1. \end{aligned}$$

Therefore (2.8) *holds and we remove* v_3. *Let* $d_1, d_2 \in \mathbb{F}$ *be such that*

$$v_4 = d_1 v_1 + d_2 v_2.$$

We have

$$\begin{aligned} v_4 &= (1,2,3,4) \\ &= d_1(1,0,0,-1)+d_2(2,1,1,0) \\ &= (d_1,0,0,-d_1)+(2d_2,d_2,d_2,0) \\ &= (d_1+2d_2,d_2,d_2,-d_1). \end{aligned}$$

Thus, we get the system

$$\begin{aligned} d_1+2d_2 &= 1 \\ d_2 &= 2 \\ d_2 &= 3 \\ -d_1 &= 4, \end{aligned}$$

which is impossible. Therefore

$$v_4 \notin span(v_1, v_2).$$

Let $e_1, e_2, e_3 \in \mathbb{F}$ *be such that*

$$v_5 = e_1 v_1 + e_2 v_2 + e_4 v_4.$$

We have

$$\begin{aligned} v_5 &= (0,1,2,3) \\ &= e_1(1,0,0,-1) + e_2(2,1,1,0) + e_3(1,2,3,4) \\ &= (e_1,0,0,-e_1) + (2e_2,e_2,e_2,0) + (e_3,2e_3,3e_3,4e_3) \\ &= (e_1+2e_2+e_3, e_2+2e_3, e_2+3e_3, -e_1+4e_3), \end{aligned}$$

whereupon we get the system

$$\begin{aligned} e_1+2e_2+e_3 &= 0 \\ e_2+2e_3 &= 1 \\ e_2+3e_3 &= 2 \\ -e_1+4e_3 &= 3, \end{aligned}$$

or

$$\begin{aligned} e_1 &= 4e_3-3 \\ e_2 &= 2-3e_3 \\ e_2 &= 1-2e_3 \\ 4e_3-3+4-6e_3+e_3 &= 0, \end{aligned}$$

or

$$
\begin{aligned}
e_3 &= 1\\
e_1 &= 1\\
e_2 &= -1\\
-e_3+1 &= 0.
\end{aligned}
$$

Consequently

$$v_5 \in span(v_1, v_2, v_4)$$

and we remove it. Now, applying Theorem 2.39, we get that the list v_1, v_2, v_4 is a basis of $\mathbb{F}^4$.

Exercise 2.42. *Consider $\mathbb{F}^2$ and*

$$
\begin{aligned}
v_1 &= (-1,1),\\
v_2 &= (2,1),\\
v_3 &= (3,2).
\end{aligned}
$$

Find a basis of $\mathbb{F}^2$.

Theorem 2.43. *Every linearly independent list of vectors in a finite-dimensional vector space can be extended to a basis of the vector space.*

Proof. Suppose that $u_1, \ldots, u_m$ is a linearly independent list in a finite-dimensional vector space V and let $w_1, \ldots, w_n$ be a basis of V. Then, the list

$$u_1, \quad \ldots, \quad u_m, \quad w_1, \quad \ldots, \quad w_n \tag{2.9}$$

spans V. Applying Theorem 2.39, by the list (2.9) we get a basis of V that contains all u's, because they are linearly independent, and some of w's. This completes the proof. □

Example 2.44. *Consider $\mathbb{F}^3$ and*

$$v_1 = (2,3,4),$$

$$v_2 = (9,6,8).$$

Let $a \in \mathbb{F}$ be such that

$$v_1 = av_2.$$

Then

$$\begin{aligned} v_1 &= (2,3,4) \\ &= a(9,6,8) \\ &= (9a,6a,8a), \end{aligned}$$

whereupon

$$\begin{aligned} 9a &= 2 \\ 6a &= 3 \\ 8a &= 4, \end{aligned}$$

or

$$\begin{aligned} a &= \frac{2}{9}, \\ a &= \frac{1}{2}. \end{aligned}$$

This is impossible. Therefore v_1 and v_2 are linearly independent. Note that

$$\begin{aligned} w_1 &= (1,0,0), \\ w_2 &= (0,1,0), \\ w_3 &= (0,0,1) \end{aligned}$$

is a basis of $\mathbb{F}^3$. Consider the list v_1, v_2, w_1, w_2, w_3. Let $b_1, b_2 \in \mathbb{F}$ be such that

$$w_1 = b_1 v_1 + b_2 v_2.$$

Then

$$\begin{aligned} w_1 &= (1,0,0) \\ &= b_1(2,3,4)+b_2(9,6,8) \\ &= (2b_1,3b_1,4b_1)+(9b_2,6b_2,8b_2) \\ &= (2b_1+9b_2,3b_1+6b_2,4b_1+8b_2), \end{aligned}$$

whereupon

$$\begin{aligned} 2b_1+9b_2 &= 1 \\ 3b_1+6b_2 &= 0 \\ 4b_1+8b_2 &= 0, \end{aligned}$$

or

$$\begin{aligned} 2b_1+9b_2 &= 1 \\ b_1 &= -2b_2, \end{aligned}$$

or

$$\begin{aligned} -4b_2+9b_2 &= 1 \\ b_1 &= -2b_2, \end{aligned}$$

or

$$\begin{aligned} b_1 &= -\frac{2}{5} \\ b_2 &= \frac{1}{5}. \end{aligned}$$

Therefore

$$w_1 \in span(v_1,v_2)$$

and we remove it. Now, consider the list v_1, v_2, w_2, w_3. *Let* $c_1, c_2 \in \mathbb{F}$ *be such that*

$$w_2 = c_1v_1 + c_2v_2.$$

Then

$$\begin{aligned} w_2 &= (0,1,0) \\ &= c_1(2,3,4) + c_2(9,6,8) \\ &= (2c_1, 3c_1, 4c_1) + (9c_2, 6c_2, 8c_2) \\ &= (2c_1 + 9c_2, 3c_1 + 6c_2, 4c_1 + 8c_2), \end{aligned}$$

whereupon

$$\begin{aligned} 2c_1 + 9c_2 &= 0 \\ 3c_1 + 6c_2 &= 1 \\ 4c_1 + 8c_2 &= 0, \end{aligned}$$

or

$$\begin{aligned} 2c_1 + 9c_2 &= 0 \\ 3c_1 + 6c_2 &= 1 \end{aligned}$$

$$4c_1 + 8c_2 = 0,$$

or

$$\begin{aligned} 2c_1 + 9c_2 &= 0 \\ 3c_1 + 6c_2 &= 1 \\ c_1 &= -2c_2, \end{aligned}$$

or

$$\begin{aligned} 2c_1+9c_2 &= 0 \\ 0 &= 1 \\ c_1 &= -2c_2, \end{aligned}$$

which is impossible. Therefore

$$w_2 \notin span(v_1, v_2).$$

Let now, $d_1, d_2, d_3 \in \mathbb{F}$ be such that

$$w_3 = d_1v_1 + d_2v_2 + d_3w_2.$$

We have

$$\begin{aligned} w_3 &= (0,0,1) \\ &= d_1(2,3,4)+d_2(9,6,8)+d_3(0,1,0) \\ &= (2d_1,3d_1,4d_1)+(9d_2,6d_2,8d_2)+(0,d_3,0) \\ &= (2d_1+9d_2,3d_1+6d_2+d_3,4d_1+8d_2), \end{aligned}$$

whereupon we get the system

$$\begin{aligned} 2d_1+9d_2 &= 0 \\ 3d_1+6d_2+d_3 &= 0 \\ 4d_1+8d_2 &= 1, \end{aligned}$$

or

$$\begin{aligned} d_1 &= -\frac{9}{2}d_2 \\ 10d_2 &= -1 \end{aligned}$$

$$3d_1 + 6d_2 + d_3 = 0,$$

or

$$\begin{aligned} d_1 &= -\frac{9}{2}d_2 \\ d_2 &= -\frac{1}{10} \\ 3d_1 + 6d_2 + d_3 &= 0, \end{aligned}$$

or

$$\begin{aligned} d_1 &= \frac{9}{20} \\ d_2 &= -\frac{1}{10} \\ \frac{27}{20} - \frac{6}{10} + d_3 &= 0, \end{aligned}$$

or

$$\begin{aligned} d_1 &= \frac{9}{20} \\ d_2 &= -\frac{1}{10} \\ d_3 &= -\frac{3}{4}. \end{aligned}$$

Consequently

$$w_3 \in span(v_1, v_2, w_2)$$

and we remove it. Now, applying Theorem 2.43, we get that the list v_1, v_2, w_2 is a basis of V.

Exercise 2.45. *Consider $\mathbb{F}^3$. Extend the list*

$$\begin{aligned} v_1 &= (-1, 2, 5), \\ v_2 &= (1, 4, 10) \end{aligned}$$

to a basis of $\mathbb{F}^3$.

Theorem 2.46. *Suppose V is a finite-dimensional vector space and U is its subspace. Then there is a subspace W of V so that*

$$V = U \oplus W.$$

Proof. Because V is finite-dimensional, by Theorem 2.33, it follows that U is finite-dimensional. Let the list

$$u_1, \ldots, u_m$$

be a basis of U. This list is linearly independent in V and then, applying Theorem 2.43, this list can be extended to a basis of V, i.e., there is a list

$$u_1, \ldots, u_m, w_1, \ldots, w_n \tag{2.10}$$

that is a basis of V. Set

$$W = \operatorname{span}(w_1, \ldots, w_n).$$

Now, we will show that

$$V = U + W. \tag{2.11}$$

Take $v \in V$ arbitrarily. Then, there are scalars $a_1, \ldots, a_m, b_1, \ldots, b_n \in \mathbb{F}$ so that

$$v = a_1 u_1 + \cdots + a_m u_m + b_1 w_1 + \cdots + b_n w_n.$$

Denote

$$\begin{aligned} u &= a_1 u_1 + \cdots + a_m u_m, \\ w &= b_1 w_1 + \cdots + b_n w_n. \end{aligned}$$

Then, $u \in U$, $w \in W$ and

$$v = u + w.$$

Because $v \in V$ was arbitrarily chosen, we get the equation (2.11). Let now, $z \in U \cap W$ be arbitrarily chosen. Then $z \in U$ and $z \in W$. Hence, there are scalars $c_1, \ldots, c_m, d_1, \ldots, d_n \in \mathbb{F}$ such that

$$\begin{aligned} z &= c_1 u_1 + \cdots + c_m u_m \\ &= d_1 w_1 + \cdots + d_n w_n. \end{aligned}$$

By the last two equations, we find

$$c_1u_1 + \cdots + c_mu_m - d_1w_1 - \cdots - d_nw_n = 0.$$

Because (2.10) is a basis of V, by the last equation, we find the system

$$\begin{aligned} c_1 &= 0 \\ &\vdots \\ c_m &= 0 \\ d_1 &= 0 \\ &\vdots \\ d_n &= 0 \end{aligned}$$

and so, $z = 0$. Because $z \in U \cap W$ was arbitrarily chosen, we conclude that

$$U \cap W = \{0\}.$$

By the last equation and the equation (2.11), applying Theorem 1.37, we arrive at

$$V = U \oplus W.$$

This completes the proof. □

Example 2.47. *Let U be a subspace of $\mathbb{R}^5$, defined by*

$$U = \{(x_1, x_2, x_3, x_4, x_5) \in \mathbb{R}^5 : x_1 = 3x_2, \quad x_3 = 7x_4\}.$$

Firstly, we will find a basis of U. Take

$$\begin{aligned} u_1 &= (3, 1, 0, 0, 0), \\ u_2 &= (0, 0, 7, 3, 0), \\ u_3 &= (0, 0, 0, 0, 1). \end{aligned}$$

Assume that there are scalars $a_1, a_2, a_3 \in \mathbb{R}$ *so that*

$$a_1u_1 + a_2u_2 + a_3u_3 = 0.$$

We have

$$\begin{aligned} 0 &= (0,0,0,0,0) \\ &= a_1(3,1,0,0,0) + a_2(0,0,7,3,0) + a_3(0,0,0,0,1) \\ &= (3a_1,a_1,0,0,0) + (0,0,7a_2,3a_2,0) + (0,0,0,0,a_3) \\ &= (3a_1,a_1,7a_2,3a_2,a_3), \end{aligned}$$

whereupon we get the system

$$\begin{aligned} 3a_1 &= 0 \\ a_1 &= 0 \\ 7a_2 &= 0 \\ 3a_2 &= 0 \\ a_3 &= 0, \end{aligned}$$

i.e.,

$$\begin{aligned} a_1 &= 0 \\ a_2 &= 0 \\ a_3 &= 0. \end{aligned}$$

Thus, the list u_1, u_2, u_3 *is linearly independent in* $\mathbb{R}^5$. *Let now,*

$$v = (3v_2, v_2, 7v_4, v_4, v_5) \in U, \quad v_2, v_4, v_5 \in \mathbb{R},$$

be arbitrarily chosen. We will search scalars $b_1, b_2, b_3 \in \mathbb{R}$ *so that*

$$v = b_1 u_1 + b_2 u_2 + b_3 u_3.$$

We have

$$\begin{aligned} v &= (3v_2, v_2, 7v_4, v_4, v_5) \\ &= b_1(3,1,0,0,0) + b_2(0,0,7,1,0) + b_3(0,0,0,0,1) \\ &= (3b_1, b_1, 0,0,0) + (0,0,7b_2, b_2, 0) + (0,0,0,0,b_3) \\ &= (3b_1, b_1, 7b_2, b_2, b_3), \end{aligned}$$

whereupon we get the system

$$\begin{aligned} 3v_2 &= 3b_1 \\ v_2 &= b_1 \\ 7v_4 &= 7b_2 \\ v_4 &= b_2 \\ v_5 &= b_3, \end{aligned}$$

or

$$\begin{aligned} b_1 &= v_2 \\ b_2 &= v_4 \\ b_3 &= v_5. \end{aligned}$$

Thus, any vector $v \in U$ *can be represented in a unique way as a linear combination of the list* u_1, u_2, u_3*. Therefore*

$$span(u_1, u_2, u_3) = V.$$

Since the list u_1, u_2, u_3 is linearly independent in U, we conclude that the list u_1, u_2, u_3 is a basis of U. Now, we will extend this basis to a basis of $\mathbb{R}^5$. Note that the standard basis of $\mathbb{R}^5$ is

$$\begin{aligned} w_1 &= (1,0,0,0,0), \\ w_2 &= (0,1,0,0,0), \\ w_3 &= (0,0,1,0,0), \\ w_4 &= (0,0,0,1,0), \\ w_5 &= (0,0,0,0,1). \end{aligned}$$

Since $u_3 = w_5$, we consider the list

$$u_1, u_2, u_3, w_1, w_2, w_3, w_4.$$

We will search scalars c_1, c_2, c_3 so that

$$w_1 = c_1 u_1 + c_2 u_2 + c_3 u_3.$$

We have

$$\begin{aligned} w_1 &= (1,0,0,0,0) \\ &= c_1(3,1,0,0,0) + c_2(0,0,7,1,0) + c_3(0,0,0,0,1) \\ &= (3c_1, c_1, 0,0,0) + (0,0,7c_2, c_2, 0) + (0,0,0,0,c_3) \\ &= (3c_1, c_1, 7c_2, c_2, c_3), \end{aligned}$$

whereupon we get the system

$$\begin{aligned} 3c_1 &= 1 \\ c_1 &= 0 \end{aligned}$$

$$\begin{aligned} 7c_2 &= 0 \\ c_2 &= 0 \\ c_3 &= 0, \end{aligned}$$

which is impossible. Thus,

$$w_1 \notin span(u_1, u_2, u_3).$$

Now, we will search scalars $d_1, d_2, d_3, d_4 \in \mathbb{R}$ *so that*

$$w_2 = d_1u_1 + d_2u_2 + d_3u_3 + d_4w_1.$$

We have

$$\begin{aligned} w_2 &= (0,1,0,0,0) \\ &= d_1(3,1,0,0,0) + d_2(0,0,7,1,0) + d_3(0,0,0,0,1) + d_4(1,0,0,0,0) \\ &= (3d_1, d_1, 0,0,0) + (0,0,7d_2, d_2, 0) + (0,0,0,0,d_3) + (d_4, 0,0,0,0) \\ &= (3d_1 + d_4, d_1, 7d_2, d_2, d_3), \end{aligned}$$

whereupon we get the system

$$\begin{aligned} 3d_1 + d_4 &= 0 \\ d_1 &= 1 \\ 7d_2 &= 0 \\ d_2 &= 0 \\ d_3 &= 0, \end{aligned}$$

or

$$d_1 = 1$$

$$\begin{aligned} d_2 &= 0 \\ d_3 &= 0, \\ d_4 &= -3. \end{aligned}$$

Thus,

$$w_2 \in span(u_1, u_2, u_3, w_1)$$

and we remove it. We will search scalars $e_1, e_2, e_3, e_4 \in \mathbb{R}$ *so that*

$$w_3 = e_1u_1 + e_2u_2 + e_3u_3 + e_4w_1.$$

We have

$$\begin{aligned} w_3 &= (0,0,1,0,0) \\ &= e_1(3,1,0,0,0) + e_2(0,0,7,1,0) + e_3(0,0,0,0,1) + e_4(1,0,0,0,0) \\ &= (3e_1, e_1, 0,0,0) + (0,0,7e_2, e_2, 0) + (0,0,0,0,e_3) + (e_4, 0,0,0,0) \\ &= (3e_1 + e_4, e_1, 7e_2, e_2, e_3), \end{aligned}$$

whereupon we get the system

$$\begin{aligned} 3e_1 + e_4 &= 0 \\ e_1 &= 0 \\ 7e_2 &= 1 \\ e_2 &= 0 \\ e_3 &= 0, \end{aligned}$$

which is impossible. So,

$$w_3 \notin span(u_1, u_2, u_3, w_1).$$

Now, we will search scalars $f_1, f_2, f_3, f_4, f_5 \in \mathbb{R}$ *so that*

$$w_4 = f_1 u_1 + f_2 u_2 + f_3 u_3 + f_4 w_1 + f_5 w_3.$$

We have

$$\begin{aligned} w_4 &= (0,0,0,1,0) \\ &= f_1(3,1,0,0,0) + f_2(0,0,7,1,0) + f_3(0,0,0,0,1) + f_4(1,0,0,0,0) \\ &\quad + f_5(0,0,1,0,0) \\ &= (3f_1, f_1, 0,0,0) + (0,0,7f_2, f_2, 0) + (0,0,0,0,f_3) + (f_4,0,0,0,0) \\ &\quad + (0,0,f_5,0,0) \\ &= (3f_1 + f_4, f_1, 7f_2 + f_5, f_2, f_3), \end{aligned}$$

whereupon we get the system

$$\begin{aligned} 3f_1 + f_4 &= 0 \\ f_1 &= 0 \\ 7f_2 + f_5 &= 0 \\ f_2 &= 1 \\ f_3 &= 0, \end{aligned}$$

or

$$\begin{aligned} f_1 &= 0 \\ f_2 &= 1 \\ f_3 &= 0, \end{aligned}$$

$$f_4 = 0,$$

$$f_5 = -7.$$

Thus,

$$w_4 \in span(u_1, u_2, u_3, w_1, w_3)$$

and we remove it. Consequently, the list

$$u_1, u_2, u_3, w_1, w_3$$

is a basis of $\mathbb{R}^5$. *Let*

$$\begin{aligned} W &= span(w_1, w_3) \\ &= \{g_1(1,0,0,0,0) + g_2(0,0,1,0,0) : g_1, g_2 \in \mathbb{R}\}. \end{aligned}$$

We have

$$U \oplus W = \mathbb{R}^5.$$

Exercise 2.48. *Let* U *be a subspace of* $\mathbb{R}^5$ *defined by*

$$U = \{(x_1, x_2, x_3, x_4, x_5) : x_1 = 2x_5, x_2 = 7x_4\}.$$

1. *Find a basis of* U.
2. *Extend the basis from 1 to a basis of* $\mathbb{R}^5$.
3. *Find a subspace* W *of* $\mathbb{R}^5$ *so that*

$$U \oplus W = \mathbb{R}^5.$$

2.7 Dimension

We will start with the following useful lemma.

Lemma 2.49. *Any two bases of a finite-dimensional vector space have the same length.*

Proof. Suppose that V is a finite-dimensional vector space and let B_1 and B_2 be two bases of V. Then B_1 is linearly independent in V and B_2 spans V. So, the length of B_1 is at most the length of B_2. Interchanging the roles of B_1 and B_2, we obtain that the length of B_2 is at most the length of B_1. Therefore the length of B_1 equals the length of B_2. This completes the proof. □

Definition 2.50. *The dimension of a finite-dimensional vector space is the length of any basis of the vector space. It is denoted by* $\dim V$.

By Lemma 2.49, it follows that the dimension of a finite-dimensional vector space does not depend on the choice of the basis.

Example 2.51. *Consider* $\mathbb{F}^n$. *The standard basis in* $\mathbb{F}^n$ *is*

$$\begin{aligned} w_1 &= (1,0,\ldots,0), \\ w_2 &= (0,1,0,\ldots,0), \\ &\vdots \\ w_n &= (0,\ldots,0,1). \end{aligned}$$

Therefore $\dim \mathbb{F}^n = n$.

Example 2.52. *Consider* $\mathscr{P}_m(\mathbb{F})$. *A basis in it is*

$$1, z, z^2, \ldots, z^m.$$

Then $\dim \mathscr{P}_m(\mathbb{F}) = m+1$.

Theorem 2.53. *If* V *is a finite-dimensional vector space and* U *is its subspace, then* $\dim U \leq \dim V$.

Proof. Note that any basis of U is linearly independent in V and a basis of V spans V. Hence and Theorem 2.32, it follows that $\dim U \leq \dim V$. This completes the proof. □

Remark 2.54. *The real vector space* $\mathbb{R}^2$ *has dimension* 2 *and the space* $\mathbb{C}$ *has dimension* 1. *On the other hand, as a set,* $\mathbb{R}^2$ *can be identified with* $\mathbb{C}$. *Thus, when we talk about the dimension of a vector space, the role played by the choice of* $\mathbb{F}$ *cannot be neglected.*

Theorem 2.55. *Suppose that* V *is finite-dimensional. Then any linearly independent list of vectors in* V *with length* $\dim V$ *is a basis of* V.

Proof. Let $\dim V = m$ and $v_1, \ldots, v_m$ be a linearly independent list in U. By Theorem 2.43, it follows that the list $v_1, \ldots, v_m$ can be extended to a basis of V. However, every basis of V has length m. So, in this case the extension is the trivial one, meaning that no elements are adjoined to $v_1, \ldots, v_m$. In other words, $v_1, \ldots, v_m$ is a basis of V. This completes the proof. □

Example 2.56. *Consider* $\mathbb{F}^2$ *and*

$$\begin{aligned} v_1 &= (1,2), \\ v_2 &= (-1,1). \end{aligned}$$

We will search a scalar $a \in \mathbb{F}$ *so that*

$$v_1 = a v_2.$$

We have

$$\begin{aligned} (1,2) &= a(-1,1) \\ &= (-a,a), \end{aligned}$$

whereupon we get the system

$$\begin{aligned} -a &= 1 \\ a &= 2, \end{aligned}$$

which is impossible. Thus, the list v_1, v_2 *is linearly independent in* $\mathbb{F}^2$ *and since its length is* 2 *and* $\dim \mathbb{F}^2 = 2$, *we conclude that the list* v_1, v_2 *is a basis of* $\mathbb{F}^2$.

Example 2.57. *Consider* $\mathscr{P}_3(\mathbb{R})$ *and the subspace*

$$U = \{p \in \mathscr{P}_3(\mathbb{R}) : p'(1) = 0\}.$$

Let also,

$$\begin{aligned} p_1(x) &= 1, \\ p_2(x) &= (x-1)^2, \\ p_3(x) &= (x-1)^3, \quad x \in \mathbb{R}. \end{aligned}$$

Then

$$p_1'(x) = 0,$$

$$\begin{aligned} p_2'(x) &= 2(x-1), \\ p_3'(x) &= 3(x-1)^2, \quad x \in \mathbb{R}, \end{aligned}$$

and

$$\begin{aligned} p_1'(1) &= 0, \\ p_2'(1) &= 0, \\ p_3'(1) &= 0. \end{aligned}$$

Therefore $p_1, p_2, p_3 \in U$*. Assume that* $a_1, a_2, a_3 \in \mathbb{F}$ *be such that*

$$a_1 p_1(x) + a_2 p_2(x) + a_3 p_3(x) = 0, \quad x \in \mathbb{R}.$$

We have

$$\begin{aligned} 0 &= a_1 + a_2(x-1)^2 + a_3(x-1)^3 \\ &= a_1 + a_2(x^2 - 2x + 1) + a_3(x^3 - 3x^2 + 3x - 1) \\ &= a_1 + a_2x^2 - 2a_2x + a_2 + a_3x^3 - 3a_3x^2 + 3a_3x - a_3 \\ &= a_3x^3 + (a_2 - 3a_3)x^2 + (3a_3 - 2a_2)x + a_1 + a_2 - a_3. \end{aligned}$$

Thus, we get the system

$$\begin{aligned} a_3 &= 0 \\ a_2 - 3a_3 &= 0 \\ 3a_3 - 2a_2 &= 0 \\ a_1 + a_2 - a_3 &= 0, \end{aligned}$$

or

$$a_1 = 0$$

$$a_2 = 0$$

$$a_3 = 0.$$

Therefore the list p_1, p_2, p_3 *is linearly independent in* U. *Hence,* $\dim U \geq 3$. *On the other hand* $\dim \mathscr{P}_3(\mathbb{F}) = 4$ *and then* $\dim U \leq 4$. *If we assume that* $\dim U = 4$, *then when we extend a basis of* U *to a basis of* $\mathscr{P}_3(\mathbb{F})$ *we will get a list with a length greater than 4. This is impossible. Consequently* $\dim U = 3$. *By Theorem 2.60, it follows that the list* 1, $(x-1)^2$, $(x-1)^3$ *is a basis of* U.

Exercise 2.58. *Consider* $\mathscr{P}_2(\mathbb{F})$ *and*

$$U = \{p \in \mathscr{P}_2(\mathbb{F}) : p(2) = 0\}.$$

1. *Find a basis of* U.
2. *Extend the basis in 1 to a basis of* $\mathscr{P}_2(\mathbb{F})$.
3. *Find a subspace* W *of* $\mathscr{P}_2(\mathbb{F})$ *so that*

$$U \oplus W = \mathscr{P}_2(\mathbb{F}).$$

Theorem 2.59. *Suppose that* V *is finite-dimensional. Then any spanning list of vectors in* V *with length* $\dim V$ *is a basis of* V.

Proof. Let $\dim V = m$ and $v_1, \ldots, v_m$ be a spanning list in U. By Theorem 2.39, it follows that the list $v_1, \ldots, v_m$ can be extended to a basis of V. However, every basis of V has length m. So, in this case the extension is the trivial one, meaning that no elements are adjoined to $v_1, \ldots, v_m$. In other words, $v_1, \ldots, v_m$ is a basis of V. This completes the proof. □

Theorem 2.60. *Suppose that* V *is finite-dimensional and* U *be its subspace so that* $\dim U = \dim V$. *Then* $U = V$.

Proof. Let $\dim U \dim V = m$ and $v_1, \ldots, v_m$ be a basis of U. By Theorem 2.43, it follows that the list $v_1, \ldots, v_m$ can be extended to a basis of V. However, every basis of V has length m. So, in this case the extension is the trivial one, meaning that no elements are adjoined to $v_1, \ldots, v_m$. In other words, $v_1, \ldots, v_m$ is a basis of V. Therefore $U = V$. This completes the proof. □

Theorem 2.61. *Let U_1 and U_2 be subspaces of a finite-dimensional vector space. Then*

$$\dim(U_1+U_2)=\dim U_1+\dim U_2-\dim(U_1\cap U_2).$$

Proof. Let $u_1,\ldots,u_m$ be a basis of $U_1\cap U_2$. Thus,

$$\dim(U_1\cap U_2)=m.$$

Since $u_1,\ldots,u_m$ is a basis of $U_1\cap U_2$, it is linearly independent in U_1 and linearly independent in U_2. Therefore it can be extended to a basis

$$u_1,\ldots,u_m,v_1,\ldots v_j \tag{2.12}$$

of U_1 and to a basis

$$u_1,\ldots,u_m,w_1,\ldots,w_k \tag{2.13}$$

of U_2. Then

$$\dim U_1=m+j$$

and

$$\dim U_2=m+k.$$

Consider the list

$$u_1,\ldots,u_m,v_1,\ldots,v_j,w_1,\ldots,w_k. \tag{2.14}$$

Note that

$$\operatorname{span}(u_1,\ldots,u_m,v_1,\ldots,v_j,w_1,\ldots,w_k)$$

contains U_1 and U_2 and thus,

$$\operatorname{span}(u_1,\ldots,u_m,v_1,\ldots,v_j,w_1,\ldots,w_k)=U_1+U_2. \tag{2.15}$$

Let $a_1,\ldots,a_m,b_1,\ldots,b_j,c_1,\ldots,c_k\in\mathscr{F}$ be such that

$$a_1u_1+\cdots+a_mu_m+b_1v_1+\cdots+b_jv_j+c_1w_1+\cdots+c_kw_k=0. \tag{2.16}$$

By the last equation, we find

$$c_1w_1+\cdots+c_kw_k=-a_1u_1-\cdots-a_mu_m-b_1v_1-\cdots-b_jv_j.$$

Since

$$-a_1u_1-\cdots-a_mu_m-b_1v_1-\cdots-b_jv_j\in U_1,$$

by the last equation, we conclude that

$$c_1w_1+\cdots+c_kw_k\in U_1.$$

Because $w_1,\ldots,w_k \in U_2$ and U_2 is a subspace, we find

$$c_1w_1+\cdots+c_kw_k \in U_2$$

and then

$$c_1w_1+\cdots+c_kw_k \in U_1\cap U_2.$$

From here, we conclude that there are $d_1,\ldots,d_m \in \mathbb{F}$ so that

$$c_1w_1+\cdots+c_kw_k = d_1u_1+\cdots+d_mu_m,$$

or

$$d_1u_1+\cdots+d_mu_m-c_1w_1-\cdots-c_kw_k=0.$$

Because the list (2.13) is a basis of U_2, the last equality is true if

$$\begin{aligned} c_1 &= 0 \\ &\vdots \\ c_k &= 0 \\ d_1 &= 0 \\ &\vdots \\ d_m &= 0. \end{aligned}$$

From here and from (2.16), we obtain

$$a_1u_1+\cdots+a_mu_m+b_1v_1+\cdots+b_jv_j=0.$$

Since the list (2.12) is a basis of U_1, it is linearly independent in U_1 and

$$\begin{aligned} a_1 &= 0 \\ &\vdots \\ a_m &= 0 \end{aligned}$$

$$\begin{aligned} b_1 &= 0 \\ &\vdots \\ b_j &= 0. \end{aligned}$$

Now, using (2.16), we conclude that the list (2.14) is linearly independent in $U_1 + U_2$. Hence and (2.15), we obtain that the list (2.14) is a basis of $U_1 + U_2$. Therefore

$$\begin{aligned} \dim(U_1 + U_2) &= m + j + k \\ &= (m + j) + (m + k) - m \\ &= \dim U_1 + \dim U_2 - \dim(U_1 \cap U_2). \end{aligned}$$

This completes the proof. □

2.8 Advanced Practical Problems

Problem 2.62. *Consider* $\mathbb{F}^4$ *and*

$$\begin{aligned} v_1 &= (0, -1, 1, -1), \\ v_2 &= (1, -2, 0, 1), \\ v_3 &= (3, 1, -1, 2), \\ v_4 &= (-1, 0, 3, 1). \end{aligned}$$

1. *Check if* v_1 *is a linear combination of* v_3 *and* v_4.
2. *Check if* v_3 *is a linear combination of* v_1 *and* v_2.
3. *Check if* v_2 *is a linear combination of* v_3 *and* v_4.
4. *Check if* v_4 *is a linear combination of* v_1, v_2 *and* v_3.

5. *Check if* v_2 *is a linear combination of* v_1, v_3 *and* v_4.

Problem 2.63. *Consider* $\mathbb{F}^2$. *Check if*

$$(-1,7) \in span((1,2),(-1,1)).$$

Problem 2.64. *Consider* $\mathbb{F}^5$. *Check if*

$$(0,0,1,0,-4) \in span((-1,0,1,1,1),(-2,0,1,3,2),(0,0,0,-1,4)).$$

Problem 2.65. *Let*

$$\begin{aligned} p_1(z) &= -12+37z+42z^2+8z^3, \\ p_2(z) &= -8-28z-22z^2+7z^3+6z^4, \\ p_3(z) &= 4+10z+5z^2+2z^3, \quad z \in \mathbb{F}. \end{aligned}$$

Find

1. $p_1(z)+p_2(z)$, $z \in \mathbb{F}$.
2. $p_1(z)+p_3(z)$, $z \in \mathbb{F}$.
3. $p_2(z)+p_3(z)$, $z \in \mathbb{F}$.
4. $p_1(z)+p_2(z)+p_3(z)$, $z \in \mathbb{F}$.
5. $-3p_1(z)$, $z \in \mathbb{F}$.
6. $4p_2(z)$, $z \in \mathbb{F}$.
7. $-p_3(z)$, $z \in \mathbb{F}$.
8. $-3p_1(z)+p_2(z)$, $z \in \mathbb{F}$.
9. $4p_2(z)-p_3(z)$, $z \in \mathbb{F}$.
10. $-3p_1(z)+4p_2(z)-p_3(z)$, $z \in \mathbb{F}$.

Problem 2.66. *Determine the degree of the following polynomials.*

1. $p(z) = 5z^3+18z^2-10z-8$, $z \in \mathbb{F}$.
2. $p(z) = 2z^3-5z^2-8z+20$, $z \in \mathbb{F}$.

3. $p(z) = 3z^3 - z^2 - 27z - 8,\ z \in \mathbb{F}.$

4. $p(z) = 3z^4 = 5z^3 - 9z^2 - 9z + 10,\ z \in \mathbb{F}.$

5. $p(z) = 2z^4 + 3z^3 - z^2 + 3z + 4,\ z \in \mathbb{F}.$

6. $p(z) = z^4 + 14z^3 + 71z^2 + 154z + 20,\ z \in \mathbb{F}.$

7. $p(z) = 3z^4 - 4z^3 - 49z^2 + 64z + 16,\ z \in \mathbb{F}.$

8. $p(z) = 2z^5 - 9z^4 + 8z^3 + 15z^2 - 28z + 12,\ z \in \mathbb{F}.$

9. $p(z) = 4z^5 - 5z^4 - 11z^3 + 23z^2 - 13z + 12,\ z \in \mathbb{F}.$

10. $p(z) = 3z^5 - 19z^4 + 9z^3 + 71z^2 - 84z + 20,\ z \in \mathbb{F}.$

Problem 2.67. *Consider $\mathbb{F}^3$. Check if the following lists of vectors are linearly independent or linearly dependent.*

1.

$$\begin{aligned} v_1 &= (3,0,4), \\ v_2 &= (2,4,4), \\ v_3 &= (-3,-4,2). \end{aligned}$$

2.

$$\begin{aligned} v_1 &= (2,3,-2), \\ v_2 &= (-2,-2,-3), \\ v_3 &= (6,7,4). \end{aligned}$$

3.

$$\begin{aligned} v_1 &= (0,1,1), \\ v_2 &= (3,2,-1), \\ v_3 &= (-6,-5,1). \end{aligned}$$

4.

$$\begin{aligned} v_1 &= (-3,2,3), \\ v_2 &= (-1,1,-1), \\ v_3 &= (0,4,1). \end{aligned}$$

5.

$$\begin{aligned} v_1 &= (-1,10,-1), \\ v_2 &= (3,-4,5), \\ v_3 &= (1,0,1). \end{aligned}$$

Problem 2.68. *Consider $\mathscr{P}_2(\mathbb{F})$. Check if the following lists of polynomials in $\mathscr{P}_2(\mathbb{F})$ are linearly independent or linearly dependent.*

1.

$$\begin{aligned} p_1(z) &= 1-2z, \\ p_2(z) &= (2z+9)^2, \\ p_3(z) &= (1-z)^2+z^2, \quad z \in \mathbb{F}. \end{aligned}$$

2.

$$\begin{aligned} p_1(z) &= 1-3z-2z^2, \\ p_2(z) &= 2-4z+z^2, \\ p_3(z) &= 2+3z+2z^2, \quad z \in \mathbb{F}. \end{aligned}$$

3.

$$p_1(z) = -1+2z+3z^2,$$

$$p_2(z) = z - 2z^2,$$

$$p_3(z) = -2 + z + 5z^2, \quad z \in \mathbb{F}.$$

4.

$$p_1(z) = -4 + 2z - z^2,$$

$$p_2(z) = 4 + z + 4z^2,$$

$$p_3(z) = -3 - 3z - 2z^2, \quad z \in \mathbb{F}.$$

5.

$$p_1(z) = -1 + 2z + z^2,$$

$$p_2(z) = -2 - z + z^2,$$

$$p_3(z) = -7 + 4z + 5z^2, \quad z \in \mathbb{F}.$$

Problem 2.69. *Let v_1, v_2, v_3, v_4 spans V. Prove that*

$$v_1 - 2v_2, \quad v_2 - 3v_3, \quad v_3 - 4v_4, \quad v_4$$

spans V.

Problem 2.70. *Let v_1, v_2, v_3, v_4, v_5 spans V. Prove that*

$$v_1 - 4v_2, \quad v_2 - 5v_3, \quad v_3 - 7v_4, \quad v_4 + 2v_5, \quad v_5$$

spans V.

Problem 2.71. *Prove that if we consider $\mathbb{C}$ as a vector space over $\mathbb{R}$, then the list*

$$(2 + 3i, 2 - 3i)$$

is linearly independent.

Problem 2.72. *Prove that if we consider $\mathbb{C}$ as a vector space over $\mathbb{C}$, then the list*

$$(3+i, 3-i)$$

is linearly dependent.

Problem 2.73. *Let v_1, v_2, v_3, v_4 be a linearly independent list in V. Prove that the list*

$$v_1 - v_2, \quad v_2 - v_3, \quad v_3 - v_4, \quad v_4$$

is linearly independent.

Problem 2.74. *Prove or give a counterexample: if $v_1, \ldots, v_m$ is a linearly independent list of vectors in V, then the list of vectors*

$$3v_1 - v_2, \quad v_2, \quad \ldots, \quad v_m$$

is linearly independent.

Problem 2.75. *Prove or give a counterexample: if $v_1, \ldots, v_m$ is a linearly independent list of vectors in V and $\lambda \in \mathbb{F}$, $\lambda \neq 0$, then the list of vectors*

$$\lambda v_1, \quad \lambda v_2, \quad \ldots, \quad \lambda v_m$$

is linearly independent.

Problem 2.76. *Prove or give a counterexample: if $v_1, \ldots, v_m$ and $w_1, \ldots, w_m$ are linearly independent, then the list of vectors*

$$v_1 + w_1, \quad v_2 + w_2, \quad \ldots, \quad v_m + w_m$$

is linearly independent.

Problem 2.77. *Let $v_1, \ldots, v_m$ be a linearly independent list of vectors in V and $w \in V$. If*

$$v_1 + w, \quad v_2 + w, \quad \ldots, v_m + w$$

is linearly dependent, prove that

$$w \in span(v_1, \ldots, v_m).$$

Problem 2.78. *Let $v_1, \ldots, v_m$ be a linearly independent list of vectors in V and $w \in V$. Prove that*

$$v_1 + w, \quad v_2 + w, \quad \ldots, v_m + w$$

is linearly independent if and only if

$$w \notin span(v_1, \ldots, v_m).$$

Problem 2.79. *Consider* $\mathbb{F}^m$ *and*

$$\begin{aligned} v_1 &= (1,0,\ldots,0), \\ &\vdots \\ v_m &= (0,0,\ldots,0,1). \end{aligned}$$

Prove that the list $v_1,\ldots,v_m$ *is a basis of* $\mathbb{F}^m$.

Problem 2.80. *Consider*

$$V = \{(x,x,y) \in \mathbb{F}^3 : x,y \in \mathbb{F}\}$$

and

$$\begin{aligned} v_1 &= (1,1,0), \\ v_2 &= (0,0,1). \end{aligned}$$

Prove that the list v_1,v_2 *is a basis of* V.

Problem 2.81. *Consider*

$$V = \{(x,y,z) \in \mathbb{F}^3 : x+y+z=0\}$$

and

$$\begin{aligned} v_1 &= (1,-1,0), \\ v_2 &= (1,0,-1). \end{aligned}$$

Prove that the list v_1,v_2 *is a basis of* V.

Problem 2.82. *Prove that the list* $1,z,\ldots,z^m$ *is a basis of* $\mathscr{P}_m(\mathbb{F})$.

Problem 2.83. *Consider* $\mathbb{F}^5$ *and*

$$\begin{aligned} v_1 &= (1,1,1,1,0), \\ v_2 &= (1,1,-1,-1,-1), \end{aligned}$$

$$v_3 = (2,2,0,0,-1),$$

$$v_4 = (1,1,5,5,2),$$

$$v_5 = (1,-1,-1,0,0).$$

Find a basis of $\mathbb{F}^5$.

Problem 2.84. *Consider* $\mathbb{F}^4$. *Extend the list*

$$v_1 = (1,-2,4),$$

$$v_2 = (-1,4,8),$$

$$v_3 = (0,-1,-2)$$

to a basis of $\mathbb{F}^4$.

Problem 2.85. *Find a basis of*

$$span(u_1,u_2,u_3),$$

where

1.

$$u_1 = (1,2,1),$$

$$u_2 = (1,1,-1),$$

$$u_3 = (1,3,3).$$

2.

$$u_1 = (2,3,-1),$$

$$u_2 = (1,1,-3),$$

$$u_3 = (1,2,2).$$

3.

$$\begin{aligned} u_1 &= (1,2,1,-2), \\ u_2 &= (2,3,2,0), \\ u_3 &= (1,2,2,-3). \end{aligned}$$

4.

$$\begin{aligned} u_1 &= (1,1,1,1), \\ u_2 &= (1,0,1,-1), \\ u_3 &= (1,3,0,-4). \end{aligned}$$

5.

$$\begin{aligned} u_1 &= (1,1,0,0), \\ u_2 &= (0,1,1,0), \\ u_3 &= (0,0,1,1). \end{aligned}$$

Problem 2.86. *Suppose that v_1, v_2, v_3, v_3 is a basis of V. Prove that*

$$v_1 + v_2, \quad v_2 + v_3, \quad v_3 + v_4, \quad v_4$$

is a basis of V.

Problem 2.87. *Suppose that U and W are subspaces of V such that*

$$U \oplus W = V.$$

Suppose also that $u_1, \ldots, u_m$ is a basis of U and $w_1, \ldots, w_n$ is a basis of W. Prove that

$$u_1, \ldots, u_m, w_1, \ldots, w_n$$

is a basis of V.

Problem 2.88. *Consider* $\mathscr{P}_4(\mathbb{F})$ *and*

$$U = \{p \in \mathscr{P}_4(\mathbb{F}) : p''(2) = 0\}.$$

1. *Find a basis of* U.
2. *Extend the basis in 1 to a basis of* $\mathscr{P}_4(\mathbb{F})$.
3. *Find a subspace* W *of* $\mathscr{P}_4(\mathbb{F})$ *so that*

$$U \oplus W = \mathscr{P}_4(\mathbb{F}).$$

Problem 2.89. *Consider* $\mathscr{P}_4(\mathbb{F})$ *and*

$$U = \{p \in \mathscr{P}_4(\mathbb{F}) : p(1) = p(2)\}.$$

1. *Find a basis of* U.
2. *Extend the basis in 1 to a basis of* $\mathscr{P}_4(\mathbb{F})$.
3. *Find a subspace* W *of* $\mathscr{P}_4(\mathbb{F})$ *so that*

$$U \oplus W = \mathscr{P}_4(\mathbb{F}).$$

Problem 2.90. *Consider* $\mathscr{P}_4(\mathbb{F})$ *and*

$$U = \{p \in \mathscr{P}_4(\mathbb{F}) : p(1) = p(2) = p(3)\}.$$

1. *Find a basis of* U.
2. *Extend the basis in 1 to a basis of* $\mathscr{P}_4(\mathbb{F})$.
3. *Find a subspace* W *of* $\mathscr{P}_4(\mathbb{F})$ *so that*

$$U \oplus W = \mathscr{P}_4(\mathbb{F}).$$

Problem 2.91. *Suppose that* U *and* W *are subspaces of* $\mathbb{R}^8$ *such that* $\dim U = 3$, $\dim W = 5$ *and* $U + W = \mathbb{R}^8$. *Prove that*

$$U \oplus W = \mathbb{R}^8.$$

Chapter 3

Linear Maps

3.1 Definition and Examples of Linear Maps

Suppose that U, V, W and Y are vector spaces.

Definition 3.1. *A linear map from V to W is a function $T: V \to W$ with the following properties:*

1. *additivity*
$$T(u+v) = T(u) + T(v)$$
for all $u, v \in V$.

2. *homogeneity*
$$T(\lambda u) = \lambda T(u)$$
for all $\lambda \in \mathbb{F}$ and for all $u \in V$. The set of all linear maps from U to V will be denoted by $\mathscr{L}(U,V)$. For a linear map T from U to V we will write $T \in \mathscr{L}(U,V)$.

Some mathematicians use the term linear transformation, which means the same as linear map. Note that for linear maps we often use the notation Tv as well as the more standard functional notation $T(v)$.

Example 3.2. *Let $T: \mathbb{R}^3 \to \mathbb{R}^2$ be defined by*

$$T(x,y,z) = (2x - y + 3z, 7x + 5y - 6z), \quad (x,y,z) \in \mathbb{R}^3.$$

We will prove that $T \in \mathscr{L}(\mathbb{R}^3, \mathbb{R}^2)$. *Take*

$$(x_1, y_1, z_1), \quad (x_2, y_2, z_2) \in \mathbb{R}^3$$

and $\lambda \in \mathbb{R}$ *arbitrarily. Then*

$$T(x_1, y_1, z_1) = (2x_1 - y_1 + 3z_1, 7x_1 + 5y_1 - 6z_1),$$

$$T(x_2, y_2, z_2) = (2x_2 - y_2 + 3z_2, 7x_2 + 5y_2 - 6z_2)$$

and

$$\begin{aligned} T(x_1, y_1, z_1) + T(x_2, y_2, z_2) &= (2x_1 - y_1 + 3z_1, 7x_1 + 5y_1 - 6z_1) \\ &\quad + (2x_2 - y_2 + 3z_2, 7x_2 + 5y_2 - 6z_2) \\ &= (2x_1 - y_1 + 3z_1 + 2x_2 - y_2 + 3z_2, 7x_1 + 5y_1 - 6z_1 + 7x_2 + 5y_2 - 6z_2) \\ &= (2(x_1 + x_2) - (y_1 + y_2) + 3(z_1 + z_2), 7(x_1 + x_2) + 5(y_1 + y_2) - 6(z_1 + z_2)), \end{aligned} \tag{3.1}$$

and

$$\begin{aligned} \lambda T(x_1, y_1, z_1) &= \lambda(2x_1 - y_1 + 3z_1, 7x_1 + 5y_1 - 6z_1) \\ &= (\lambda(2x_1 - y_1 + 3z_1), \lambda(7x_1 + 5y_1 - 6z_1)) \qquad (3.2) \\ &= (2\lambda x_1 - \lambda y_1 + 3\lambda z_1, 7\lambda x_1 + 5\lambda y_1 - 6\lambda z_1). \end{aligned}$$

Next,

$$(x_1, y_1, z_1) + (x_2, y_2, z_2) = (x_1 + x_2, y_1 + y_2, z_1 + z_2)$$

and

$$\begin{aligned} T((x_1, y_1, z_1) + (x_2, y_2, z_2)) &= T(x_1 + x_2, y_1 + y_2, z_1 + z_2) \\ &= (2(x_1 + x_2) - (y_1 + y_2) + 3(z_1 + z_2), 7(x_1 + x_2) + 5(y_1 + y_2) - 6(z_1 + z_2)). \end{aligned}$$

By the last equation and (3.1), *we find*

$$T((x_1, y_1, z_1) + (x_2, y_2, z_2)) = T(x_1, y_1, z_1) + T(x_2, y_2, z_2). \tag{3.3}$$

Moreover,

$$\lambda(x_1, y_1, z_1) = (\lambda x_1, \lambda y_1, \lambda z_1)$$

and

$$\begin{aligned} T(\lambda(x_1,y_1,z_1)) &= T(\lambda x_1,\lambda y_1,\lambda z_1) \\ &= (2(\lambda x_1)-\lambda y_1+3(\lambda z_1),7(\lambda x_1)+5(\lambda y_1)-6(\lambda z_1)) \\ &= (2\lambda x_1-\lambda y_1+3\lambda z_1,7\lambda x_1+5\lambda y_1-6\lambda z_1). \end{aligned}$$

From the last equation and (3.2)*, we find*

$$T(\lambda(x_1,y_1,z_1)) = \lambda T(x_1,y_1,z_1).$$

Hence and (3.3)*, we conclude that* $T \in \mathscr{L}(\mathbb{R}^3,\mathbb{R}^2)$.

Example 3.3. *Let* $T : \mathscr{P}(\mathbb{F}) \to \mathscr{P}(\mathbb{F})$ *be defined by*

$$Tp(x) = x^2 p(x), \quad x \in \mathbb{R}, \quad p \in \mathscr{P}(\mathbb{F}).$$

We will prove that $T \in \mathscr{L}(\mathscr{P}(\mathbb{F}),\mathscr{P}(\mathbb{F}))$*. Take* $p_1, p_2 \in \mathscr{P}(\mathbb{F})$ *and* $\lambda \in \mathbb{F}$ *arbitrarily. Then, using that* $\mathscr{P}(\mathbb{F})$ *is a vector space, we have that* $p_1 + p_2 \in \mathscr{P}(\mathbb{F})$ *and* $\lambda p_1 \in \mathscr{P}(\mathbb{F})$*. We have*

$$\begin{aligned} Tp_1(x) &= x^2 p_1(x), \\ Tp_2(x) &= x^2 p_2(x), \quad x \in \mathbb{R}, \end{aligned}$$

whereupon

$$\begin{aligned} Tp_1(x)+Tp_2(x) &= x^2 p_1(x)+x^2 p_2(x) \\ &= x^2(p_1(x)+p_2(x)), \quad x \in \mathbb{R}. \end{aligned} \tag{3.4}$$

Note that

$$T(p_1(x)+p_2(x)) = x^2(p_1(x)+p_2(x)), \quad x \in \mathbb{R}.$$

By the last equation and (3.4)*, we find*

$$T(p_1(x)+p_2(x)) = Tp_1(x)+Tp_2(x), \quad x \in \mathbb{R}. \tag{3.5}$$

Next,

$$\lambda Tp_1(x) = \lambda x^2 p_1(x), \quad x \in \mathbb{R}, \tag{3.6}$$

and

$$\begin{aligned} T(\lambda p_1(x)) &= x^2(\lambda p_1(x)) \\ &= \lambda x^2 p_1(x), \quad x \in \mathbb{R}. \end{aligned}$$

From the last equation and (3.6)*, we obtain*

$$T(\lambda p_1(x)) = \lambda T p_1(x), \quad x \in \mathbb{R}.$$

Hence and (3.5)*, we conclude that* $T \in \mathscr{L}(\mathscr{P}(\mathbb{F}), \mathscr{P}(\mathbb{F}))$.

Example 3.4. *Let* $T : \mathbb{F}^\infty \to \mathbb{F}^\infty$ *be defined by*

$$T(x_1, x_2, x_3, \dots) = (x_2, x_3, \dots), \quad (x_1, x_2, x_3, \dots) \in \mathbb{F}^\infty.$$

We will prove that $T \in \mathscr{L}(\mathbb{F}^\infty, \mathbb{F}^\infty)$*. Take*

$$y = (y_1, y_2, y_3, \dots), \quad z = (z_1, z_2, z_3, \dots) \in \mathbb{F}^\infty$$

and $\lambda \in \mathbb{F}$ *arbitrarily. Then*

$$\begin{aligned} y + z &= (y_1, y_2, y_3, \dots) + (z_1, z_2, z_3, \dots) \\ &= (y_1 + z_1, y_2 + z_2, y_3 + z_3, \dots), \\ \lambda y &= \lambda(y_1, y_2, y_3, \dots) \\ &= (\lambda y_1, \lambda y_2, \lambda y_3, \dots), \\ Ty &= (y_2, y_3, \dots), \\ Tz &= (z_2, z_3, \dots). \end{aligned}$$

Hence,

$$\begin{aligned} Ty + Tz &= (y_2, y_3, \dots) + (z_2, z_3, \dots) \\ &= (y_2 + z_2, y_3 + z_3, \dots) \end{aligned} \tag{3.7}$$

and

$$T(y + z) = (y_2 + z_2, y_3 + z_3, \dots).$$

By the last equation and (3.7), *we get*

$$T(y+z) = Ty + Tz. \tag{3.8}$$

Next,

$$\begin{aligned} \lambda Ty &= \lambda(y_2, y_3, \ldots) \\ &= (\lambda y_2, \lambda y_3, \ldots) \end{aligned} \tag{3.9}$$

and

$$T(\lambda y) = (\lambda y_2, \lambda y_3, \ldots).$$

From the last equation and (3.9), *we arrive at*

$$T(\lambda y) = \lambda y.$$

Hence and (3.7), *we conclude that* $T \in \mathscr{L}(\mathbb{F}^\infty, \mathbb{F}^\infty)$.

Exercise 3.5. *Let* $T : \mathbb{R}^3 \to \mathbb{R}^3$ *be defined by*

1. $T(x,y,z) = (6x - 5y - 4z, 3x - 2y - z, y + 2z)$, $(x,y,z) \in \mathbb{R}^3$.
2. $T(x,y,z) = (x - 5y - 4z, 3x - 2y - z, y)$, $(x,y,z) \in \mathbb{R}^3$.
3. $T(x,y,z) = (5x - 4y - 3z, 2x - y, y + z)$, $(x,y,z) \in \mathbb{R}^3$.
4. $T(x,y,z) = (x + y + z, z, y)$, $(x,y,z) \in \mathbb{R}^3$.
5. $T(x,y,z) = (x + 5y + 2z, 2x - y, 7y + 2z)$, $(x,y,z) \in \mathbb{R}^3$.

Check if $T \in \mathscr{L}(\mathbb{R}^3, \mathbb{R}^3)$.

Theorem 3.6. *Suppose that* $v_1, \ldots, v_m$ *is a basis of* V *and* $w_1, \ldots, w_m \in W$. *Then there exists a unique linear map* $T : V \to W$ *such that*

$$Tv_j = w_j.$$

Proof. Since $v_1, \ldots, v_m$ is a basis of V, any element of V is represented in a unique way as a linear combination of $v_1, \ldots, v_m$. For arbitrary constants $c_1, \ldots, c_m$, define the map

$$T(c_1 v_1 + \cdots + c_m v_m) = c_1 w_1 + \cdots + c_m w_m. \tag{3.10}$$

Since $c_1 v_1 + \cdots + c_m v_m \in V$ and W is a subspace, we have that $c_1 w_1 + \cdots + c_m w_m \in W$ and thus, $T : V \to W$. For any $j \in \{1, \ldots, m\}$, take $c_j = 1$ and

$$c_1 = \ldots = c_{j-1} = c_{j+1} = \ldots c_m = 0,$$

and we get

$$Tv_j = w_j.$$

Note that the map T is uniquely determined by the equality (3.10). Now, we will prove that $T : V \to W$ is a linear map. Let $u, z \in V$ be such that

$$\begin{aligned} u &= a_1v_1 + \cdots + a_mv_m, \\ z &= b_1v_1 + \cdots + b_mv_m, \end{aligned}$$

and $\lambda \in \mathbb{F}$. Then

$$\begin{aligned} u+v &= a_1v_1 + \cdots + a_mv_m \\ &\quad +b_1v_1 + \cdots + b_mv_m \\ &= (a_1+b_1)v_1 + \cdots + (a_m+b_m)v_m, \\ \lambda u &= \lambda(a_1v_1 + \cdots + a_mv_m) \\ &= \lambda a_1v_1 + \cdots + \lambda a_mv_m. \end{aligned}$$

Hence,

$$\begin{aligned} Tu &= T(a_1v_1 + \cdots + a_mv_m) \\ &= a_1w_1 + \cdots + a_mw_m, \\ Tv &= T(b_1v_1 + \cdots + b_mv_m) \\ &= b_1w_1 + \cdots + b_mw_m, \\ Tu+Tv &= a_1w_1 + \cdots + a_mw_m \\ &\quad +b_1w_1 + \cdots + b_mw_m \\ &= (a_1+b_1)w_1 + \cdots + (a_m+b_m)w_m \end{aligned}$$

$$= T((a_1+b_1)v_1+\cdots+(a_m+b_m)v_m)$$

$$= T(u+v),$$

and

$$\begin{aligned}\lambda Tu &= \lambda T(a_1v_1+\cdots+a_mv_m)\\ &= \lambda(a_1w_1+\cdots+a_mw_m)\\ &= (\lambda a_1)w_1+\cdots+(\lambda a_m)w_m\\ &= T((\lambda a_1)v_1+\cdots+(\lambda a_m)v_m)\\ &= T(\lambda(a_1v_1+\cdots+a_mv_m))\\ &= T(\lambda u).\end{aligned}$$

Consequently $T \in \mathscr{L}(V,W)$. This completes the proof. □

3.2 Algebraic Properties of $\mathscr{L}(V,W)$

In this section, we will define some basic operations in $\mathscr{L}(V,W)$.

Definition 3.7. *Let $S,T \in \mathscr{L}(V,W)$ and $\lambda \in \mathbb{F}$. The sum $S+T$ and the product λT are the maps from V to W defined by*

$$\begin{aligned}(S+T)(v) &= Sv+Tv,\\ (\lambda T)v &= \lambda Tv, \quad v \in V.\end{aligned}$$

Since W is a vector space and $S,T \in \mathscr{L}(V,W)$, we have $Sv+Tv \in W$ and $\lambda Tv \in W$.

Example 3.8. *Let $S,T : \mathbb{R}^3 \to \mathbb{R}$ be defined by*

$$S(x,y,z) = x+y+z,$$

$$T(x,y,z) = 2x - y - 2z, \quad (x,y,z) \in \mathbb{R}^3.$$

We will find

$$(S+T)(x,y,z), \quad (2T)(x,y,z), \quad (x,y,z) \in \mathbb{R}^3.$$

We have

$$\begin{aligned} (S+T)(x,y,z) &= S(x,y,z) + T(x,y,z) \\ &= x + y + z + 2x - y - 2z \\ &= 3x - z, \quad (x,y,z) \in \mathbb{R}^3, \end{aligned}$$

and

$$\begin{aligned} (2T)(x,y,z) &= 2T(x,y,z) \\ &= 2(2x - y - 2z) \\ &= 4x - 2y - 4z, \quad (x,y,z) \in \mathbb{R}^3. \end{aligned}$$

Exercise 3.9. *Let $S, T : \mathbb{R}^4 \to \mathbb{R}^2$ be defined by*

$$\begin{aligned} S(x,y,z,t) &= (x - y + z - t, y), \\ T(x,y,z,t) &= (-2x - y + z + 3t, x + y), \quad (x,y,z,t) \in \mathbb{R}^4. \end{aligned}$$

Find

1. $(S+T)(x,y,z,t)$, $(x,y,z,t) \in \mathbb{R}^4$.
2. $(3S)(x,y,z,t)$, $(x,y,z,t) \in \mathbb{R}^4$.
3. $(2T)(x,y,z,t)$, $(x,y,z,t) \in \mathbb{R}^4$.

Definition 3.10. *Let $S, T \in \mathscr{L}(V,W)$. For $a, b \in \mathbb{F}$, define*

$$(aS + bT)v = aSv + bTv,$$

$$(aS-bT)v = aSv+(-b)Tv, \quad v \in V.$$

Since W is a vector space, we have $aSv \pm bTv \in W$, $v \in V$, *and thus,* $aS \pm bT : V \to W$.

Example 3.11. *Let* $S,T : \mathbb{R}^2 \to \mathbb{R}$ *be defined by*

$$S(x,y) = 2x-y,$$

$$T(x,y) = x+y, \quad (x,y) \in \mathbb{R}^2.$$

We have

$$\begin{aligned}(S+T)(x,y) &= S(x,y)+T(x,y)\\ &= 2x-y+x+y\\ &= 3x, \quad (x,y) \in \mathbb{R}^2,\end{aligned}$$

and

$$\begin{aligned}(3S)(x,y) &= 3S(x,y)\\ &= 3(2x-y)\\ &= 6x-3y,\\ (2T)(x,y) &= 2T(x,y)\\ &= 2(x+y)\\ &= 2x+2y, \quad (x,y) \in \mathbb{R}^2.\end{aligned}$$

Hence,

$$\begin{aligned}(3S+2T)(x,y) &= (3S)(x,y)+(2T)(x,y)\\ &= 6x-3y+2x+2y\end{aligned}$$

$$= \quad 8x - y, \quad (x,y) \in \mathbb{R}^2.$$

Next,

$$\begin{aligned}(-(2T))(x,y) &= (-1)(2T)(x,y)\\ &= (-1)(2x+2y)\\ &= -2x-2y, \quad (x,y) \in \mathbb{R}^2,\end{aligned}$$

and

$$\begin{aligned}(3S-2T)(x,y) &= (3S)(x,y)+(-(2T))(x,y)\\ &= 6x-3y-2x-y\\ &= 4x-4y, \quad (x,y) \in \mathbb{R}^2.\end{aligned}$$

Exercise 3.12. *Let $S,T : \mathbb{R}^3 \to \mathbb{R}$ be defined by*

$$\begin{aligned}S(x,y,z) &= 3x-y+z,\\ T(x,y,z) &= x+y-4z, \quad (x,y,z) \in \mathbb{R}^3.\end{aligned}$$

Find

1. $(S+T)(x,y,z)$, $(x,y,z) \in \mathbb{R}^3$.
2. $(2S)(x,y,z)$, $(x,y,z) \in \mathbb{R}^3$.
3. $(4T)(x,y,z)$, $(x,y,z) \in \mathbb{R}^3$.
4. $(S-T)(x,y,z)$, $(x,y,z) \in \mathbb{R}^3$.
5. $(2S-4T)(x,y,z)$, $(x,y,z) \in \mathbb{R}^3$.

Theorem 3.13. *Let $S,T \in \mathscr{L}(V,W)$. For any $a,b \in \mathbb{F}$, we have*

$$aS+bT, \quad aS-bT \in \mathscr{L}(V,W).$$

Proof. Let $v_1, v_2 \in V$ and $c \in \mathbb{F}$ be arbitrarily chosen. Then

$$\begin{aligned} S(v_1+v_2) &= Sv_1+Sv_2, \\ T(v_1+v_2) &= Tv_1+Tv_2, \\ S(cv_1) &= cSv_1, \\ T(cv_1) &= cTv_1. \end{aligned}$$

Hence,

$$\begin{aligned} (aS+bT)(v_1+v_2) &= aS(v_1+v_2)+bT(v_1+v_2) \\ &= a(Sv_1+Sv_2)+b(Tv_1+Tv_2) \\ &= aSv_1+aSv_2+bTv_1+bTv_2 \\ &= (aSv_1+bTv_1)+(aSv_2+bTv_2) \\ &= (aS+bT)v_1+(aS+bT)v_2 \end{aligned}$$

and

$$\begin{aligned} (aS+bT)(cv_1) &= aS(cv_1)+bT(cv_1) \\ &= a(cSv_1)+b(cTv_1) \\ &= acSv_1+bcTv_1 \\ &= caSv_1+cbTv_1 \\ &= c(aSv_1+bTv_1) \\ &= c(aS+bT)v_1. \end{aligned}$$

Therefore $aS+bT \in \mathscr{L}(V,W)$. Next,

$$\begin{aligned}(aS-bT)(v_1+v_2) &= aS(v_1+v_2)-bT(v_1+v_2)\\ &= a(Sv_1+Sv_2)-b(Tv_1+Tv_2)\\ &= aSv_1+aSv_2-bTv_1-bTv_2\\ &= (aSv_1-bTv_1)+(aSv_2-bTv_2)\\ &= (aS-bT)v_1+(aS-bT)v_2\end{aligned}$$

and

$$\begin{aligned}(aS-bT)(cv_1) &= aS(cv_1)-bT(cv_1)\\ &= a(cSv_1)-b(cTv_1)\\ &= acSv_1-bcTv_1\\ &= caSv_1-cbTv_1\\ &= c(aSv_1-bTv_1)\\ &= c(aS-bT)v_1.\end{aligned}$$

Consequently $aS-bT \in \mathscr{L}(V,W)$. This completes the proof. □

Definition 3.14. *Define the zero linear map from V to W as follows:*

$$Ov = 0, \quad v \in V.$$

Definition 3.15. *Define the identity linear map from V to W as follows:*

$$Iv = v, \quad v \in V.$$

Theorem 3.16. *$\mathscr{L}(V,W)$ is a vector space.*

Proof. Suppose that $S,T,K \in \mathscr{L}(V,W)$, $a,b \in \mathbb{F}$ and $v \in V$ be arbitrarily chosen and fixed. Then, we have the following.

1.
$$\begin{aligned}(S+T)v &= Sv+Tv\\ &= Tv+Sv\\ &= (T+S)v,\end{aligned}$$
i.e.,
$$S+T=T+S.$$

2.
$$\begin{aligned}(S+(T+K))v &= Sv+(T+K)v\\ &= Sv+(Tv+Kv)\\ &= (Sv+Tv)+Kv\\ &= (S+T)v+Kv\\ &= ((S+T)+K)v,\end{aligned}$$
i.e.,
$$S+(T+K)=(S+T)+K.$$

3.
$$\begin{aligned}(a(S+T))v &= a(S+T)v\\ &= a(Sv+Tv)\\ &= aSv+aTv\\ &= (aS)v+(aT)v\end{aligned}$$

$$= (aS+aT)v,$$

i.e.,

$$a(S+T) = aS+aT.$$

4.

$$\begin{aligned}((a+b)S)v &= (a+b)Sv \\ &= aSv+bSv \\ &= (aS)v+(bS)v \\ &= (aS+bS)v,\end{aligned}$$

i.e.,

$$(a+b)S = aS+bS.$$

5.

$$\begin{aligned}(S+(-S))v &= Sv+(-S)v \\ &= Sv-Sv \\ &= 0 \\ &= Ov,\end{aligned}$$

i.e.,

$$S+(-S) = O.$$

6.

$$\begin{aligned}(S+O)v &= Sv+Ov \\ &= Sv+0\end{aligned}$$

$$= \quad Sv,$$

i.e.,

$$S+O=S.$$

7.

$$\begin{aligned}(1S)v &= 1Sv \\ &= Sv,\end{aligned}$$

i.e.,

$$1S=S.$$

Therefore $\mathscr{L}(V,W)$ is a vector space. This completes the proof. □

Definition 3.17. *Let $T\in\mathscr{L}(V,U)$ and $S\in\mathscr{L}(U,W)$. Then the product $ST\in\mathscr{L}(V,W)$ is defined by*

$$(ST)u=S(Tu), \quad u\in V.$$

Theorem 3.18. *Let $T\in\mathscr{L}(V,U)$ and $S\in\mathscr{L}(U,W)$. Then $ST\in\mathscr{L}(V,W)$.*

Proof. Let $v_1,v_2\in V$ and $a\in\mathbb{F}$ be arbitrarily chosen. Then

$$\begin{aligned}T(v_1+v_2) &= Tv_1+Tv_2, \\ S(Tv_1+Tv_2) &= S(Tv_1)+S(Tv_2)\end{aligned}$$

and

$$\begin{aligned}T(av_1) &= aTv_1, \\ S(aTv_1) &= aS(Tv_1).\end{aligned}$$

Hence,

$$\begin{aligned}(ST)(v_1+v_2) &= S(T(v_1+v_2)) \\ &= S(Tv_1+Tv_2)\end{aligned}$$

$$\begin{aligned} &= S(Tv_1)+S(Tv_2) \\ &= (ST)v_1+(ST)v_2 \end{aligned}$$

and

$$\begin{aligned} (ST)(av_1) &= S(T(av_1)) \\ &= S(aTv_1) \\ &= aS(Tv_1) \\ &= a(ST)(v_1). \end{aligned}$$

Consequently $ST \in \mathscr{L}(V,W)$. This completes the proof. □

Example 3.19. *Let $S, T : \mathbb{R} \to \mathbb{R}$ be defined by*

$$\begin{aligned} Sx &= 2x+3, \\ Tx &= x-4, \quad x \in \mathbb{R}. \end{aligned}$$

Then

$$\begin{aligned} (ST)x &= S(Tx) \\ &= S(x-4) \\ &= 2(x-4)+3 \\ &= 2x-8+3 \\ &= 2x-5, \quad x \in \mathbb{R}, \end{aligned}$$

and

$$(TS)x = T(Sx)$$

$$
\begin{aligned}
&= T(2x+3)\\
&= 2x+3-4\\
&= 2x-1, \quad x\in\mathbb{R}.
\end{aligned}
$$

Remark 3.20. *By the above example, it follows that in the general case we have*

$$ST \neq TS.$$

Example 3.21. *Let $T:\mathbb{R}^3\to\mathbb{R}^2$ and $S:\mathbb{R}^2\to\mathbb{R}^2$ be defined by*

$$
\begin{aligned}
T(x,y,z) &= (x+2z, y-3z), \quad (x,y,z)\in\mathbb{R}^3,\\
S(x,y) &= (2x-y, x+y), \quad (x,y)\in\mathbb{R}^2.
\end{aligned}
$$

We will find

$$(ST)(x,y,z), \quad (x,y,z)\in\mathbb{R}^3.$$

We have

$$
\begin{aligned}
(ST)(x,y,z) &= S(T(x,y,z))\\
&= S(x+2z, y-3z)\\
&= (2(x+2z)-(y-3z), x+2z+y-3z)\\
&= (2x+4z-y+3z, x+y-z)\\
&= (2x-y+7z, x+y-z), \quad (x,y,z)\in\mathbb{R}^3.
\end{aligned}
$$

Exercise 3.22. *Let $\mathbb{T}:\mathbb{R}^4\to\mathbb{R}^4$ and $S:\mathbb{R}^4\to\mathbb{R}^2$ be defined by*

$$
\begin{aligned}
T(x,y,z,t) &= (x-1-2y+3z, y+z, z-t, 3t),\\
S(x,y,z,t) &= (2x-3y, y-2t), \quad (x,y,z,t)\in\mathbb{R}^4.
\end{aligned}
$$

Find

1. $(ST)(x,y,z,t)$, $(x,y,z,t) \in \mathbb{R}^4$.

2. $(2T)(x,y,z,t)$, $(x,y,z,t) \in \mathbb{R}^4$.

3. $(ST-4T)(x,y,z,t)$, $(x,y,z,t) \in \mathbb{R}^4$.

4. $((3S)T)(x,y,z,t)$, $(x,y,z,t) \in \mathbb{R}^4$.

5. $(S(-4T))(x,y,z,t)$, $(x,y,z,t) \in \mathbb{R}^4$.

Theorem 3.23. *Let* $T_1 \in \mathscr{L}(U,V)$, $T_2 \in \mathscr{L}(W,U)$, $T_3 \in \mathscr{L}(X,W)$. *Then*

$$(T_1T_2)T_3 = T_1(T_2T_3). \tag{3.11}$$

Proof. Let $v \in X$ be arbitrarily chosen. Then, we have the following.

$$\begin{aligned}(T_1T_2)T_3v &= (T_1T_2)(T_3v)\\ &= T_1(T_2(T_3v))\\ &= T_1((T_2T_3)v)\\ &= T_1(T_2T_3)v.\end{aligned}$$

Because $v \in X$ was arbitrarily chosen, we get (3.11). This completes the proof. □

Theorem 3.24. . *Let* $T \in \mathscr{L}(V,W)$. *Then*

$$IT = TI = T, \tag{3.12}$$

where the first identity is on W *and the second identity is on* V.

Proof. Let I be the identity on W. Take $v \in V$ arbitrarily. Then

$$\begin{aligned}ITv &= I(Tv)\\ &= Tv.\end{aligned}$$

Because $v \in V$ was arbitrarily chosen, we get

$$IT = T. \tag{3.13}$$

Now, suppose that I is the identity on V. Let $u \in V$ be arbitrarily chosen and fixed. Then

$$\begin{aligned} TIu &= T(Iu) \\ &= Tu. \end{aligned}$$

Because $u \in V$ was arbitrarily chosen, we find

$$TI = T.$$

By the last equality and by (3.13), we find (3.12). This completes the proof. □

Theorem 3.25. *Let $T_1, T_2 \in \mathscr{L}(U,V)$, $S \in \mathscr{L}(V,W)$. Then*

$$S(T_1 + T_2) = ST_1 + ST_2. \tag{3.14}$$

Proof. Let $u \in V$ be arbitrarily chosen. Then

$$\begin{aligned} S(T_1 + T_2)u &= S((T_1 + T_2)u) \\ &= S(T_1 u + T_2 u) \\ &= S(T_1 u) + S(T_2 u) \\ &= ST_1 u + ST_2 u. \end{aligned}$$

Because $u \in U$ was arbitrarily chosen, we get (3.14). This completes the proof. □

Theorem 3.26. *Let $T \in \mathscr{L}(U,V)$ and $S_1, S_2 \in \mathscr{L}(V,W)$. Then*

$$(S_1 + S_2)T = S_1 T + S_2 T. \tag{3.15}$$

Proof. Let $u \in U$ be arbitrarily chosen and fixed. Then

$$\begin{aligned} (S_1 + S_2)Tu &= (S_1 + S_2)(Tu) \\ &= S_1(Tu) + S_2(Tu) \end{aligned}$$

$$= \quad S_1Tu + S_2Tu.$$

Because $u \in U$ was arbitrarily chosen, we get (3.15). This completes the proof. □

Theorem 3.27. *Let $T \in \mathscr{L}(V,W)$. Then*

$$T0 = 0. \tag{3.16}$$

Proof. Note that

$$0 = 0 + 0.$$

Since $T : V \to W$ is a linear map, we get

$$\begin{aligned} T0 &= T(0+0) \\ &= T0 + T0, \end{aligned}$$

whereupon we obtain (3.16). This completes the proof. □

3.3 Kernels and Ranges

Definition 3.28. *Let $T \in \mathscr{L}(V,W)$. The kernel of T, denoted by* $\ker T$, *is the subset of V consisting of those vectors in V that T maps to* 0. *In other words,*

$$\ker T = \{v \in V : Tv = 0\}.$$

Some mathematicians use the term null space instead of kernel.

Example 3.29. *Let $\phi : \mathbb{R}^3 \to \mathbb{R}$ be defined by*

$$\phi(x,y,z) = x + 2y + 3z, \quad (x,y,z) \in \mathbb{R}^3.$$

Then

$$\ker \phi = \{(x,y,z) : x + 2y + 3z = 0\}.$$

Define

$$\begin{aligned} v_1 &= (-2,1,0), \\ v_2 &= (-3,0,1). \end{aligned}$$

We will show that the list v_1, v_2 *is a basis of* $\ker\phi$. *Let* $a \in \mathbb{R}$ *be such that*

$$v_1 = av_2.$$

Then

$$\begin{aligned}(-2,1,0) &= a(-3,0,1)\\ &= (-3a,0,a),\end{aligned}$$

whereupon we get the system

$$\begin{aligned}-3a &= -2\\ 1 &= 0\\ a &= 0,\end{aligned}$$

which is impossible. Thus, the list v_1, v_2 *is linearly independent in* $\mathbb{R}^3$. *In particular, the list* v_1, v_2 *is linearly independent in* $\ker\phi$. *Let now,* $u \in \ker\phi$ *be arbitrarily chosen. Then*

$$u = (-2u_2 - 3u_3, u_2, u_3), \quad u_2, u_3 \in \mathbb{R}.$$

Let also, $b_1, b_2 \in \mathbb{R}$ *be such that*

$$u = b_1 v_1 + b_2 v_2.$$

Then

$$\begin{aligned}(-2u_2 - 3u_3, u_2, u_3) &= b_1(-2,1,0) + b_2(-3,0,1)\\ &= (-2b_1, b_1, 0) + (-3b_2, 0, b_2)\\ &= (-2b_1 - 3b_2, b_1, b_2),\end{aligned}$$

whereupon we get the system

$$-2u_2 - 3u_3 = -2b_1 - 3b_2$$

$$u_2 = b_1$$

$$u_3 = b_2.$$

Therefore

$$span(v_1, v_2) = \ker\phi.$$

Hence, using that the list v_1, v_2 is linearly independent in $\ker\phi$*, we conclude that the list v_1, v_2 is a basis of* $\ker\phi$*.*

Example 3.30. *Let $\phi : \mathbb{R}^2 \to \mathbb{R}$ be defined by*

$$\phi(x,y) = 2x - 4y + 6, \quad (x,y) \in \mathbb{R}^2.$$

Then

$$\begin{aligned} \phi(x,y) &= 0 \iff \\ 2x - 4y + 6 &= 0 \iff \\ 2x &= 4y - 6 \iff \\ x &= 2y - 3, \quad (x,y) \in \mathbb{R}^2. \end{aligned}$$

Therefore

$$\ker\phi = \{(2y-3, y) : y \in \mathbb{R}\}.$$

Example 3.31. *Let $\phi : \mathbb{R}^3 \to \mathbb{R}^2$ be defined by*

$$\phi(x,y,z) = (x - 2y, y - z), \quad (x,y,z) \in \mathbb{R}^2.$$

Then

$$\begin{aligned} \phi(x,y,z) &= 0 \iff \\ (x - 2y, y - z) &= (0,0) \iff \end{aligned}$$

$$x - 2y = 0$$

$$y - z = 0$$

or

$$x = 2y$$

$$y = z.$$

Therefore

$$\ker \phi = \{(2y, y, y) : y \in \mathbb{R}\}.$$

Exercise 3.32. *Let $\phi : \mathbb{R}^3 \to \mathbb{R}^2$ be defined by*

$$\phi(x, y, z) = (x^2 - y^2, x + y + z, z), \quad (x, y, z) \in \mathbb{R}^3.$$

Find $\ker \phi$.

Theorem 3.33. *Let $T \in \mathscr{L}(V, W)$. Then T is injective if and only if* $\ker T = \{0\}$.

Proof.

1. Let T be injective. By Theorem 3.27, it follows that $\{0\} \subseteq \ker T$. Suppose that $u \in \ker T$. Then

$$Tu = 0.$$

Since $T0 = 0$ and T is injective, we conclude that $u = 0$ and $\ker T = \{0\}$.

2. Let $\ker T = \{0\}$. Take $u, v \in V$ and suppose that

$$Tu = Tv.$$

Since T is linear, by the last equation, we find

$$T(u - v) = 0,$$

whereupon $u - v \in \ker T$ and $u = v$. Therefore T is injective. This completes the proof.

□

Since any $T \in \mathscr{L}(V, W)$ can be considered as a function, we have

$$\text{range} T = \{Tv : v \in V\}.$$

If $T \in \mathscr{L}(V, W)$ is surjective, then $\text{range} T = W$.

Theorem 3.34. *Let $T \in \mathscr{L}(V,W)$. Then rangeT is a subspace of W.*

Proof. Since $T0 = 0$, we get $0 \in \text{range}T$. Let now, $v_1, v_2 \in V$ and $a \in \mathbb{F}$. By the equations

$$T(v_1 + v_2) = Tv_1 + Tv_2,$$

$$T(av_1) = aTv_1,$$

it follows that range$T$ is closed under addition and scalar multiplication. Therefore range$T$ is a subspace of W. This completes the proof. □

Theorem 3.35 (Fundamental Theorem of Linear Maps)**.** *Suppose that V is finite-dimensional and $T \in \mathscr{L}(V,W)$. Then rangeT is finite-dimensional and*

$$dimV = dim\ker T + dimrangeT.$$

Proof. Let $u_1, \ldots, u_m$ be a basis of $\ker T$. Then it is linearly independent in V. By Theorem 2.43, it follows that it can be extended to a basis of V

$$u_1, \ldots, u_m, v_1, \ldots, v_n. \tag{3.17}$$

Thus,

$$\dim V = m + n.$$

Consider the list

$$Tv_1, \ldots, Tv_n. \tag{3.18}$$

Let $v \in V$ be arbitrarily chosen. Since the list (3.17) is a basis of V, we have that

$$v = a_1u_1 + \cdots + a_mu_m + b_1v_1 + \cdots + b_nv_n,$$

where $a_1, \ldots, a_m, b_1, \ldots, b_n \in \mathbb{F}$. Note that

$$Tu_j = 0, \quad j \in \{1, \ldots, m\},$$

because $u_1, \ldots, u_m$ is a basis of $\ker T$ and $u_j \in \ker T$ for any $j \in \{1, \ldots, m\}$. Hence,

$$\begin{aligned} Tv &= T(a_1u_1 + \cdots + a_mu_m + b_1v_1 + \cdots + b_nv_n) \\ &= a_1Tu_1 + \cdots + a_mTu_m + b_1Tv_1 + \cdots + b_nTv_n \end{aligned}$$

$$= \quad b_1 T v_1 + \cdots + b_n T v_n.$$

Therefore

$$\operatorname{span}(Tv_1, \ldots, Tv_n) = \operatorname{range} T. \tag{3.19}$$

Let now, $c_1, \ldots, c_n \in \mathbb{F}$ be such that

$$c_1 T v_1 + \cdots + c_n T v_n = 0.$$

Then

$$0 = T(c_1 v_1 + \cdots + c_n v_n).$$

Hence,

$$c_1 v_1 + \cdots + c_n v_n \in \ker T.$$

Since the list $u_1, \ldots, u_m$ is a basis of $\ker T$, there are $d_1, \ldots, d_m \in \mathbb{F}$ so that

$$c_1 v_1 + \cdots + c_n v_n = d_1 u_1 + \cdots + d_m u_m.$$

Now, using that the list (3.17) is a basis of V, we get that it is linearly independent. Then

$$\begin{aligned} c_1 &= 0 \\ &\vdots \\ c_n &= 0 \\ d_1 &= 0 \\ &\vdots \\ d_m &= 0. \end{aligned}$$

From here, we conclude that (3.18) is linearly independent. Hence, applying (3.19), we find that (3.18) is a basis of $\operatorname{range} T$. Therefore

$$\dim \operatorname{range} T = n$$

and

$$\begin{aligned}\dim V &= m+n \\ &= \dim\ker T + \dim\operatorname{range} T.\end{aligned}$$

This completes the proof. □

Theorem 3.36. *Suppose V and W are finite-dimensional vector spaces such that* $\dim V > \dim W$. *Then no linear map from V to W is injective.*

Proof. Let $T \in \mathscr{L}(V,W)$ be arbitrarily chosen and fixed. By Theorem 3.35, we find

$$\begin{aligned}\dim\ker T &= \dim V - \dim\operatorname{range} T \\ &\geq \dim V - \dim W \\ &> 0.\end{aligned}$$

This means $\ker T \neq \{0\}$. Now, applying Theorem 3.33, we conclude that T is not injective. This completes the proof. □

Theorem 3.37. *Suppose V and W are finite-dimensional vector spaces and* $\dim V < \dim W$. *Then no linear map from V to W is surjective.*

Proof. Let $T \in \mathscr{L}(V,W)$ be arbitrarily chosen. Then, applying Theorem 3.35, we obtain

$$\begin{aligned}\dim\operatorname{range} T &= \dim V - \dim\ker T \\ &\leq \dim V \\ &< \dim W.\end{aligned}$$

This means that range$T$ cannot equal W. Thus, T is not surjective. This completes the proof. □

3.4 Applications to Systems of Linear Equations

Suppose that $a_{jk} \in \mathbb{F}$, $j \in \{1,\ldots,m\}$, $k \in \{1,\ldots,n\}$. Consider the homogeneous system

$$\begin{aligned} \sum_{k=1}^{n} a_{1k}x_k &= 0 \\ &\vdots \\ \sum_{k=1}^{n} a_{mk}x_k &= 0. \end{aligned} \tag{3.20}$$

Remark 3.38. *Homogeneous, in this context, means that the constant term on the right hand side of each equation above is* 0*.*

Obviously

$$\begin{aligned} x_1 &= 0 \\ &\vdots \\ x_n &= 0 \end{aligned}$$

is a solution to the system (3.20). The question here is whether any other solutions exist. Define the map $T : \mathbb{F}^n \to \mathbb{F}^m$ by

$$T(x_1,\ldots,x_n) = \left(\sum_{k=1}^{n} a_{1k}x_k, \ldots, \sum_{k=1}^{b} a_{mk}x_k \right), \quad (x_1,\ldots,x_n) \in \mathbb{F}^n. \tag{3.21}$$

The equation

$$T(x_1,\ldots,x_n) = (0,\ldots,0), \quad (x_1,\ldots,x_n) \in \mathbb{F}^n$$

is the same as the homogeneous system (3.20).

Theorem 3.39. *A homogeneous system of linear equations with more variables than equations has nonzero solutions.*

Proof. Note that $T \in \mathscr{L}(\mathbb{F}^n, \mathbb{F}^m)$. In this case, we have $n > m$. Applying Theorem 3.36, we get that T is not injective. This completes the proof. $\square$

Let $n > m$. Then, to find a nonzero solution to the system (3.20), we introduce $n-m$ parameters. Set

$$\begin{aligned} x_{m+1} &= p_1 \\ &\vdots \\ x_n &= p_{n-m}. \end{aligned}$$

Then, we find a representation of the list $x_1, \ldots, x_m$ via the list $p_1, \ldots, p_{n-m}$.

Example 3.40. *Consider the system*

$$\begin{aligned} x_1 - 2x_2 + x_3 &= 0 \\ x_2 - 2x_3 &= 0. \end{aligned}$$

Here, we have three variables and two equations. By Theorem 3.39, it follows that the considered system has a nonzero solution. Set

$$x_3 = p.$$

Then

$$\begin{aligned} x_1 &= 2x_2 - p \\ x_2 &= 2p \end{aligned}$$

and

$$\begin{aligned} x_1 &= 4p - p \\ x_2 &= 2p, \end{aligned}$$

or

$$\begin{aligned} x_1 &= 3p \\ x_2 &= 2p. \end{aligned}$$

Take $p = 1$ and we find that

$$x_1 = 3$$

$$x_2 = 2$$

is a solution to the considered system.

Example 3.41. *Consider the system*

$$x_1 - x_2 + x_3 - x_4 = 0$$

$$x_1 + x_2 - x_3 = 0$$

$$x_2 - 2x_3 - 3x_4 = 0.$$

Here, we have four variables and three equations. By Theorem 3.39, it follows that the considered system has a nonzero solution. Set

$$x_4 = p.$$

Then

$$x_1 - x_2 + x_3 = p$$

$$x_1 + x_2 - x_3 = 0$$

$$x_2 - 2x_3 = 3p,$$

whereupon

$$2x_1 = p$$

$$x_1 + x_2 - x_3 = 0$$

$$x_2 - 2x_3 = 3p,$$

or

$$x_1 = \frac{p}{2}$$

$$\begin{aligned} x_2 - x_3 &= -\frac{p}{2} \\ x_2 - 2x_3 &= 3p, \end{aligned}$$

or

$$\begin{aligned} x_1 &= \frac{p}{2} \\ x_3 &= -\frac{7p}{2} \\ x_2 - x_3 &= -\frac{p}{2}, \end{aligned}$$

or

$$\begin{aligned} x_1 &= \frac{p}{2} \\ x_2 &= -4p \\ x_3 &= -\frac{7p}{2}. \end{aligned}$$

Take $p = 2$. Then

$$\begin{aligned} x_1 &= 1 \\ x_2 &= -8 \\ x_3 &= -7 \end{aligned}$$

is a solution of the considered system.

Example 3.42. *Consider the system*

$$x_1 - 2x_2 + x_3 - 4x_4 = 0.$$

Here, we have four variables and one equation. By Theorem 3.39, it follows that the considered system has a nonzero solution. Set

$$x_2 = p,$$

$$x_3 = q,$$

$$x_4 = r.$$

Then

$$x_1 = 2p - q + 4r.$$

Take

$$p = 1,$$

$$q = 2,$$

$$r = 1.$$

Then

$$x_1 = 4$$

$$x_2 = 1$$

$$x_3 = 2$$

$$x_4 = 1$$

is a solution of the considered system.

Exercise 3.43. *Find a nonzero solution to the following systems.*

1.

$$x_1 - 2x_2 = 0.$$

2.

$$\begin{aligned} x_1 - x_2 - x_3 &= 0 \\ x_1 + 2x_2 + 3x_3 &= 0. \end{aligned}$$

3.

$$\begin{aligned} 2x_1+3x_2-4x_3 &= 0 \\ x_1-4x_2-5x_3 &= 0. \end{aligned}$$

4.

$$\begin{aligned} x_1+x_2-2x_3-3x_4 &= 0 \\ x_1+2x_2-3x_3-x_4 &= 0 \\ x_2+2x_3+3x_4 &= 0. \end{aligned}$$

5.

$$\begin{aligned} 2x_1-x_2+x_3-x_4 &= 0 \\ x_2-2x_3+3x_4 &= 0 \\ 3x_1+x_3 &= 0. \end{aligned}$$

Suppose that $c_1,\ldots,c_m \in \mathbb{F}$. Consider the inhomogeneous system

$$\begin{aligned} \sum_{k=1}^{n} a_{1k}x_k &= c_1 \\ &\vdots \\ \sum_{k=1}^{n} a_{mk}x_k &= c_m. \end{aligned} \tag{3.22}$$

The question here is whether there is a choice of $c_1,\ldots,c_m \in \mathbb{F}$ so that the system (3.22) has no solutions. Note that the equation

$$T(x_1,\ldots,x_n) = (c_1,\ldots,c_m), \quad (x_1,\ldots,x_n) \in \mathbb{F}^n,$$

is the same as the homogeneous system (3.22).

Theorem 3.44. *An inhomogeneous system of linear equations with more equations than variables has no solutions for some choice of the constant terms.*

Proof. Since $T \in \mathscr{L}(\mathbb{F}^n, \mathbb{F}^m)$, $n < m$, applying Theorem 3.37, we have that T is not surjective. This completes the proof. $\square$

Remark 3.45. *Our results for homogeneous systems with more variables than equations and inhomogeneous systems with more equations than variables are often proved using Gaussian elimination. The abstract approach taken here leads to cleaner proofs.*

Example 3.46. *Consider the system*

$$\begin{aligned} 2x_1 + 3x_2 &= 5 \\ x_1 - x_2 &= 2 \\ x_1 + 4x_2 &= 2. \end{aligned}$$

We have

$$\begin{aligned} 2x_1 + 3x_2 &= 5 \\ x_1 &= 2 + x_2 \\ x_1 &= 2 - 4x_2. \end{aligned}$$

By the last two equations, we find

$$2 + x_2 = 2 - 4x_2,$$

whereupon

$$x_2 = 0$$

and

$$x_1 = 2.$$

We put them in the first equation of the considered system and we obtain

$$4 + 0 = 5,$$

which is impossible.

Example 3.47. *Consider the system*

$$\begin{aligned} 2x_1+3x_2 &= 5 \\ x_1-x_2 &= 2 \\ x_1+4x_2 &= 3. \end{aligned}$$

We have

$$\begin{aligned} 2x_1+3x_2 &= 5 \\ x_1 &= 2+x_2 \\ x_1 &= 3-4x_2. \end{aligned}$$

By the last two equations, we find

$$2+x_2 = 3-4x_2,$$

whereupon

$$5x_2 = 1,$$

or

$$x_2 = \frac{1}{5}$$

and

$$\begin{aligned} x_1 &= 2+\frac{1}{5} \\ &= \frac{11}{5}. \end{aligned}$$

We put them in the first equation of the given system and we arrive at

$$\frac{22}{5}+\frac{3}{5} = 5.$$

Therefore

$$x_1 = \frac{11}{5}$$

$$x_2 = \frac{1}{5}$$

is a solution of the considered system.

Example 3.48. *We will investigate the system*

$$\begin{aligned} ax_1 + bx_2 &= a \\ (a-2)x_1 + x_2 &= 3 \\ x_1 + x_2 &= 1, \end{aligned}$$

where $a, b \in \mathbb{R}$ are parameters. We have

$$\begin{aligned} ax_1 + bx_2 &= a \\ x_2 &= 3 - (a-2)x_1 \\ x_2 &= 1 - x_1. \end{aligned}$$

By the last two equations, we get

$$1 - x_1 = 3 - (a-2)x_1,$$

or

$$(a-3)x_1 = 2.$$

1. *Let $a = 3$. Then the last equation has no solutions and from here, the considered system has no any solutions.*

2. *Let $a \neq 3$. Then*

$$x_1 = \frac{2}{a-3}$$

and

$$\begin{aligned} x_2 &= 1 - x_1 \\ &= 1 - \frac{2}{a-3} \end{aligned}$$

$$= \frac{a-3-2}{a-3}$$

$$= \frac{a-5}{a-3}.$$

We put them in the first equation of the given system and we find

$$\frac{2a}{a-3} + \frac{b(a-5)}{a-3} = a,$$

whereupon

$$2a + b(a-5) = a(a-3),$$

or

$$2a + b(a-5) = a^2 - 3a,$$

or

$$b(a-5) = a^2 - 5a,$$

or

$$b(a-5) = a(a-5).$$

We have the following cases.

(a) *If* $a = 5$, *then*

$$x_1 = 1$$

$$x_2 = 0$$

is a solution of the considered system for any $b \in \mathbb{R}$.

(b) *Let* $a \neq 5$. *Then* $b = a$. *Therefore for* $a \neq 3, 5$ *and* $a = b$ *the considered system has a solution*

$$x_1 = \frac{2}{a-3}$$

$$x_2 = \frac{a-5}{a-3}.$$

Exercise 3.49. *Investigate the following system*

$$x_1 + 2x_2 = 3$$

$$\begin{aligned} ax_1 - 4x_2 &= -6 \\ x_1 + x_2 &= 1, \end{aligned}$$

where $a \in \mathbb{R}$ is a parameter.

3.5 Matrices. Representing a Linear Map by a Matrix

Suppose that the list $v_1, \dots, v_n$ is a basis of V and the list $w_1, \dots, w_m$ is a basis of W. Let $T \in \mathscr{L}(V, W)$.

Definition 3.50. *Let m and n denote positive integers. A $m \times n$ matrix A is a rectangular array of elements of $\mathbb{F}$ with m rows and n columns*

$$A = \begin{pmatrix} a_{11} & \cdots & a_{1n} \\ \vdots & \vdots & \vdots \\ a_{m1} & \cdots & a_{mn} \end{pmatrix}.$$

The notation a_{jk} denotes the entry in row j and column k of the matrix A, shortly jk entry of A.

Example 3.51. *We have that*

$$A = \begin{pmatrix} -1 & 1 & 2 & 3i \\ 0 & -1 & 3 & 4 \end{pmatrix}$$

is a 2×4 matrix with

$$\begin{aligned} a_{23} &= 3, \\ a_{14} &= 3i. \end{aligned}$$

Example 3.52. *We have that*

$$A = \begin{pmatrix} 1 & -1 & 0 & 2 \\ -1 & 1 & 2 & 3 \\ -3 & 1 & 0 & 1 \\ 0 & 1 & -2 & 1 \end{pmatrix}$$

is a 4×4 *matrix with*

$$a_{23} = 2,$$

$$a_{14} = 2.$$

Because $T\in\mathscr{L}(V,W)$, there are $a_{jk}\in\mathbb{F}$, $j\in\{1,\dots,m\}$, $k\in\{1,\dots,n\}$, such that

$$\begin{aligned} Tv_1 &= a_{11}w_1 + a_{21}w_2 + \cdots + a_{m1}w_m \\ Tv_2 &= a_{12}w_1 + a_{22}w_2 + \cdots + a_{m2}w_m \\ &\vdots \\ Tv_n &= a_{1n}w_1 + a_{2n}w_2 + \cdots + a_{mn}w_m. \end{aligned} \tag{3.23}$$

Definition 3.53. *The matrix T with respect the bases* $v_1,\dots,v_n$ *and* $w_1,\dots,w_m$ *is the matrix* $\mathscr{M}(T)$ *whose entries* a_{jk}, $j\in\{1,\dots,m\}$, $k\in\{1,\dots,n\}$, *are determined by* (3.23). *If the bases are not clear from the context, then we will use the notation*

$$\mathscr{M}(T,(v_1,\dots,v_n),(w_1,\dots,w_m)).$$

Example 3.54. *Let* $T\in\mathscr{L}(\mathbb{F}^2,\mathbb{F}^3)$ *be defined by*

$$T(x,y) = (x+y, x-y, 2x+3y), \quad (x,y)\in\mathbb{F}^2.$$

We have that

$$v_1 = (1,0),$$

$$v_2 = (0,1)$$

is a basis of $\mathbb{F}^2$ *and*

$$w_1 = (1,0,0),$$

$$w_2 = (0,1,0),$$

$$w_3 = (0,0,1)$$

is a basis of $\mathbb{F}^3$. *We compute*

$$\begin{aligned} Tv_1 &= (1,1,2) \\ &= w_1+w_2+2w_3 \end{aligned}$$

and

$$\begin{aligned} Tv_2 &= (1,-1,3) \\ &= w_1-w_2+3w_2. \end{aligned}$$

Therefore the matrix of T *is*

$$\mathscr{M}(T)=\begin{pmatrix} 1 & 1 \\ 1 & -1 \\ 2 & 3 \end{pmatrix}.$$

Example 3.55. *Let* $T\in\mathscr{L}(\mathbb{F}^3,\mathbb{F}^3)$ *be defined by*

$$T(x,y,z)=(x+y+z,2x-y-z,2y-3z),\quad (x,y,z)\in\mathbb{R}^3.$$

Here

$$\begin{aligned} v_1 &= (1,0,0), \\ v_2 &= (0,1,0), \\ v_3 &= (0,0,1) \end{aligned}$$

is a basis of $\mathbb{F}^3$. *Then*

$$\begin{aligned} Tv_1 &= (1,2,0) \\ &= v_1+2v_2, \\ Tv_2 &= (1,-1,2) \\ &= v_1-v_2+2v_3, \end{aligned}$$

$$\begin{aligned} Tv_3 &= (1,-1,-3) \\ &= v_1 - v_2 - 3v_3. \end{aligned}$$

Therefore

$$\mathscr{M}(T) = \begin{pmatrix} 1 & 1 & 1 \\ 2 & -1 & -1 \\ 0 & 2 & -3 \end{pmatrix}.$$

Example 3.56. *Let $V, W \subset \mathbb{F}^2$, v_1, v_2 be a basis of V and*

$$\begin{aligned} w_1 &= (-1,1), \\ w_2 &= (2,-1) \end{aligned}$$

be a basis of W. Suppose that $T \in \mathscr{L}(V,W)$ be defined by

$$T(x,y) = (x-y, x+y), \quad (x,y) \in \mathbb{F}^2,$$

and

$$\mathscr{M}(T) = \begin{pmatrix} 1 & -1 \\ -1 & 2 \end{pmatrix}.$$

We will find the basis v_1, v_2. Let

$$\begin{aligned} v_1 &= (v_{11}, v_{12}), \\ v_2 &= (v_{21}, v_{22}). \end{aligned}$$

Then

$$\begin{aligned} Tv_1 &= (v_{11} - v_{12}, v_{11} + v_{12}), \\ Tv_2 &= (v_{21} - v_{22}, v_{21} + v_{22}) \end{aligned}$$

and

$$\begin{aligned} Tv_1 &= w_1 - w_2 \\ &= (-1,1) - (2,-1) \end{aligned}$$

$$
\begin{aligned}
&= (-1-2,1+1)\\
&= (-3,2),\\
Tv_2 &= -w_1+2w_2\\
&= -(-1,1)+2(2,-1)\\
&= (1,-1)+(4,-2)\\
&= (1+4,-1-2)\\
&= (5,-3).
\end{aligned}
$$

Therefore

$$
\begin{aligned}
(v_{11}-v_{12},v_{11}+v_{12}) &= (-3,2),\\
(v_{21}-v_{22},v_{21}+v_{22}) &= (5,-3).
\end{aligned}
$$

By the last system, we find

$$
\begin{aligned}
v_{11}-v_{12} &= -3\\
v_{11}+v_{12} &= 2,
\end{aligned}
$$

whereupon

$$
\begin{aligned}
2v_{11} &= -1\\
2v_{12} &= 5,
\end{aligned}
$$

or

$$
v_{11} = -\frac{1}{2}
$$

$$v_{12} = \frac{5}{2}.$$

Moreover,

$$v_{21} - v_{22} = 5$$

$$v_{21} + v_{22} = -3.$$

Hence,

$$2v_{21} = 2$$

$$2v_{22} = -8,$$

or

$$v_{21} = 1$$

$$v_{22} = -4.$$

Thus,

$$v_1 = \left(-\frac{1}{2}, \frac{5}{2}\right),$$

$$v_2 = (1, -4).$$

Exercise 3.57. *Let* $T \in \mathscr{L}(\mathbb{R}^4, \mathbb{R}^3)$ *be given by*

$$T(x,y,z,t) = (x - 2y + z + t, y + z, 2y - 3z + 4t), \quad (x,y,z,t) \in \mathbb{R}^4.$$

Find $\mathscr{M}(T)$.

3.6 Addition and Scalar Multiplication of Matrices

Definition 3.58. *The sum of two matrices of the same size is the matrix obtained by adding the corresponding entries in the matrices*

$$\begin{pmatrix} a_{11} & \cdots & a_{1n} \\ \vdots & \vdots & \vdots \\ a_{m1} & \cdots & a_{mn} \end{pmatrix} + \begin{pmatrix} b_{11} & \cdots & b_{1n} \\ \vdots & \vdots & \vdots \\ b_{m1} & \cdots & b_{mn} \end{pmatrix}$$

$$= \begin{pmatrix} a_{11}+b_{11} & \cdots & a_{1n}+b_{1n} \\ \vdots & \vdots & \vdots \\ a_{m1}+b_{m1} & \cdots & a_{mn}+b_{mn} \end{pmatrix}.$$

In other words,

$$(A+B)_{jk} = a_{jk}+b_{jk}, \quad j \in \{1,\ldots,m\}, \quad k \in \{1,\ldots,n\}.$$

Example 3.59. *We have*

$$\begin{pmatrix} -1 & 2 & 3 & 4 \\ -3 & 2 & 1 & 0 \\ 0 & 0 & 1 & 2 \end{pmatrix} + \begin{pmatrix} 2 & 1 & -1 & 3 \\ 1 & -2 & 1 & 2 \\ -1 & 1 & 1 & -1 \end{pmatrix}$$

$$= \begin{pmatrix} -1+2 & 2+1 & 3+(-1) & 4+3 \\ -3+1 & 2+(-2) & 1+1 & 0+2 \\ 0+(-1) & 0+1 & 1+1 & 2+(-1) \end{pmatrix}$$

$$= \begin{pmatrix} 1 & 3 & 2 & 7 \\ -2 & 0 & 2 & 2 \\ -1 & 1 & 2 & 1 \end{pmatrix}.$$

Exercise 3.60. *Find $A+B$, where*

$$A = \begin{pmatrix} -3 & 1 & 2 & 5 & 7 & 8 \\ 3 & 4 & 2 & -1 & 3 & 4 \\ -1 & 2 & 3 & -4 & 0 & 1 \end{pmatrix},$$

$$B = \begin{pmatrix} -1 & 0 & -2 & -5 & -11 & 9 \\ -3 & 1 & -2 & 1 & -4 & 7 \\ 1 & 0 & 2 & 0 & 11 & 0 \end{pmatrix}.$$

Theorem 3.61. *Let $S, T \in \mathscr{L}(V,W)$. Then*

$$\mathscr{M}(S+T) = \mathscr{M}(S) + \mathscr{M}(T).$$

Proof. Let $v_1,\ldots,v_n$ be a basis of V and $w_1,\ldots,w_m$ be a basis of W. Let also,

$$Sv_1 = a_{11}w_1 + \cdots + a_{1m}w_m$$

$$\vdots$$

$$Sv_n = a_{n1}w_1 + \cdots + a_{nm}w_m$$

and

$$\begin{aligned} Tv_1 &= b_{11}w_1 + \cdots + b_{1m}w_m \\ &\vdots \\ Tv_n &= b_{n1}w_1 + \cdots + b_{nm}w_m, \end{aligned}$$

where $a_{jk}, b_{jk} \in \mathbb{F}$, $j \in \{1, \ldots, n\}$, $k \in \{1, \ldots, m\}$. Then

$$\mathscr{M}(S) = \begin{pmatrix} a_{11} & \cdots & a_{1m} \\ \vdots & \vdots & \vdots \\ a_{n1} & \cdots & a_{nm} \end{pmatrix}$$

and

$$\mathscr{M}(T) = \begin{pmatrix} b_{11} & \cdots & b_{1m} \\ \vdots & \vdots & \vdots \\ b_{n1} & \cdots & b_{nm} \end{pmatrix}.$$

Moreover,

$$\begin{aligned} (S+T)v_1 &= Sv_1 + Tv_1 \\ &= a_{11}w_1 + \cdots + a_{1m}w_m + b_{11}w_1 + \cdots + b_{1m}w_m \\ &= (a_{11} + b_{11})w_1 + \cdots + (a_{1m} + b_{1m})w_m \\ &\vdots \\ (S+T)v_n &= Sv_n + Tv_n \\ &= a_{n1}w_1 + \cdots + a_{nm}w_m + b_{n1}w_1 + \cdots + b_{nm}w_m \end{aligned}$$

$$= \quad (a_{n1}+b_{n1})w_1+\cdots+(a_{nm}+b_{nm})w_m.$$

Therefore

$$\begin{aligned}
\mathscr{M}(S+T) &= \begin{pmatrix} a_{11}+b_{11} & \cdots & a_{1m}+b_{1m} \\ \vdots & \vdots & \vdots \\ a_{n1}+b_{n1} & \cdots & a_{nm}+b_{nm} \end{pmatrix} \\
&= \begin{pmatrix} a_{11} & \cdots & a_{1m} \\ \vdots & \vdots & \vdots \\ a_{n1} & \cdots & a_{nm} \end{pmatrix} + \begin{pmatrix} b_{11} & \cdots & b_{1m} \\ \vdots & \vdots & \vdots \\ b_{n1} & \cdots & b_{nm} \end{pmatrix} \\
&= \mathscr{M}(S)+\mathscr{M}(T).
\end{aligned}$$

This completes the proof. □

Definition 3.62. *The product of a scalar and a matrix is the matrix obtained by multiplying each entry in the matrix by the scalar*

$$\lambda \begin{pmatrix} a_{11} & \cdots & a_{1m} \\ \vdots & \vdots & \vdots \\ a_{n1} & \cdots & a_{nm} \end{pmatrix} = \begin{pmatrix} \lambda a_{11} & \cdots & \lambda a_{1m} \\ \vdots & \vdots & \vdots \\ \lambda a_{n1} & \cdots & \lambda a_{nm} \end{pmatrix}.$$

In other words,

$$(\lambda A)_{jk} = \lambda(A_{jk}), \quad j \in \{1,\ldots,n\}, \quad k \in \{1,\ldots,m\}.$$

Example 3.63. *Let*

$$A = \begin{pmatrix} -1 & 2 & 1 & 3 \\ 0 & 1 & 4 & 2 \\ -3 & 1 & 2 & -1 \end{pmatrix}.$$

We will find 2A. We have

$$\begin{aligned}
2A &= \begin{pmatrix} 2\cdot(-1) & 2\cdot 2 & 2\cdot 1 & 2\cdot 3 \\ 2\cdot 0 & 2\cdot 1 & 2\cdot 4 & 2\cdot 2 \\ 2\cdot(-3) & 2\cdot 1 & 2\cdot 2 & 2\cdot(-1) \end{pmatrix} \\
&= \begin{pmatrix} -2 & 4 & 2 & 6 \\ 0 & 2 & 8 & 4 \\ -6 & 2 & 4 & -2 \end{pmatrix}.
\end{aligned}$$

Exercise 3.64. *Let*

$$A = \begin{pmatrix} -4 & 0 & \frac{1}{2} & 3 & 5 \\ 2 & 1 & -3 & 4 & \frac{5}{3} \\ 3 & -1 & \frac{7}{8} & \frac{4}{5} & -1 \end{pmatrix}.$$

Find

1. $2A$.
2. $(-4)A$.
3. $5A$.
4. $\frac{1}{2}A$.
5. $\frac{7}{3}A$.

Definition 3.65. *Let*

$$A = \begin{pmatrix} a_{11} & \cdots & a_{1m} \\ \vdots & \vdots & \vdots \\ a_{n1} & \cdots & a_{nm} \end{pmatrix}$$

and

$$B = \begin{pmatrix} b_{11} & \cdots & b_{1m} \\ \vdots & \vdots & \vdots \\ b_{n1} & \cdots & b_{nm} \end{pmatrix}.$$

Let also, $a, b \in \mathbb{F}$. Define $aA + bB$ by

$$aA + bB = \begin{pmatrix} aa_{11} + bb_{11} & \cdots & aa_{1m} + bb_{1m} \\ \vdots & \vdots & \vdots \\ aa_{n1} + bb_{n1} & \cdots & aa_{nm} + bb_{nm} \end{pmatrix}.$$

Define $aA - bB$ by

$$aA - bB = aA + (-b)B.$$

Example 3.66. *Let*

$$A = \begin{pmatrix} 2 & -1 & 0 \\ 3 & 4 & -2 \\ -3 & 1 & 5 \end{pmatrix},$$

$$B = \begin{pmatrix} 3 & 1 & 2 \\ -2 & 1 & 3 \\ 0 & 2 & -4 \end{pmatrix}.$$

We will find

$$4A - 5B.$$

We have

$$\begin{aligned} 4A &= \begin{pmatrix} 4\cdot 2 & 4\cdot(-1) & 4\cdot 0 \\ 4\cdot 3 & 4\cdot 4 & 4\cdot(-2) \\ 4\cdot(-3) & 4\cdot 1 & 4\cdot 5 \end{pmatrix} \\ &= \begin{pmatrix} 8 & -4 & 0 \\ 12 & 16 & -8 \\ -12 & 4 & 20 \end{pmatrix} \end{aligned}$$

and

$$\begin{aligned} (-5)B &= \begin{pmatrix} (-5)\cdot 3 & (-5)\cdot 1 & (-5)\cdot 2 \\ (-5)\cdot(-2) & (-5)\cdot 1 & (-5)\cdot 3 \\ (-5)\cdot 0 & (-5)\cdot 2 & (-5)\cdot(-4) \end{pmatrix} \\ &= \begin{pmatrix} -15 & -5 & -10 \\ 10 & -5 & -15 \\ 0 & -10 & 20 \end{pmatrix}. \end{aligned}$$

Therefore

$$\begin{aligned} 4A - 5B &= 4A + (-5)B \\ &= \begin{pmatrix} 8 & -4 & 0 \\ 12 & 16 & -8 \\ -12 & 4 & 20 \end{pmatrix} + \begin{pmatrix} -15 & -5 & -10 \\ 10 & -5 & -15 \\ 0 & -10 & 20 \end{pmatrix} \\ &= \begin{pmatrix} 8+(-15) & -4+(-5) & 0+(-10) \\ 12+10 & 16+(-5) & -8+(-15) \\ -12+0 & 4+(-10) & 20+20 \end{pmatrix} \end{aligned}$$

$$= \begin{pmatrix} -7 & -9 & -10 \\ 22 & 11 & -23 \\ -12 & -6 & 40 \end{pmatrix}.$$

Exercise 3.67. *Let*

$$A = \begin{pmatrix} -2 & 3 & 1 \\ 5 & 4 & 0 \\ 2 & -1 & -5 \end{pmatrix},$$

$$B = \begin{pmatrix} 1 & -2 & -3 \\ 0 & -3 & 1 \\ 4 & -4 & 5 \end{pmatrix}.$$

Find

1. $2A$.
2. $3B$.
3. $A - B$.
4. $-3A$.
5. $4A - 2B$.

Theorem 3.68. *Let $T \in \mathscr{L}(V,W)$ and $\lambda \in \mathbb{F}$. Then*

$$\mathscr{M}(\lambda T) = \lambda \mathscr{M}(T).$$

Proof. Let $v_1, \ldots, v_n$ be a basis of V and $w_1, \ldots, w_m$ be a basis of W. Let also,

$$\begin{aligned} Tv_1 &= a_{11}w_1 + \cdots + a_{1m}w_m \\ &\vdots \\ Tv_n &= a_{n1}w_1 + \cdots + a_{nm}w_m, \end{aligned}$$

i.e.,

$$\mathscr{M}(T) = \begin{pmatrix} a_{11} & \cdots & a_{1m} \\ \vdots & \vdots & \vdots \\ a_{n1} & \cdots & a_{nm} \end{pmatrix}.$$

Then

$$
\begin{aligned}
(\lambda T)v_1 &= \lambda T v_1 \\
&= \lambda(a_{11}w_1 + \cdots + a_{1m}w_m) \\
&= (\lambda a_{11})w_1 + \cdots + (\lambda a_{1m})w_m \\
&\vdots \\
(\lambda T)v_n &= \lambda T v_n \\
&= \lambda(a_{n1}w_1 + \cdots + a_{nm}w_m) \\
&= (\lambda a_{n1})w_1 + \cdots + (\lambda a_{nm})w_m.
\end{aligned}
$$

Therefore

$$
\begin{aligned}
\mathscr{M}(\lambda T) &= \begin{pmatrix} \lambda a_{11} & \cdots & \lambda a_{1m} \\ \vdots & \vdots & \vdots \\ \lambda a_{n1} & \cdots & \lambda a_{nm} \end{pmatrix} \\
&= \lambda \begin{pmatrix} a_{11} & \cdots & a_{1n} \\ \vdots & \vdots & \vdots \\ a_{n1} & \cdots & a_{nm} \end{pmatrix} \\
&= \lambda \mathscr{M}(T).
\end{aligned}
$$

This completes the proof. □

Definition 3.69. *For m and n positive integers, the set of all $m \times n$ matrices with entries in $\mathbb{F}$ is denoted by $\mathbb{F}^{m,n}$.*

Definition 3.70. *Define the zero matrix*

$$
O = \begin{pmatrix} 0 & \cdots & 0 \\ \vdots & \vdots & \vdots \\ 0 & \cdots & 0 \end{pmatrix},
$$

i.e., the zero matrix is an $m \times n$ matrix whose entries are zero.

Theorem 3.71. *The set $\mathbb{F}^{m,n}$ is a vector space.*

Proof. Let

$$
\begin{aligned}
A &= \begin{pmatrix} a_{11} & \cdots & a_{1n} \\ \vdots & \vdots & \vdots \\ a_{m1} & \cdots & a_{mn} \end{pmatrix} \in \mathbb{F}^{m,n}, \\
B &= \begin{pmatrix} b_{11} & \cdots & b_{1n} \\ \vdots & \vdots & \vdots \\ b_{m1} & \cdots & b_{mn} \end{pmatrix} \in \mathbb{F}^{m,n}, \\
C &= \begin{pmatrix} c_{11} & \cdots & c_{1n} \\ \vdots & \vdots & \vdots \\ c_{m1} & \cdots & c_{mn} \end{pmatrix} \in \mathbb{F}^{m,n}
\end{aligned}
$$

and $a, b \in \mathbb{F}$ be arbitrarily chosen. Then, we have the following.

1.

$$
\begin{aligned}
A+B &= \begin{pmatrix} a_{11} & \cdots & a_{1n} \\ \vdots & \vdots & \vdots \\ a_{m1} & \cdots & a_{mn} \end{pmatrix} + \begin{pmatrix} b_{11} & \cdots & b_{1n} \\ \vdots & \vdots & \vdots \\ b_{m1} & \cdots & b_{mn} \end{pmatrix} \\
&= \begin{pmatrix} a_{11}+b_{11} & \cdots & a_{1n}+b_{1n} \\ \vdots & \vdots & \vdots \\ a_{m1}+b_{m1} & \cdots & a_{mn}+b_{mn} \end{pmatrix} \\
&= \begin{pmatrix} b_{11}+a_{11} & \cdots & b_{1n}+a_{1n} \\ \vdots & \vdots & \vdots \\ b_{m1}+a_{m1} & \cdots & b_{mn}+a_{mn} \end{pmatrix} \\
&= \begin{pmatrix} b_{11} & \cdots & b_{1n} \\ \vdots & \vdots & \vdots \\ b_{m1} & \cdots & b_{mn} \end{pmatrix} + \begin{pmatrix} a_{11} & \cdots & a_{1n} \\ \vdots & \vdots & \vdots \\ a_{m1} & \cdots & a_{mn} \end{pmatrix} \\
&= B+A,
\end{aligned}
$$

i.e.,

$$A+B = B+A.$$

2.

$$
\begin{aligned}
A+(B+C) &= \begin{pmatrix} a_{11} & \cdots & a_{1n} \\ \vdots & \vdots & \vdots \\ a_{m1} & \cdots & a_{mn} \end{pmatrix} \\
&\quad + \left(\begin{pmatrix} b_{11} & \cdots & b_{1n} \\ \vdots & \vdots & \vdots \\ b_{m1} & \cdots & b_{mn} \end{pmatrix} + \begin{pmatrix} c_{11} & \cdots & c_{1n} \\ \vdots & \vdots & \vdots \\ c_{m1} & \cdots & c_{mn} \end{pmatrix} \right) \\
&= \begin{pmatrix} a_{11} & \cdots & a_{1n} \\ \vdots & \vdots & \vdots \\ a_{m1} & \cdots & a_{mn} \end{pmatrix} + \begin{pmatrix} b_{11}+c_{11} & \cdots & b_{1n}+c_{1n} \\ \vdots & \vdots & \vdots \\ b_{m1}+c_{m1} & \cdots & b_{mn}+c_{mn} \end{pmatrix} \\
&= \begin{pmatrix} a_{11}+(b_{11}+c_{11}) & \cdots & a_{1n}+(b_{1n}+c_{1n}) \\ \vdots & \vdots & \vdots \\ a_{m1}+(b_{m1}+c_{m1}) & \cdots & a_{mn}+(b_{mn}+c_{mn}) \end{pmatrix} \\
&= \begin{pmatrix} (a_{11}+b_{11})+c_{11} & \cdots & (a_{1n}+b_{1n})+c_{1n} \\ \vdots & \vdots & \vdots \\ (a_{m1}+b_{m1})+c_{m1} & \cdots & (a_{mn}+b_{mn})+c_{mn} \end{pmatrix} \\
&= \begin{pmatrix} a_{11}+b_{11} & \cdots & a_{1n}+b_{1n} \\ \vdots & \vdots & \vdots \\ a_{m1}+b_{m1} & \cdots & a_{mn}+b_{mn} \end{pmatrix} + \begin{pmatrix} c_{11} & \cdots & c_{1n} \\ \vdots & \vdots & \vdots \\ c_{m1} & \cdots & c_{mn} \end{pmatrix} \\
&= \left(\begin{pmatrix} a_{11} & \cdots & a_{1n} \\ \vdots & \vdots & \vdots \\ a_{m1} & \cdots & a_{mn} \end{pmatrix} + \begin{pmatrix} b_{11} & \cdots & b_{1n} \\ \vdots & \vdots & \vdots \\ b_{m1} & \cdots & b_{mn} \end{pmatrix} \right) \\
&\quad + \begin{pmatrix} c_{11} & \cdots & c_{1n} \\ \vdots & \vdots & \vdots \\ c_{m1} & \cdots & c_{mn} \end{pmatrix} \\
&= (A+B)+C,
\end{aligned}
$$

i.e.,

$$A+(B+C) = (A+B)+C.$$

3. Let

$$-A = \begin{pmatrix} -a_{11} & \cdots & -a_{1n} \\ \vdots & \vdots & \vdots \\ -a_{m1} & \cdots & -a_{mn} \end{pmatrix}.$$

Then

$$\begin{aligned} A+(-A) &= \begin{pmatrix} a_{11} & \cdots & a_{1n} \\ \vdots & \vdots & \vdots \\ a_{m1} & \cdots & a_{mn} \end{pmatrix} + \begin{pmatrix} -a_{11} & \cdots & -a_{1n} \\ \vdots & \vdots & \vdots \\ -a_{m1} & \cdots & -a_{mn} \end{pmatrix} \\ &= \begin{pmatrix} a_{11}+(-a_{11}) & \cdots & a_{1n}+(-a_{1n}) \\ \vdots & \vdots & \vdots \\ a_{m1}+(-a_{m1}) & \cdots & a_{mn}+(-a_{mn}) \end{pmatrix} \\ &= \begin{pmatrix} 0 & \cdots & 0 \\ \vdots & \vdots & \vdots \\ 0 & \cdots & 0 \end{pmatrix} \\ &= O, \end{aligned}$$

i.e.,

$$A+(-A) = O.$$

4.

$$\begin{aligned} A+O &= \begin{pmatrix} a_{11} & \cdots & a_{1n} \\ \vdots & \vdots & \vdots \\ a_{m1} & \cdots & a_{mn} \end{pmatrix} + \begin{pmatrix} 0 & \cdots & 0 \\ \vdots & \vdots & \vdots \\ 0 & \cdots & 0 \end{pmatrix} \\ &= \begin{pmatrix} a_{11}+0 & \cdots & a_{1n}+0 \\ \vdots & \vdots & \vdots \\ a_{m1}+0 & \cdots & a_{mn}+0 \end{pmatrix} \\ &= \begin{pmatrix} a_{11} & \cdots & a_{1n} \\ \vdots & \vdots & \vdots \\ a_{m1} & \cdots & a_{mn} \end{pmatrix} \\ &= A, \end{aligned}$$

i.e.,

$$A+O = A.$$

5.

$$\begin{aligned}
1A &= 1\begin{pmatrix} a_{11} & \cdots & a_{1n} \\ \vdots & \vdots & \vdots \\ a_{m1} & \cdots & a_{mn} \end{pmatrix} \\
&= \begin{pmatrix} 1\cdot a_{11} & \cdots & 1\cdot a_{1n} \\ \vdots & \vdots & \vdots \\ 1\cdot a_{m1} & \cdots & 1\cdot a_{mn} \end{pmatrix} \\
&= \begin{pmatrix} a_{11} & \cdots & a_{1n} \\ \vdots & \vdots & \vdots \\ a_{m1} & \cdots & a_{mn} \end{pmatrix} \\
&= A,
\end{aligned}$$

i.e.,

$$1A = A.$$

6.

$$\begin{aligned}
a(A+B) &= a\left(\begin{pmatrix} a_{11} & \cdots & a_{1n} \\ \vdots & \vdots & \vdots \\ a_{m1} & \cdots & a_{mn} \end{pmatrix} + \begin{pmatrix} b_{11} & \cdots & b_{1n} \\ \vdots & \vdots & \vdots \\ b_{m1} & \cdots & b_{mn} \end{pmatrix}\right) \\
&= a\begin{pmatrix} a_{11}+b_{11} & \cdots & a_{1n}+b_{1n} \\ \vdots & \vdots & \vdots \\ a_{m1}+b_{m1} & \cdots & a_{mn}+b_{mn} \end{pmatrix} \\
&= \begin{pmatrix} a(a_{11}+b_{11}) & \cdots & a(a_{1n}+b_{1n}) \\ \vdots & \vdots & \vdots \\ a(a_{m1}+b_{m1}) & \cdots & a(a_{mn}+b_{mn}) \end{pmatrix} \\
&= \begin{pmatrix} aa_{11}+ab_{11} & \cdots & aa_{1n}+ab_{1n} \\ \vdots & \vdots & \vdots \\ aa_{m1}+ab_{m1} & \cdots & aa_{mn}+ab_{mn} \end{pmatrix} \\
&= \begin{pmatrix} aa_{11} & \cdots & aa_{1n} \\ \vdots & \vdots & \vdots \\ aa_{m1} & \cdots & aa_{mn} \end{pmatrix} + \begin{pmatrix} ab_{11} & \cdots & ab_{1n} \\ \vdots & \vdots & \vdots \\ ab_{m1} & \cdots & ab_{mn} \end{pmatrix}
\end{aligned}$$

$$= \quad aA + aB,$$

i.e.,

$$a(A+B) = aA + aB.$$

7.

$$\begin{aligned}
(a+b)A &= (a+b)\begin{pmatrix} a_{11} & \cdots & a_{1n} \\ \vdots & \vdots & \vdots \\ a_{m1} & \cdots & a_{mn} \end{pmatrix} \\
&= \begin{pmatrix} (a+b)a_{11} & \cdots & (a+b)a_{1n} \\ \vdots & \vdots & \vdots \\ (a+b)a_{m1} & \cdots & (a+b)a_{mn} \end{pmatrix} \\
&= \begin{pmatrix} aa_{11}+ba_{11} & \cdots & aa_{1n}+ba_{1n} \\ \vdots & \vdots & \vdots \\ aa_{m1}+ba_{m1} & \cdots & aa_{mn}+ba_{mn} \end{pmatrix} \\
&= \begin{pmatrix} aa_{11} & \cdots & aa_{1n} \\ \vdots & \vdots & \vdots \\ aa_{m1} & \cdots & aa_{mn} \end{pmatrix} + \begin{pmatrix} ba_{11} & \cdots & ba_{1n} \\ \vdots & \vdots & \vdots \\ ba_{m1} & \cdots & ba_{mn} \end{pmatrix} \\
&= aA + bA,
\end{aligned}$$

i.e.,

$$(a+b)A = aA + bA.$$

Therefore $\mathbb{F}^{m,n}$ is a vector space. This completes the proof.

□

Theorem 3.72. *We have*

$$\dim \mathbb{F}^{m,n} = mn. \tag{3.24}$$

Proof. Let A^{jk}, $j \in \{1,\ldots,m\}$, $k \in \{1,\ldots,n\}$, be a $m \times n$ matrix that have 0 in all entries except for a 1 in one entry. Let $a_{ij} \in \mathbb{F}$, $j \in \{1,\ldots,m\}$, $k \in \{1,\ldots,n\}$, be such that

$$O \quad = \quad a_{11}A^{11} + \cdots + a_{1n}A^{1n}$$

$$\vdots$$

$$+a_{m1}A^{m1}+\cdots+a_{mn}A^{mn}$$

$$= \begin{pmatrix} a_{11} & \cdots & a_{1n} \\ \vdots & \vdots & \vdots \\ a_{m1} & \cdots & a_{mn} \end{pmatrix},$$

whereupon

$$a_{jk}=0, \quad j\in\{1,\ldots,m\}, \quad k\in\{1,\ldots,n\}.$$

Therefore, the list

$$A^{jk}, \quad j\in\{1,\ldots,m\}, \quad k\in\{1,\ldots,n\}, \tag{3.25}$$

is linearly independent in $\mathbb{F}^{m,n}$. Next, let

$$B = \begin{pmatrix} b_{11} & \cdots & b_{1n} \\ \vdots & \vdots & \vdots \\ b_{m1} & \cdots & b_{mn} \end{pmatrix} \in \mathbb{F}^{m,n}$$

be arbitrarily chosen. Then, we have the following

$$B = b_{11}A^{11}+\cdots+b_{1n}A^{1n}$$

$$\vdots$$

$$+b_{m1}A^{m1}+\cdots+b_{mn}A^{mn}.$$

Assume that there are c_{jk}, $j\in\{1,\ldots,m\}$, $k\in\{1,\ldots,n\}$, so that

$$B = c_{11}A^{11}+\cdots+c_{1n}A^{1n}$$

$$\vdots$$

$$+c_{m1}A^{m1}+\cdots+c_{mn}A^{mn}.$$

Therefore

$$O = (b_{11}-c_{11})A^{11}+\cdots+(b_{1n}-c_{1n})A^{1n}$$

$$
\begin{aligned}
&\vdots \\
&+(b_{m1}-c_{m1})A^{m1}+\cdots+(b_{mn}-c_{mn})A^{mn} \\
&= \begin{pmatrix} b_{11}-c_{11} & \cdots & b_{1n}-c_{1n} \\ \vdots & \vdots & \vdots \\ b_{m1}-c_{m1} & \cdots & b_{mn}-c_{mn} \end{pmatrix}.
\end{aligned}
$$

Hence,

$$
b_{jk}=c_{jk}, \quad j\in\{1,\ldots,m\}, \quad k\in\{1,\ldots,n\}.
$$

Thus, the list (3.25) spans $\mathbb{F}^{m,n}$ and it is a basis of $\mathbb{F}^{m,n}$. Since the list (3.25) has mn elements, we conclude that (3.24) holds. This completes the proof.

□

3.7 Matrix Multiplication

Let $v_1,\ldots,v_n$ be a basis of V, $w_1,\ldots,w_m$ be a basis of W and $u_1,\ldots,u_p$ be a basis of U. Consider the linear maps $T\in\mathscr{L}(U,V)$ and $S\in\mathscr{L}(V,W)$. Suppose that $\mathscr{M}(S)=A$ and $\mathscr{M}(T)=C$. Then, for any $k\in\{1,\ldots,p\}$, we have

$$
\begin{aligned}
(ST)u_k &= S(Tu_k) \\
&= S\left(\sum_{r=1}^{k} C_{rk}v_r\right) \\
&= \sum_{r=1}^{n} S(C_{rk}v_r) \\
&= \sum_{r=1}^{n} C_{rk}Sv_k \\
&= \sum_{r=1}^{n} C_{rk}\sum_{j=1}^{m} A_{jr}w_j
\end{aligned}
$$

$$= \sum_{r=1}^{n} \sum_{j=1}^{m} C_{rk} A_{jr} w_j$$

$$= \sum_{j=1}^{m} \left(\sum_{r=1}^{n} C_{rk} A_{jr} \right) w_j.$$

Thus, $\mathcal{M}(ST)$ is the $m \times p$ matrix whose jk entry is

$$\sum_{r=1}^{n} C_{rk} A_{jr}. \tag{3.26}$$

Definition 3.73. *Let $A \in \mathbb{F}^{m,n}$, $C \in \mathbb{F}^{n,p}$. Then AC is the matrix defined to be the $m \times p$ matrix whose jk entry is given by* (3.26)*. In other words, the jk entry of AC is computed by taking row j of A and column k of C, multiplying together corresponding entries, and then summing.*

By the above definition, it follows that

$$\mathcal{M}(ST) = \mathcal{M}(S)\mathcal{M}(T).$$

Remark 3.74. *Let $A \in \mathbb{F}^{n,p}$ and $B \in \mathbb{F}^{q,m}$. If $p \neq q$, then the multiplication AB does not exist.*

Example 3.75. *Let*

$$A = \begin{pmatrix} 3 & -2 \\ 5 & -4 \end{pmatrix},$$

$$B = \begin{pmatrix} 3 & 4 \\ 2 & 5 \end{pmatrix}.$$

Then

$$AB = \begin{pmatrix} 3 & -2 \\ 5 & -4 \end{pmatrix} \begin{pmatrix} 3 & 4 \\ 2 & 5 \end{pmatrix}$$

$$= \begin{pmatrix} 3 \cdot 3 + (-2) \cdot 2 & 3 \cdot 4 + (-2) \cdot 5 \\ 5 \cdot 3 + (-4) \cdot 2 & 5 \cdot 4 + (-4) \cdot 5 \end{pmatrix}$$

$$= \begin{pmatrix} 5 & 2 \\ 7 & 0 \end{pmatrix}$$

and

$$BA = \begin{pmatrix} 3 & 4 \\ 2 & 5 \end{pmatrix} \begin{pmatrix} 3 & -2 \\ 5 & -4 \end{pmatrix}$$

$$= \begin{pmatrix} 3\cdot3+4\cdot5 & 3\cdot(-2)+4\cdot(-4) \\ 2\cdot3+5\cdot5 & 2\cdot(-2)+5\cdot(-4) \end{pmatrix}$$

$$= \begin{pmatrix} 29 & -22 \\ 31 & -24 \end{pmatrix}.$$

Remark 3.76. *By the last example, it follows that in the general case we have*

$$AB \neq BA.$$

Example 3.77. *Let*

$$A = \begin{pmatrix} 1 & 2 & 3 \\ 1 & 0 & -1 \end{pmatrix},$$

$$B = \begin{pmatrix} 3 & 4 & 5 \\ 6 & 0 & -2 \\ 7 & 1 & 8 \end{pmatrix}.$$

Then

$$AB = \begin{pmatrix} 1 & 2 & 3 \\ 1 & 0 & -1 \end{pmatrix} \begin{pmatrix} 3 & 4 & 5 \\ 6 & 0 & -2 \\ 7 & 1 & 8 \end{pmatrix}$$

$$= \left(\begin{matrix} 1\cdot3+2\cdot6+3\cdot7 & 1\cdot4+2\cdot0+3\cdot1 \\ 1\cdot3+0\cdot6+(-1)\cdot7 & 1\cdot4+0\cdot0+(-1)\cdot1 \end{matrix} \right.$$

$$\left. \begin{matrix} 1\cdot5+2\cdot(-2)+3\cdot1 \\ 1\cdot5+0\cdot(-2)+(-1)\cdot8 \end{matrix} \right)$$

$$= \begin{pmatrix} 36 & 7 & 4 \\ -4 & 3 & -3 \end{pmatrix}.$$

Note that the multiplication BA does not exist.

Example 3.78. *Let*

$$A = \begin{pmatrix} 2 & 0 & 3 \\ -1 & 2 & 1 \end{pmatrix},$$

$$B = \begin{pmatrix} -4 \\ -3 \\ 5 \end{pmatrix},$$

$$C = \begin{pmatrix} 2 \\ -3 \end{pmatrix}.$$

We will find

$$2AB - 3C.$$

We have

$$\begin{aligned} AB &= \begin{pmatrix} 2 & 0 & 3 \\ -1 & 2 & 1 \end{pmatrix} \begin{pmatrix} -4 \\ -3 \\ 5 \end{pmatrix} \\ &= \begin{pmatrix} 2\cdot(-4)+0\cdot(-3)+3\cdot 5 \\ -1\cdot(-4)+2\cdot(-3)+1\cdot 5 \end{pmatrix} \\ &= \begin{pmatrix} 7 \\ 3 \end{pmatrix}, \\ 2AB &= 2\begin{pmatrix} 7 \\ 3 \end{pmatrix} \\ &= \begin{pmatrix} 2\cdot 7 \\ 2\cdot 3 \end{pmatrix} \\ &= \begin{pmatrix} 14 \\ 6 \end{pmatrix}, \\ 3C &= 3\begin{pmatrix} 2 \\ -3 \end{pmatrix} \\ &= \begin{pmatrix} 3\cdot 2 \\ 3\cdot(-3) \end{pmatrix} \\ &= \begin{pmatrix} 6 \\ -9 \end{pmatrix}. \end{aligned}$$

Therefore

$$\begin{aligned} 2AB - 3C &= \begin{pmatrix} 14 \\ 6 \end{pmatrix} - \begin{pmatrix} 6 \\ -9 \end{pmatrix} \\ &= \begin{pmatrix} 14-6 \\ 6+9 \end{pmatrix} \end{aligned}$$

$$= \begin{pmatrix} 8 \\ 15 \end{pmatrix}.$$

Exercise 3.79. *Find AB and BA, provided they exist, where*

1.

$$A = \begin{pmatrix} 3 & 5 & -1 \\ 2 & -2 & 0 \end{pmatrix},$$

$$B = \begin{pmatrix} 2 & 4 \\ -3 & 0 \\ 5 & 1 \end{pmatrix}.$$

2.

$$A = \begin{pmatrix} -2 & 3 & 1 \\ 5 & 4 & 0 \\ 2 & -1 & -5 \end{pmatrix},$$

$$B = \begin{pmatrix} 1 & -2 & -3 \\ 0 & -3 & 1 \\ 4 & -4 & 5 \end{pmatrix}.$$

3.

$$A = \begin{pmatrix} 1 & -3 & 2 \\ 3 & -4 & 1 \\ 2 & -5 & 3 \end{pmatrix},$$

$$B = \begin{pmatrix} 2 & 5 & 6 \\ 1 & 2 & 5 \\ 1 & 3 & 2 \end{pmatrix}.$$

$$A = \begin{pmatrix} 5 & 8 & -4 \\ 6 & 9 & -5 \\ 4 & 7 & -3 \end{pmatrix},$$

$$B = \begin{pmatrix} 3 & 2 & 5 \\ 4 & -1 & 3 \\ 9 & 6 & 5 \end{pmatrix}.$$

4.

$$A = \begin{pmatrix} 2 & -1 & 3 & -4 \\ 3 & -2 & 4 & -3 \\ 5 & -3 & -2 & 1 \\ 3 & -3 & -1 & 2 \end{pmatrix},$$

$$B = \begin{pmatrix} 7 & 8 & 6 & 9 \\ 5 & 7 & 4 & 5 \\ 3 & 4 & 5 & 6 \\ 2 & 1 & 1 & 2 \end{pmatrix}.$$

Definition 3.80. *Let $A \in \mathbb{F}^{n,p}$, $B \in \mathbb{F}^{p,q}$, $C \in \mathbb{F}^{q,m}$. Then, we define*

$$ABC = A(BC). \tag{3.27}$$

Remark 3.81. *For $A \in \mathbb{F}^{n,p}$, $B \in \mathbb{F}^{p,q}$, $C \in \mathbb{F}^{q,m}$, we can define*

$$ABC = (AB)C. \tag{3.28}$$

In fact, (3.27) *and* (3.28) *are equivalently. Really, we have the following.*

$$\begin{aligned} (BC)_{jk} &= \sum_{r=1}^{q} B_{jr} C_{rk}, \quad j \in \{1,\ldots,p\}, \quad k \in \{1,\ldots,m\}, \\ (A(BC))_{jk} &= \sum_{r=1}^{p} A_{jr} (BC)_{rk} \\ &= \sum_{r=1}^{p} A_{jr} \left(\sum_{l=1}^{q} B_{rl} C_{lk} \right) \\ &= \sum_{r=1}^{p} \sum_{l=1}^{q} A_{jr} B_{rl} C_{lk}, \quad j \in \{1,\ldots,n\}, \quad k \in \{1,\ldots,m\}, \end{aligned}$$

and

$$\begin{aligned} (AB)_{jk} &= \sum_{r=1}^{p} A_{jr} B_{rk}, \quad j \in \{1,\ldots,n\}, \quad k \in \{1,\ldots,q\}, \\ ((AB)C)_{jk} &= \sum_{l=1}^{q} (AB)_{jl} C_{lk} \end{aligned}$$

$$\begin{aligned}
&= \sum_{l=1}^{q}\left(\sum_{r=1}^{p} A_{jr}B_{rl}\right)C_{lk}\\
&= \sum_{l=1}^{q}\sum_{r=1}^{p} A_{jr}B_{rl}C_{lk}\\
&= \sum_{r=1}^{q}\sum_{l=1}^{q} A_{jr}B_{rl}C_{lk}, \quad j \in \{1,\ldots,n\}, \quad k \in \{1,\ldots,m\}.
\end{aligned}$$

Hence,

$$(A(BC))_{jk} = ((AB)C)_{jk}, \quad j \in \{1,\ldots,n\}, \quad k \in \{1,\ldots,m\}.$$

Therefore

$$A(BC) = (AB)C$$

and (3.27) *and* (3.28) *are equivalently.*

Example 3.82. *Let*

$$\begin{aligned}
A &= \begin{pmatrix} 0 & 2 & -1 \\ -2 & -1 & 2 \\ 3 & -2 & -1 \end{pmatrix},\\
B &= \begin{pmatrix} 70 & 34 & -107 \\ 52 & 26 & -68 \\ 101 & 50 & -140 \end{pmatrix},\\
C &= \begin{pmatrix} 27 & -18 & 10 \\ -46 & 31 & -17 \\ 3 & -2 & 1 \end{pmatrix}.
\end{aligned}$$

We will find

$$ABC.$$

We have

$$\begin{aligned}
AB &= \begin{pmatrix} 0 & 2 & -1 \\ -2 & -1 & 2 \\ 3 & -2 & -1 \end{pmatrix}\begin{pmatrix} 70 & 34 & -107 \\ 52 & 26 & -68 \\ 101 & 50 & -140 \end{pmatrix}\\
&= \left(\begin{matrix} 0\cdot 70+2\cdot 52+(-1)\cdot 101 & 0\cdot 34+2\cdot 26+(-1)\cdot 50 \\ (-2)\cdot 70+(-1)\cdot 52+2\cdot 101 & (-2)\cdot 34+(-1)\cdot 26+2\cdot 50 \\ 3\cdot 70+(-2)\cdot 52+(-1)\cdot 101 & 3\cdot 34+(-2)\cdot 26+(-1)\cdot 50 \end{matrix}\right.
\end{aligned}$$

$$\left.\begin{matrix} 0\cdot(-107)+2\cdot(-68)+(-1)\cdot(-140) \\ (-2)\cdot(-107)+(-1)\cdot(-68)+2\cdot(-140) \\ 3\cdot(-107)+(-2)\cdot(-68)+(-1)\cdot(-140) \end{matrix}\right)$$

$$= \begin{pmatrix} 3 & 2 & 4 \\ 10 & 6 & 2 \\ 5 & 0 & -45 \end{pmatrix}$$

and

$$ABC = (AB)C$$

$$= \begin{pmatrix} 3 & 2 & 4 \\ 10 & 6 & 2 \\ 5 & 0 & -45 \end{pmatrix} \begin{pmatrix} 27 & -18 & 10 \\ -46 & 31 & -17 \\ 3 & -2 & 1 \end{pmatrix}$$

$$= \left(\begin{matrix} 3\cdot 27+2\cdot(-46)+4\cdot 3 & 3\cdot(-18)+2\cdot 31+4\cdot(-2) \\ 10\cdot 27+6\cdot(-46)+2\cdot 3 & 10\cdot(-18)+6\cdot 31+2\cdot(-2) \\ 5\cdot 27+0\cdot(-46)+(-45)\cdot 3 & 5\cdot(-18)+0\cdot 31+(-45)\cdot(-2) \end{matrix}\right.$$

$$\left.\begin{matrix} 3\cdot 10+2\cdot(-17)+4\cdot 1 \\ 10\cdot 10+6\cdot(-17)+2\cdot 1 \\ 5\cdot 10+0\cdot(-17)+(-45)\cdot 1 \end{matrix}\right)$$

$$= \begin{pmatrix} 1 & 0 & 0 \\ 0 & 2 & 0 \\ 0 & 0 & 5 \end{pmatrix}.$$

Exercise 3.83. *Find ABC, where*

$$A = \begin{pmatrix} 4 & 3 \\ 7 & 5 \end{pmatrix},$$

$$B = \begin{pmatrix} -28 & 93 \\ 38 & -126 \end{pmatrix},$$

$$C = \begin{pmatrix} 7 & 3 \\ 2 & 1 \end{pmatrix}.$$

Definition 3.84. *Let* $A_1 \in \mathbb{F}^{p_1,p_2}$, $A_2 \in \mathbb{F}^{p_2,p_3}$, …, $A_n \in \mathbb{F}^{p_n,p_{n+1}}$. *Define* $A_1A_2\ldots A_n$ *as follows:*

$$A_1A_2\ldots A_n = (\ldots(((A_1A_2)A_3)A_4)\ldots A_{n-1})A_n.$$

Example 3.85. *Let A, B, C be as in Example 3.82 and*

$$D = \begin{pmatrix} -3 & -1 & 2 \\ 1 & 1 & 0 \\ -1 & 1 & 1 \end{pmatrix}.$$

We will find

$$ABCD.$$

Using the computations in Example 3.82, we get

$$\begin{aligned} ABCD &= ((AB)C)D \\ &= (ABC)D \\ &= \begin{pmatrix} 1 & 0 & 0 \\ 0 & 2 & 0 \\ 0 & 0 & 5 \end{pmatrix} \begin{pmatrix} -3 & -1 & 2 \\ 1 & 1 & 0 \\ -1 & 1 & 1 \end{pmatrix} \\ &= \begin{pmatrix} 1\cdot(-3)+0\cdot 1+0\cdot(-1) & 1\cdot(-1)+0\cdot 1+0\cdot 1 & 1\cdot 2+0\cdot 0+0\cdot 1 \\ 0\cdot(-3)+2\cdot 1+0\cdot(-1) & 0\cdot(-1)+2\cdot 1+0\cdot 1 & 0\cdot 2+2\cdot 0+0\cdot 1 \\ 0\cdot(-3)+0\cdot 1+5\cdot(-1) & 0\cdot(-1)+0\cdot 1+5\cdot 1 & 0\cdot 2+0\cdot 0+5\cdot 1 \end{pmatrix} \\ &= \begin{pmatrix} -3 & -1 & 2 \\ 2 & 2 & 0 \\ -5 & 5 & 5 \end{pmatrix}. \end{aligned}$$

Exercise 3.86. *Find ABCD, where*

$$\begin{aligned} A &= \begin{pmatrix} -1 & 1 & 2 \\ 2 & 0 & -2 \\ -3 & 1 & 1 \end{pmatrix}, \\ B &= \begin{pmatrix} 1 & 2 & 3 \\ -5 & -4 & -6 \\ 1 & -1 & -1 \end{pmatrix}, \\ C &= \begin{pmatrix} -3 & 0 & 0 \\ 0 & 1 & 0 \\ 0 & 0 & -2 \end{pmatrix}, \end{aligned}$$

$$D = \begin{pmatrix} -1 & 0 & 1 \\ -3 & 1 & -2 \\ -4 & -1 & 1 \end{pmatrix}.$$

Definition 3.87. *A matrix for which the number of the rows equals the number of the columns is said to be quadratic.*

Example 3.88. *Let*

$$A = \begin{pmatrix} -3 & 1 & 2 & 0 \\ 1 & 2 & -3 & 1 \\ -1 & 1 & 4 & 2 \\ 4 & 0 & -2 & -3 \end{pmatrix}.$$

Here the number of the rows is 4 *and the number of the columns is* 4. *Therefore A is a quadratic matrix.*

Example 3.89. *Let*

$$A = \begin{pmatrix} -1 & 7 & 8 & 9 & 0 & 1 \\ 1 & 2 & 2 & 4 & 6 & 8 \\ -1 & 3 & 3 & 2 & 2 & 0 \end{pmatrix}.$$

Here the number of the rows is 3 *and the number of the columns is* 6. *Therefore A is not a quadratic matrix.*

Exercise 3.90. *Determine which matrices are quadratic.*

1.

$$A = \begin{pmatrix} -1 & 7 & 3 & 4 & 8 & 0 & 1 \end{pmatrix}.$$

2.

$$A = \begin{pmatrix} -1 & 0 & 1 & 3 & 2 \\ -2 & -3 & 0 & 0 & 2 \\ 1 & 0 & 0 & 0 & 4 \\ 0 & 3 & 5 & 101 & 32 \\ 2 & 1 & -1 & -4 & 2 \end{pmatrix}.$$

3.

$$A = \begin{pmatrix} 0 \\ -1 \\ 1 \\ 2 \\ -2 \end{pmatrix}.$$

4.
$$A = \begin{pmatrix} -1 & 1 \\ 3 & 2 \end{pmatrix}.$$

5.
$$A = \begin{pmatrix} -1 & 3 & 1 \\ 1 & 0 & 4 \\ -2 & 1 & 2 \end{pmatrix}.$$

Definition 3.91. *For a quadratic matrix $A \in \mathbb{F}^{m,m}$, define*

$$\begin{aligned} A^2 &= AA, \\ A^3 &= A^2A = AA^2, \\ &\vdots \\ A^m &= A^{m-1}A = AA^{m-1}, \quad m \in \mathbb{N}. \end{aligned}$$

Example 3.92. *Let*
$$A = \begin{pmatrix} 1 & -2 \\ 3 & -4 \end{pmatrix}.$$

We will find A^3. We have

$$\begin{aligned} A^2 &= AA \\ &= \begin{pmatrix} 1 & -2 \\ 3 & -4 \end{pmatrix} \begin{pmatrix} 1 & -2 \\ 3 & -4 \end{pmatrix} \\ &= \begin{pmatrix} 1\cdot 1 + (-2)\cdot 3 & 1\cdot(-2) + (-2)\cdot(-4) \\ 3\cdot 1 + (-4)\cdot 3 & 3\cdot(-2) + (-4)\cdot(-4) \end{pmatrix} \\ &= \begin{pmatrix} -5 & 6 \\ -9 & 10 \end{pmatrix} \end{aligned}$$

and

$$A^3 = A^2A$$

$$
\begin{aligned}
&= \begin{pmatrix} -5 & 6 \\ -9 & 10 \end{pmatrix} \begin{pmatrix} 1 & -2 \\ 3 & -4 \end{pmatrix} \\
&= \begin{pmatrix} -5\cdot 1+6\cdot 3 & -5\cdot(-2)+6\cdot(-4) \\ -9\cdot 1+10\cdot 3 & -9\cdot(-2)+10\cdot(-4) \end{pmatrix} \\
&= \begin{pmatrix} 13 & -14 \\ 21 & -22 \end{pmatrix}
\end{aligned}
$$

.

Exercise 3.93. *Let*

$$A = \begin{pmatrix} 4 & -1 \\ 5 & -2 \end{pmatrix}.$$

Find A^5.

Definition 3.94. *Let* $A \in \mathbb{F}^{m,n}$ *and* $j \in \{1,\dots,m\}$, $k \in \{1,\dots,n\}$. *With* $A_{j,}$ *we will denote the* $1\times n$ *matrix consisting of the row* j *of* A. *With* $A_{,k}$ *we will denote the* $m \times 1$ *matrix consisting of the column* k *of* A.

Example 3.95. *Let*

$$A = \begin{pmatrix} 8 & 4 & 5 \\ 1 & 9 & 7 \end{pmatrix}.$$

Then

$$
\begin{aligned}
A_{1,} &= \begin{pmatrix} 8 & 4 & 5 \end{pmatrix}, \\
A_{2,} &= \begin{pmatrix} 1 & 9 & 7 \end{pmatrix}, \\
A_{,1} &= \begin{pmatrix} 8 \\ 1 \end{pmatrix}, \\
A_{,2} &= \begin{pmatrix} 4 \\ 9 \end{pmatrix}, \\
A_{,3} &= \begin{pmatrix} 5 \\ 7 \end{pmatrix}.
\end{aligned}
$$

Exercise 3.96. *Let*

$$A = \begin{pmatrix} -1 & 0 & 3 & 5 & 6 \\ 2 & 3 & -1 & 4 & 5 \\ 0 & 1 & 2 & 3 & 4 \end{pmatrix}.$$

Find

1. $A_{2,\cdot}$.
2. $A_{3,\cdot}$.
3. $A_{,3}$.
4. $A_{,5}$.
5. $2A_{2,} - 3A_{3,}$.
6. $3A_{,3} + 4A_{,5}$.
7. $A_{,3}A_{2,}$.
8. $A_{,5}A_{3,}$.
9. $2A_{,3}A_{2,}$.
10. $4A_{,5}A_{3,}$.

Theorem 3.97. *Let $A \in \mathbb{F}^{m,n}$, $C \in \mathbb{F}^{n,p}$. Then*

$$(AC)_{jk} = A_{j,}C_{,k}, \quad j \in \{1,\ldots,m\}, \quad k \in \{1,\ldots,p\}.$$

Proof. We have

$$\begin{aligned} (AC)_{jk} &= \sum_{r=1}^{n} A_{jr}C_{rk} \\ &= A_{j,}C_{,k}, \quad j \in \{1,\ldots,m\}, \quad k \in \{1,\ldots,p\}. \end{aligned}$$

This completes the proof. □

Theorem 3.98. *Let $A \in \mathbb{F}^{m,n}$, $C \in \mathbb{F}^{n,p}$. Then*

$$(AC)_{jk} = AC_{,k}, \quad k \in \{1,\ldots,p\}. \tag{3.29}$$

Proof. We have

$$(AC)_{jk} = \sum_{r=1}^{n} A_{jr}C_{rk}, \quad j \in \{1,\ldots,m\}, \quad k \in \{1,\ldots,n\}.$$

Then

$$(AC)_{jk} = \begin{pmatrix} \sum_{r=1}^{n} A_{1r}C_{rk} \\ \sum_{r=1}^{n} A_{2r}C_{rk} \\ \vdots \\ \sum_{r=1}^{n} A_{mr}C_{rk} \end{pmatrix}. \qquad (3.30)$$

Next,

$$C_{,k} = \begin{pmatrix} C_{1k} \\ C_{2k} \\ \vdots \\ C_{nk} \end{pmatrix}$$

and

$$\begin{aligned} AC_{,k} &= \begin{pmatrix} A_{11} & \cdots & A_{1n} \\ A_{21} & \cdots & A_{2n} \\ \vdots & \vdots & \vdots \\ A_{m1} & \cdots & A_{mn} \end{pmatrix} \begin{pmatrix} C_{1k} \\ C_{2k} \\ \vdots \\ C_{nk} \end{pmatrix} \\ &= \begin{pmatrix} \sum_{r=1}^{n} A_{1r}C_{rk} \\ \sum_{r=1}^{n} A_{2r}C_{rk} \\ \vdots \\ \sum_{r=1}^{n} A_{mr}C_{rk} \end{pmatrix}. \end{aligned}$$

By the last equation and (3.30), it follows (3.29). This completes the proof.
□

Theorem 3.99. *Let* $A \in \mathbb{F}^{m,n}$, $C \in \mathbb{F}^{n,1}$,

$$A = \begin{pmatrix} A_{11} & \cdots & A_{1n} \\ A_{21} & \cdots & A_{2n} \\ \vdots & \vdots & \vdots \\ A_{m1} & \cdots & A_{mn} \end{pmatrix},$$

$$C = \begin{pmatrix} C_1 \\ C_2 \\ \vdots \\ C_n \end{pmatrix}.$$

Then

$$AC = C_1 A_{,1} + \cdots + C_n A_{,n}. \tag{3.31}$$

Proof. We have

$$\begin{aligned} AC &= \begin{pmatrix} A_{11} & \cdots & A_{1n} \\ A_{21} & \cdots & A_{2n} \\ \vdots & \vdots & \vdots \\ A_{m1} & \cdots & A_{mn} \end{pmatrix} \begin{pmatrix} C_1 \\ C_2 \\ \vdots \\ C_n \end{pmatrix} \\ &= \begin{pmatrix} \sum_{r=1}^{n} A_{1r} C_r \\ \sum_{r=1}^{n} A_{2r} C_r \\ \vdots \\ \sum_{r=1}^{n} A_{mr} C_r \end{pmatrix}. \end{aligned} \tag{3.32}$$

Next,

$$\begin{aligned} C_1 A_{,1} + C_2 A_{,2} + \cdots + C_n A_{,n} &= C_1 \begin{pmatrix} A_{11} \\ A_{21} \\ \vdots \\ A_{m1} \end{pmatrix} + C_2 \begin{pmatrix} A_{12} \\ A_{22} \\ \vdots \\ A_{m2} \end{pmatrix} \\ &\quad + \cdots + C_n \begin{pmatrix} A_{1n} \\ A_{2n} \\ \vdots \\ A_{mn} \end{pmatrix} \\ &= \begin{pmatrix} C_1 A_{11} \\ C_1 A_{21} \\ \vdots \\ C_1 A_{n1} \end{pmatrix} + \begin{pmatrix} C_2 A_{12} \\ C_2 A_{22} \\ \vdots \\ C_2 A_{m2} \end{pmatrix} \end{aligned}$$

$$+\cdots+\begin{pmatrix} C_nA_{1n} \\ C_nA_{2n} \\ \vdots \\ C_nA_{mn} \end{pmatrix}$$

$$= \begin{pmatrix} \sum_{r=1}^{n} A_{1r}C_r \\ \sum_{r=1}^{n} A_{2r}C_r \\ \vdots \\ \sum_{r=1}^{n} A_{mr}C_r \end{pmatrix}.$$

By the last equation and (3.32), we get (3.31). This completes the proof. □

3.8 Advanced Practical Problems

Problem 3.100. *Let $T : \mathbb{R}^3 \to \mathbb{R}^3$ be defined by*

1. $T(x,y,z) = (4x-3y-2z, x, x+3z)$, $(x,y,z) \in \mathbb{R}^3$.
2. $T(x,y,z) = (x-4z, x, x)$, $(x,y,z) \in \mathbb{R}^3$.
3. $T(x,y,z) = (x, 2y, 3z)$, $(x,y,z) \in \mathbb{R}^3$.
4. $T(x,y,z) = (y+z, 2x+z, 3x-y+z)$, $(x,y,z) \in \mathbb{R}^3$.
5. $T(x,y,z) = (x-y+z, z, y)$, $(x,y,z) \in \mathbb{R}^3$.

Check if $T \in \mathscr{L}(\mathbb{R}^3, \mathbb{R}^3)$.

Problem 3.101. *Let $S, T : \mathbb{R}^4 \to \mathbb{R}^3$ be defined by*

$$\begin{aligned} S(x,y,z,t) &= (2x-2y+z+t, 2x-y, z+t), \\ T(x,y,z,t) &= (x+y-z, x-2y+z+t, z+2t) \quad (x,y,z,t) \in \mathbb{R}^4. \end{aligned}$$

Find

1. $(S+T)(x,y,z,t)$, $(x,y,z,t) \in \mathbb{R}^4$.

2. $(3S)(x,y,z,t)$, $(x,y,z,t) \in \mathbb{R}^4$.
3. $(2T)(x,y,z,t)$, $(x,y,z,t) \in \mathbb{R}^4$.
4. $(S-T)(x,y,z,t)$, $(x,y,z,t) \in \mathbb{R}^4$.
5. $(3S-5T)(x,y,z,t)$, $(x,y,z,t) \in \mathbb{R}^4$.

Problem 3.102. *Let $T : \mathbb{R}^5 \to \mathbb{R}^4$ and $S : \mathbb{R}^4 \to \mathbb{R}^3$ be defined by*

$$T(x,y,z,t,w) = (x, 3y+z+t, z+2t+w, t-3w), \quad (x,y,z,t,w) \in \mathbb{R}^5,$$

$$S(x,y,z,t) = (x+y+z, y-2z, y+z+t), \quad (x,y,z,t) \in \mathbb{R}^4.$$

Find

1. $(ST)(x,y,z,t,w)$, $(x,y,z,t,w) \in \mathbb{R}^5$.
2. $(2T)(x,y,z,t,w)$, $(x,y,z,t,w) \in \mathbb{R}^5$.
3. $(ST-2T)(x,y,z,t,w)$, $(x,y,z,t,w) \in \mathbb{R}^5$.
4. $((3S)T)(x,y,z,t,w)$, $(x,y,z,t,w) \in \mathbb{R}^5$.
5. $((2S)T-4T)(x,y,z,t,w)$, $(x,y,z,t,w) \in \mathbb{R}^5$.

Problem 3.103. *Let $\phi : \mathbb{R}^3 \to \mathbb{R}^2$ be defined by*

$$\phi(x,y,z) = (x-y, y^2-z^2, x+z), \quad (x,y,z) \in \mathbb{R}^3.$$

Find $\ker\phi$.

Problem 3.104. *Find a nonzero solution to the following systems.*

1.
$$x_1 - 5x_2 + 4x_3 = 0.$$

2.
$$x_1 - 3x_2 + 4x_3 + 8x_4 = 0.$$

3.
$$\begin{aligned} x_1 + 2x_3 + x_4 + x_5 &= 0 \\ x_3 + x_4 + x_5 &= 0 \\ -x_1 + x_2 + 4x_4 &= 0. \end{aligned}$$

4.

$$\begin{aligned} x_1+x_2-3x_3 &= 0 \\ x_2-x_3+x_4 &= 0. \end{aligned}$$

5.

$$\begin{aligned} x_1+x_2 &= 0 \\ x_2+x_3+x_4+x_5+x_6 &= 0. \end{aligned}$$

Problem 3.105. *Investigate the system*

$$\begin{aligned} x_1+ax_2 &= 1 \\ 2x_1+4x_2 &= 2 \\ bx_1+4x_2 &= 2, \end{aligned}$$

where $a, b \in \mathbb{R}$ are parameters.

Problem 3.106. *Let $T \in \mathscr{L}(\mathbb{R}^5, \mathbb{R}^3)$ be given by*

$$T(x,y,z,t,w) = (x-y+3z+4t-w, y+z-w, 2y-z+t+4w),$$

$(x,y,z,t,w) \in \mathbb{R}^5$. *Find $\mathscr{M}(T)$.*

Problem 3.107. *Find $A+B$, where*

$$A = \begin{pmatrix} -4 & 1 & 2 & 3 & -11 & 12 & 15 \\ -3 & 0 & 0 & 1 & 2 & 3 & 5 \\ -4 & -1 & -3 & 1 & 2 & -1 & 3 \end{pmatrix},$$

$$B = \begin{pmatrix} -3 & 0 & 1 & 2 & 11 & -15 & 1 \\ -4 & 1 & 2 & 3 & -5 & 6 & 7 \\ -3 & -1 & 2 & -1 & 4 & -3 & 1 \end{pmatrix}.$$

Problem 3.108. *Let*

$$A = \begin{pmatrix} -3 & 1 & -\frac{1}{2} & \frac{1}{3} & \frac{1}{4} \\ 0 & \frac{1}{5} & -1 & 2 & -\frac{1}{3} \\ -\frac{1}{4} & 1 & \frac{2}{3} & 1 & 2 \end{pmatrix}.$$

Find

1. $3A$.
2. $(-2)A$.
3. $4A$.
4. $\frac{1}{3}A$.
5. $\frac{5}{4}A$.

Problem 3.109. *Let*

$$A = \begin{pmatrix} 1 & -2 & 5 & 3 \\ 2 & 0 & -3 & 1 \\ 5 & -1 & 0 & 4 \end{pmatrix},$$

$$B = \begin{pmatrix} 0 & 2 & 7 & -5 \\ -8 & 1 & 3 & 0 \\ 4 & 2 & -2 & 5 \end{pmatrix}.$$

Find

1. $3A$.
2. $2B$.
3. $2A+4B$.
4. $-3B$.
5. $2A-5B$.

Problem 3.110. *Find AB and BA, provided they exist, where*

1.

$$A = \begin{pmatrix} 5 & 7 & -3 & -4 \\ 7 & 6 & -4 & -5 \\ 6 & 4 & -3 & -2 \\ 8 & 5 & -6 & -1 \end{pmatrix},$$

$$B = \begin{pmatrix} 1 & 2 & 3 & 4 \\ 2 & 3 & 4 & 5 \\ 1 & 3 & 5 & 7 \\ 2 & 4 & 6 & 8 \end{pmatrix}.$$

2.

$$A = \begin{pmatrix} 2 & -3 \\ 4 & -6 \end{pmatrix},$$

$$B = \begin{pmatrix} 9 & -6 \\ 6 & -4 \end{pmatrix}.$$

3.

$$A = \begin{pmatrix} 5 & 2 & -2 & -3 \\ 6 & 4 & -3 & 5 \\ 9 & 2 & -3 & 4 \\ 7 & 6 & -4 & 7 \end{pmatrix},$$

$$B = \begin{pmatrix} 2 & 2 & 2 & 2 \\ -1 & -5 & 3 & 11 \\ 16 & 24 & 8 & -8 \\ 8 & 16 & 0 & -16 \end{pmatrix}.$$

4.

$$A = \begin{pmatrix} 0 & 2 & -1 \\ -2 & -1 & 2 \\ 3 & -2 & -1 \end{pmatrix},$$

$$B = \begin{pmatrix} 70 & 34 & -107 \\ 52 & 26 & -68 \\ 101 & 50 & -140 \end{pmatrix}.$$

5.

$$A = \begin{pmatrix} 1 & 1 & 1 & -1 \\ -5 & -3 & -4 & 4 \\ 5 & 1 & 4 & -3 \\ -16 & -11 & -15 & 14 \end{pmatrix},$$

$$B = \begin{pmatrix} 7 & -2 & 3 & 4 \\ 11 & 0 & 3 & 4 \\ 5 & 4 & 3 & 0 \\ 22 & 2 & 9 & 8 \end{pmatrix}.$$

Problem 3.111. *Find ABC, where*

$$A = \begin{pmatrix} 1 & 3 & 1 \\ 2 & 2 & 1 \\ 3 & 4 & 2 \end{pmatrix},$$

$$B = \begin{pmatrix} 1 & 0 & 0 \\ 0 & 2 & 0 \\ 0 & 0 & 1 \end{pmatrix},$$

$$C = \begin{pmatrix} 0 & 2 & -1 \\ -1 & 1 & -1 \\ -2 & -5 & 4 \end{pmatrix}.$$

Problem 3.112. *Find ABCD, where*

$$A = \begin{pmatrix} 1 & 0 & -4 \\ 3 & 1 & 2 \\ 2 & -2 & 7 \end{pmatrix},$$

$$B = \begin{pmatrix} 1 & 1 & -3 \\ 2 & 3 & 5 \\ -1 & 1 & 1 \end{pmatrix},$$

$$C = \begin{pmatrix} 1 & 1 & 0 \\ 1 & 1 & 1 \\ 0 & 0 & 2 \end{pmatrix},$$

$$D = \begin{pmatrix} 4 & -1 & -1 \\ 1 & 10 & 2 \\ 0 & 1 & 3 \end{pmatrix}.$$

Problem 3.113. *Determine which matrices are quadratic.*

1.

$$A = \begin{pmatrix} -1 & 7 & 3 & 4 & 8 & 0 \\ 0 & -3 & 2 & 1 & 4 & -5 \end{pmatrix}.$$

2.

$$A = \begin{pmatrix} 2 & 1 & 2 & 0 \\ 1 & -1 & 1 & 1 \\ 1 & -4 & 0 & 2 \\ 2 & 3 & 5 & 10 \end{pmatrix}.$$

3.

$$A = \begin{pmatrix} 7 & 0 \\ -3 & 1 \\ 4 & 0 \\ 1 & -3 \\ 2 & 2 \\ 0 & -4 \end{pmatrix}.$$

4.
$$A = \begin{pmatrix} -1 & 1 & 0 \\ 3 & 2 & 6 \\ 0 & -1 & 3 \end{pmatrix}.$$

5.
$$A = \begin{pmatrix} 1 & 0 & 2 \\ -3 & 2 & 1 \\ -1 & 3 & 0 \end{pmatrix}.$$

Problem 3.114. *Let*
$$A = \begin{pmatrix} 2 & -1 \\ 3 & -2 \end{pmatrix}.$$
Find A^5.

Problem 3.115. *Let*
$$A = \begin{pmatrix} -2 & 1 & 0 & 0 & 3 & 5 \\ 0 & -1 & 1 & 0 & 1 & 2 \\ 3 & 4 & 5 & -1 & 2 & 1 \end{pmatrix}.$$
Find

1. $A_{1,\cdot}$.
2. $A_{3,\cdot}$.
3. $A_{,2}$.
4. $A_{,6}$.
5. $A_{1,} + 3A_{3,}$.
6. $4A_{,2} - A_{,6}$.
7. $A_{,2}A_{1,}$.
8. $A_{,6}A_{3,}$.
9. $3A_{,2}A_{1,} - A_{,6}A_{3,}$.
10. $A_{,2}A_{1,} + 4A_{,6}A_{3,}$.

Problem 3.116. *Let* $A \in \mathbb{F}^{m,n}$, $C \in \mathbb{F}^{n,p}$. *Prove that*
$$(AC)_{j,} = A_{j,}C, \quad j \in \{1, \ldots, m\}.$$

Problem 3.117. *Let $C \in \mathbb{F}^{n,p}$ and*

$$a = (a_1, \dots, a_n) \in \mathbb{F}^{1,n}.$$

Prove that

$$aC = a_1 C_{1,\cdot} + \cdots + a_n C_{n,\cdot}.$$

Problem 3.118. *Suppose that V and W are finite-dimensional and $T \in \mathscr{L}(V,W)$. Prove that* $\dim range T = 1$ *if and only if there are a basis of V and a basis of W such that with respect to these bases all entries of $\mathscr{M}(T)$ equal* 1.

Problem 3.119. *Let $A, B \in \mathbb{F}^{n,n}$ and $AB = BA$. Prove that*

$$\begin{aligned} (A+B)^2 &= A^2 + 2AB + B^2, \\ (A-B)^2 &= A^2 - 2AB + B^2. \end{aligned}$$

Chapter 4

Invertibility and Isomorphic Vector Spaces

Suppose that U, V and W are vector spaces.

4.1 Invertible Linear Maps

We begin this section by defining the notions of invertible and inverse in the context of the linear maps.

Definition 4.1. *A linear map $T \in \mathscr{L}(V,W)$ is called invertible if there exists a linear map $S \in \mathscr{L}(W,V)$ such that ST equals the identity map on V and TS equals the identity map on W. The linear map S is called an inverse of the linear map T.*

Theorem 4.2. *An invertible linear map has a unique inverse.*

Proof. Suppose that $T \in \mathscr{L}(V,W)$ is an invertible linear map. Let S_1 and S_2 be inverses of T. Then

$$\begin{aligned} S_1 &= S_1 I \\ &= S_1(TS_2) \\ &= (S_1T)S_2 \end{aligned}$$

$$= IS_2$$

$$= S_2,$$

i.e.,

$$S_1 = S_2.$$

This completes the proof. □

If $T \in \mathscr{L}(V,W)$ is an invertible linear map, then its inverse is denoted by T^{-1}. In other words, if $T \in \mathscr{L}(V,W)$ is an invertible linear map, then there exists an unique element T^{-1} of $\mathscr{L}(W,V)$ such that

$$T^{-1}T = I \quad \text{and} \quad TT^{-1} = I.$$

Theorem 4.3. *A linear map is invertible if and only if it is injective and surjective.*

Proof. Let $T \in \mathscr{L}(V,W)$.

1. Assume that T is invertible. Take $u, v \in V$ arbitrarily. Suppose that

   $$Tu = Tv.$$

   Then

   $$\begin{aligned} u &= T^{-1}(Tu) \\ &= T^{-1}(Tv) \\ &= v. \end{aligned}$$

   Hence, we conclude that T is injective. Now, we take $w \in W$ arbitrarily. Then

   $$w = T(T^{-1}w).$$

   This shows that w is in the range of T. Because $w \in W$ was arbitrarily chosen, we conclude that

   $$\text{range}T = W.$$

   Therefore T is surjective.

2. Suppose that T is injective and surjective. For each $w \in W$, define Sw to be the unique element of V such that

$$T(Sw) = w.$$

Note that Sw exists and it is unique because T is injective and surjective. Then TS is the identity map on W. Let now, $w, v \in V$ be arbitrarily chosen. Then

$$\begin{aligned} T((ST)v) &= (TS)(Tv) \\ &= I(Tv) \\ &= Tv. \end{aligned}$$

This equality implies that

$$(ST)v = v$$

because T is injective. Because $v \in V$ was arbitrarily chosen, we conclude that ST equals the identity map on V. Now, we will show that S is linear. Let $w_1, w_2 \in W$ and $a \in \mathbb{F}$ be arbitrarily chosen. Then

$$\begin{aligned} T(Sw_1 + Sw_2) &= T(Sw_1) + T(Sw_2) \\ &= w_1 + w_2 \\ &= S(w_1 + w_2), \end{aligned}$$

from where

$$S(w_1 + w_2) = Sw_1 + Sw_2.$$

Next,

$$\begin{aligned} T(aSw_1) &= aT(Sw_1) \\ &= aw_1 \\ &= S(aw_1). \end{aligned}$$

Therefore

$$S(aw_1) = aSw_1.$$

This completes the proof.

□

Theorem 4.4. *Let $T \in \mathscr{L}(U,V)$ and $S \in \mathscr{L}(V,W)$ be invertible linear maps. Then $ST \in \mathscr{L}(U,W)$ is invertible and*

$$(ST)^{-1} = T^{-1}S^{-1}.$$

Proof. Firstly, we will prove that $ST \in \mathscr{L}(U,W)$. Let $u_1, u_2 \in U$ and $\lambda \in \mathbb{F}$ be arbitrarily chosen. Then $Tu_1, Tu_2, \lambda Tu_1 \in V$ and

$$\begin{aligned}(ST)(u_1+u_2) &= S(T(u_1+u_2)) \\ &= S(Tu_1+Tu_2) \\ &= S(Tu_1)+S(Tu_2) \\ &= (ST)(u_1)+(ST)(u_2),\end{aligned}$$

and

$$\begin{aligned}(ST)(\lambda u_1) &= S(T(\lambda u_1)) \\ &= S(\lambda Tu_1) \\ &= \lambda S(Tu_1) \\ &= \lambda(ST)(u_1).\end{aligned}$$

Therefore $ST \in \mathscr{L}(U,W)$. Now, we will prove that $ST \in \mathscr{L}(U,W)$ is invertible. Since $T \in \mathscr{L}(U,V)$ and $S \in \mathscr{L}(V,W)$ are invertible, they are injective and surjective. Let $w \in W$ be arbitrarily chosen and fixed. Since $S \in \mathscr{L}(V,W)$ is surjective, there is a $v \in V$ so that

$$Sv = w.$$

Because $T \in \mathscr{L}(U,V)$ is surjective and $v \in V$, there is an $u \in U$ so that

$$Tu = v.$$

Consequently

$$\begin{aligned} w &= Sv \\ &= STu. \end{aligned}$$

Hence, we conclude that $ST \in \mathscr{L}(U,W)$ is surjective. Let now, $z, y \in U$. Assume that

$$STy = STz.$$

Because $S \in \mathscr{L}(V,W)$ is injective, by the last equation we find

$$Ty = Tz.$$

Since $T \in \mathscr{L}(U,V)$, by the last equation, we conclude that $y = z$ and then $ST \in \mathscr{L}(U,W)$ is injective. Now, using that $ST \in \mathscr{L}(U,W)$ is surjective and injective, we obtain that $ST \in \mathscr{L}(U,W)$ is invertible. Moreover, we have

$$\begin{aligned} STT^{-1}S^{-1} &= S((TT^{-1})S^{-1}) \\ &= S(IS^{-1}) \\ &= SS^{-1} \\ &= I. \end{aligned}$$

Consequently

$$(ST)^{-1} = T^{-1}S^{-1}.$$

This completes the proof. □

Remark 4.5. *Let V be finite-dimensional and* $\dim V = m > 1$. *Let $v_1, \ldots, v_m$ be a basis of V. With A we will denote the subset of $\mathscr{L}(V)$ containing all noninvertible linear operators on V. Take $T_1, T_2 \in \mathscr{L}(V)$ so that*

$$T_1 v_1 = v_1$$

$$T_1 v_j = 0, \quad j \in \{2, \ldots, m\},$$

and

$$T_2 v_1 = v_1$$

$$T_2 v_j = 0, \quad j \in \{2, \ldots, m\}.$$

Then T_1 and T_2 are noninvertible and $T_1, T_2 \in A$. On the other hand, we have

$$(T_1 + T_2) v_j = v_j, \quad j \in \{1, \ldots, m\}.$$

Then $T_1 + T_2$ is invertible and $T_1 + T_2 \notin A$. Therefore the set A is not closed under the operation addition and A is not a subspace of $\mathscr{L}(V)$.

Remark 4.6. *Let $T \in \mathscr{L}(V)$ be invertible and $\mathscr{M}(T)$ be its matrix. Then the matrix of the linear map T^{-1} will be denoted by $\mathscr{M}(T)^{-1}$.*

4.2 Isomorphic Vector Spaces

Definition 4.7. *An isomorphism is an invertible linear map.*

Definition 4.8. *Two vector spaces are called isomorphic if there is an isomorphism from one vector space onto the other vector space.*

Theorem 4.9. *Two finite-dimensional vector spaces over $\mathbb{F}$ are isomorphic if and only if they have the same dimension.*

Proof. 1. Let V and W be isomorphic finite-dimensional vector spaces. Then there is an isomorphism T from V onto W. Since T is invertible, it is injective and surjective. Hence, $\ker T = \{0\}$ and $\mathrm{range} T = W$. Thus,

$$\dim \ker T = 0.$$

By the fundamental theorem of linear maps, we get

$$\begin{aligned} \dim V &= \dim \ker T + \dim \mathrm{range} T \\ &= \dim \mathrm{range} T \\ &= \dim W. \end{aligned}$$

2. Let V and W be vector spaces with the same dimension n. Suppose that $v_1,\dots,v_n$ is a basis of V and $w_1,\dots,w_n$ is a basis of W. For any scalars $c_1,\dots,c_n \in \mathbb{F}$, define the map

$$T(c_1v_1+\cdots+c_nv_n) = c_1w_1+\cdots+c_nw_n.$$

By Theorem 3.6 and its proof, it follows that $T \in \mathscr{L}(V,W)$ is well defined. Let $w \in W$ be arbitrarily chosen. Then there are scalars $a_1,\dots,a_n \in \mathbb{F}$ so that

$$w = a_1w_1+\cdots+a_nw_n.$$

Take

$$v = a_1v_1+\cdots+a_nv_n.$$

Then, by the definition of T, it follows that

$$\begin{aligned} T(a_1v_1+\cdots+a_nv_n) &= a_1w_1+\cdots+a_nw_n \\ &= w. \end{aligned}$$

Because $w \in W$ was arbitrarily chosen, we conclude that $T \in \mathscr{L}(V,W)$ is surjective. Let now, $v \in \ker T$ be arbitrarily chosen. Then there are scalars $b_1,\dots,b_n$ so that

$$v = b_1v_1+\cdots+b_nv_n.$$

Hence,

$$\begin{aligned} 0 &= Tv \\ &= T(b_1v_1+\cdots+b_nv_n) \\ &= b_1w_1+\cdots+b_nw_n. \end{aligned}$$

By the last equality, using that $w_1,\dots,w_n$ is a basis of W, we obtain

$$\begin{aligned} b_1 &= 0 \\ &\vdots \end{aligned}$$

$$b_n = 0,$$

i.e., $v = 0$. Now, applying Theorem 3.33, we find that $T \in \mathscr{L}(V,W)$ is injective. Because $T \in \mathscr{L}(V,W)$ is injective and surjective, using Theorem 4.3, we conclude that T is invertible. Therefore V and W are isomorphic. This completes the proof.

□

Theorem 4.10. *Suppose that $v_1, \ldots, v_n$ is a basis of V, $w_1, \ldots, w_m$ is a basis of W. Then $\mathscr{L}(V,W)$ and $\mathbb{F}^{m,n}$ are isomorphic.*

Proof. Note that for each $T \in \mathscr{L}(V,W)$ there is a matrix $\mathscr{M}(T) \in \mathbb{F}^{m,n}$. Consider $\mathscr{M}$ as a function from $\mathscr{L}(V,W)$ to $\mathbb{F}^{m,n}$. By Theorem 3.68, it follows that $\mathscr{M} \in \mathscr{L}(\mathscr{L}(V,W), \mathbb{F}^{m,n})$. Let $A \in \mathbb{F}^{m,n}$ be arbitrarily chosen. Take $T \in \mathscr{L}(V,W)$ so that

$$Tv_k = \sum_{j=1}^{m} a_{jk} w_j,$$

where a_{jk}, $j \in \{1, \ldots, m\}$, $k \in \{1, \ldots, n\}$, are the elements of the matrix A. Then

$$\mathscr{M}(T) = A.$$

Because $A \in \mathbb{F}^{m,n}$ was arbitrarily chosen, we conclude that $\mathscr{M}$ is surjective. Let now, $S \in \mathscr{L}(V,W) \cap \ker \mathscr{M}$. Then

$$\mathscr{M}(S) = 0$$

and

$$Sv_k = 0, \quad k \in \{1, \ldots, n\}.$$

Since $v_1, \ldots, v_n$ is a basis of V, we get $S = O$. Because $S \in \mathscr{L}(V,W) \cap \ker \mathscr{M}$ was arbitrarily chosen, we get $\ker \mathscr{M} = \{O\}$. Therefore $\mathscr{M}$ is injective. Hence, using that it is surjective, we conclude that $\mathscr{M}$ is invertible. Now, applying Theorem 4.3, we conclude that $\mathscr{M}$ is an isomorphism between $\mathscr{L}(V,W)$ and $\mathbb{F}^{m,n}$. This completes the proof. □

Theorem 4.11. *Suppose that V and W are finite-dimensional. Then $\mathscr{L}(V,W)$ is finite-dimensional and*

$$\dim \mathscr{L}(V,W) = \dim V \dim W.$$

Proof. Let $v_1, \ldots, v_n$ be a basis of V and $w_1, \ldots, w_m$ be a basis of W. Then

$$\begin{aligned} \dim V &= n, \\ \dim W &= m. \end{aligned}$$

By Theorem 4.10, it follows that $\mathscr{L}(V,W)$ and $\mathbb{F}^{m,n}$ are isomorphic. Hence, applying Theorem 4.9, we get

$$\begin{aligned} \dim \mathscr{L}(V,W) &= \dim \mathbb{F}^{m,n} \\ &= mn \\ &= \dim V \dim W. \end{aligned}$$

This completes the proof. □

4.3 Matrix of a Vector

Previously, we defined a matrix of a linear map. Now, we will define the matrix of a vector.

Definition 4.12. *Suppose that $v \in V$ and $v_1, \ldots, v_n$ is a basis of V. The matrix of v with respect to this basis is the $n \times 1$ matrix*

$$\mathscr{M}(v) = \begin{pmatrix} c_1 \\ \vdots \\ c_n \end{pmatrix},$$

where $c_1, \ldots, c_n$ are scalars so that

$$v = c_1 v_1 + \cdots + c_n v_n.$$

Note that the matrix $\mathscr{M}(v)$ depends on $v_1, \ldots, v_n$, as well as on v. However, the basis should be clear from the context and thus it is not included in the notation.

Example 4.13. *Consider $\mathbb{R}^3$. Let*

$$v_1 = \begin{pmatrix} 2 \\ -1 \\ -1 \end{pmatrix}, \quad v_2 = \begin{pmatrix} 3 \\ 1 \\ 1 \end{pmatrix}, \quad v_3 = \begin{pmatrix} -2 \\ -1 \\ -2 \end{pmatrix}$$

and

$$v = \begin{pmatrix} -2 \\ 2 \\ -2 \end{pmatrix}.$$

We will find the matrix of v with respect to the basis v_1, v_2, v_3. Firstly, we will prove that v_1, v_2, v_3 is a basis of $\mathbb{R}^3$. Let $a_1, a_2, a_3 \in \mathbb{R}$ be such that

$$a_1 v_1 + a_2 v_2 + a_3 v_3 = 0.$$

Then

$$\begin{aligned} \begin{pmatrix} 0 \\ 0 \\ 0 \end{pmatrix} &= a_1 \begin{pmatrix} 2 \\ -1 \\ -1 \end{pmatrix} + a_2 \begin{pmatrix} 3 \\ 1 \\ 1 \end{pmatrix} + a_3 \begin{pmatrix} -2 \\ -1 \\ -2 \end{pmatrix} \\ &= \begin{pmatrix} 2a_1 \\ -a_1 \\ -a_1 \end{pmatrix} + \begin{pmatrix} 3a_2 \\ a_2 \\ a_2 \end{pmatrix} + \begin{pmatrix} -2a_3 \\ -a_3 \\ -2a_3 \end{pmatrix} \\ &= \begin{pmatrix} 2a_1 + 3a_2 - 2a_3 \\ -a_1 + a_2 - a_3 \\ -a_1 + a_2 - 2a_3 \end{pmatrix}. \end{aligned}$$

Hence, we get the system

$$\begin{aligned} 2a_1 + 3a_2 - 2a_3 &= 0 \\ -a_1 + a_2 - a_3 &= 0 \\ -a_1 + a_2 - 2a_3 &= 0. \end{aligned} \tag{4.1}$$

By the second and third equations of the last system, we get

$$\begin{aligned} 0 &= -a_1 + a_2 - 2a_3 \\ &= -a_1 + a_2 - a_3 - a_3 \\ &= -a_3, \end{aligned}$$

i.e.,

$$a_3 = 0.$$

By the second equation of the system (4.1), *we find*

$$a_1 = a_2$$

and then, using the first equation of (4.1), *we get*

$$\begin{aligned} 0 &= 2a_1 + 3a_2 - 2a_3 \\ &= 2a_1 + 3a_1 \\ &= 5a_1, \end{aligned}$$

i.e.,

$$a_1 = a_2 = 0.$$

Therefore v_1, v_2 *and* v_3 *are linearly independent. Let now,*

$$w = \begin{pmatrix} w_1 \\ w_2 \\ w_3 \end{pmatrix} \in \mathbb{R}^3$$

be arbitrarily chosen. We will search $b_1, b_2, b_3 \in \mathbb{R}$ *so that*

$$w = b_1 v_1 + b_2 v_2 + b_3 v_3.$$

Then

$$\begin{aligned} \begin{pmatrix} w_1 \\ w_2 \\ w_3 \end{pmatrix} &= b_1 \begin{pmatrix} 2 \\ -1 \\ -1 \end{pmatrix} + b_2 \begin{pmatrix} 3 \\ 1 \\ 1 \end{pmatrix} + b_3 \begin{pmatrix} -2 \\ -1 \\ -2 \end{pmatrix} \\ &= \begin{pmatrix} 2b_1 \\ -b_1 \\ -b_1 \end{pmatrix} + \begin{pmatrix} 3b_2 \\ b_2 \\ b_2 \end{pmatrix} + \begin{pmatrix} -2b_3 \\ -b_3 \\ -2b_3 \end{pmatrix} \\ &= \begin{pmatrix} 2b_1 + 3b_2 - 2b_3 \\ -b_1 + b_2 - b_3 \\ -b_1 + b_2 - 2b_3 \end{pmatrix}. \end{aligned}$$

From here, we find the system

$$\begin{aligned} 2b_1 + 3b_2 - 2b_3 &= w_1 \\ -b_1 + b_2 - b_3 &= w_2 \\ -b_1 + b_2 - 2b_3 &= w_3. \end{aligned} \tag{4.2}$$

By the last two equations of the system (4.2), *we obtain*

$$\begin{aligned} w_3 &= -b_1+b_2-2b_3 \\ &= -b_1+b_2-b_3-b_3 \\ &= w_2-b_3, \end{aligned}$$

whereupon

$$b_3 = w_2 - w_3.$$

Now, employing the second equation of the system (4.2), *we get*

$$-b_1+b_2-(w_2-w_3) = w_2,$$

whereupon

$$\begin{aligned} -b_1+b_2 &= w_2+w_2-w_3 \\ &= 2w_2-w_3 \end{aligned}$$

and

$$b_2 = b_1+2w_2-w_3.$$

Hence, using the first equation of (4.2), *we arrive at the equation*

$$2b_1-3(b_1+2w_2-w_3)-2(w_2-w_3) = w_1,$$

or

$$2b_1-3b_1-6w_2+3w_3-2w_2+2w_3 = w_1,$$

or

$$-b_1-8w_2+5w_3 = w_1,$$

or

$$b_1 = -w_1-8w_2+5w_3,$$

and

$$\begin{aligned} b_2 &= b_1+2w_2-w_3 \\ &= -w_1-8w_2+5w_3+2w_2-w_1 \end{aligned}$$

$$= \quad -w_1 - 6w_2 + 4w_3.$$

Consequently

$$\begin{aligned} b_1 &= -w_1 - 8w_2 + 5w_3 \\ b_2 &= -w_1 - 6w_2 + 4w_3 \qquad (4.3) \\ b_3 &= w_2 - w_3. \end{aligned}$$

This shows that any vector $w \in \mathbb{R}^3$ can be represented as a linear combination of the vectors v_1, v_2, v_3. Consequently, v_1, v_2, v_3 is a basis of $\mathbb{R}^3$. For the given vector v, the system (4.3) *takes the form*

$$\begin{aligned} b_1 &= -v^1 + 8v^2 + 5v^3 \\ b_2 &= -v^1 - 6v^2 + 4v^3 \\ b_3 &= v^2 - v^3, \end{aligned}$$

or

$$\begin{aligned} b_1 &= -(-2) - 8 \cdot 2 + 5 \cdot (-2) \\ b_2 &= -(-2) - 6 \cdot 2 + 4 \cdot (-2) \\ b_3 &= 2 - (-2), \end{aligned}$$

or

$$\begin{aligned} b_1 &= 2 - 16 - 10 \\ b_2 &= 2 - 12 - 8 \\ b_3 &= 2 + 2, \end{aligned}$$

or

$$b_1 = -24$$

$$b_2 = -18$$

$$b_3 = 4.$$

Therefore

$$\mathcal{M}(v) = \begin{pmatrix} -24 \\ -18 \\ 4 \end{pmatrix}.$$

Exercise 4.14. *Prove that v_1, v_2, v_3 is a basis of $\mathbb{R}^3$ and find $\mathcal{M}(w)$, where*

1.

$$v_1 = \begin{pmatrix} -3 \\ 1 \\ 3 \end{pmatrix}, \quad v_2 = \begin{pmatrix} -2 \\ -1 \\ 1 \end{pmatrix}, \quad v_3 = \begin{pmatrix} -2 \\ -1 \\ -1 \end{pmatrix}, \quad w = \begin{pmatrix} 1 \\ -2 \\ 3 \end{pmatrix}.$$

2.

$$v_1 = \begin{pmatrix} -3 \\ 1 \\ 2 \end{pmatrix}, \quad v_2 = \begin{pmatrix} 1 \\ 1 \\ 3 \end{pmatrix}, \quad v_3 = \begin{pmatrix} -2 \\ -2 \\ -1 \end{pmatrix}, \quad w = \begin{pmatrix} 2 \\ -1 \\ 1 \end{pmatrix}.$$

If $T \in \mathcal{L}(V,W)$ and $v_1, \ldots, v_n$ is a basis of V and $w_1, \ldots, w_m$ is a basis of W, then

$$\mathcal{M}(T)_{,k} = \mathcal{M}(v_k).$$

Theorem 4.15. *Suppose that $T \in \mathcal{L}(V,W)$ and $v \in V$. Suppose that $v_1, \ldots, v_n$ is a basis of V and $w_1, \ldots, w_m$ is a basis of W. Then*

$$\mathcal{M}(Tv) = \mathcal{M}(T)\mathcal{M}(v).$$

Proof. Let

$$v = c_1 v_1 + \cdots + c_n v_n,$$

where $c_1, \ldots, c_n \in \mathbb{F}$. Then

$$\mathcal{M}(v) = \begin{pmatrix} c_1 \\ \vdots \\ c_n \end{pmatrix}$$

and

$$Tv = T(c_1 v_1 + \cdots + c_n v_n)$$

$$
\begin{aligned}
&= T(c_1v_1)+\cdots+T(c_nv_n) \\
&= c_1Tv_1+\cdots+c_nTv_n.
\end{aligned}
$$

Hence,

$$
\begin{aligned}
\mathcal{M}(Tv) &= \mathcal{M}(c_1Tv_1+\cdots+c_nTv_n) \\
&= \mathcal{M}(c_1Tv_1)+\cdots+\mathcal{M}(c_nTv_n) \\
&= c_1\mathcal{M}(Tv_1)+\cdots+c_n\mathcal{M}(Tv_n) \\
&= c_1\mathcal{M}(T)_{,1}+\cdots+c_n\mathcal{M}(T)_{,n} \\
&= \mathcal{M}(T)\mathcal{M}(v).
\end{aligned}
$$

This completes the proof. □

4.4 Operators

Definition 4.16. *A linear map from a vector space to itself is called an operator. The set of all operators on V will be denoted by $\mathcal{L}(V)$.*

Theorem 4.17. *Suppose that V is a finite-dimensional vector space and $T \in \mathcal{L}(V)$. Then the following are equivalent.*

1. *T is invertible.*
2. *T is injective.*
3. *T is surjective.*

Proof. 1. Let T be invertible. Then T is injective and so, $1 \Longrightarrow 2$.

2. Let T be injective. Then $\ker T = \{0\}$. Now, applying the fundamental theorem of linear maps, we get

$$
\dim \operatorname{range} T = \dim V - \dim \ker T
$$

$$= \dim V - 0$$

$$= \dim V.$$

Therefore $\text{range} T = V$ and T is surjective. So, $2 \Longrightarrow 3$. Because T is injective and surjective, by Theorem 4.3, we get that T is invertible. Thus, $2 \Longrightarrow 1$.

3. Suppose that T is surjective. Then $\text{range} T = V$ and by the fundamental theorem of linear maps, we get

$$\begin{aligned} \dim \ker T &= \dim V - \dim \text{range} T \\ &= \dim V - \dim V \\ &= 0. \end{aligned}$$

Therefore $\ker T = \{0\}$ and T is injective. So, $3 \Longrightarrow 2$. This completes the proof.

□

4.5 Product of Vector Spaces

Definition 4.18. *Suppose that V_1, …, V_m are vector spaces. Then the product $V_1 \times \ldots \times V_m$ is defined by*

$$V_1 \times \ldots \times V_m = \{(v_1, \ldots, v_m) : v_1 \in V_1, \ldots, v_m \in V_m\}.$$

Addition in $V_1 \times \ldots \times V_m$ is defined by

$$(v_1, \ldots, v_m) + (u_1, \ldots, u_m) = (v_1 + u_1, \ldots, v_m + u_m), \tag{4.4}$$

where $(v_1, \ldots, v_m)$, $(u_1, \ldots, u_m) \in V_1 \times \ldots \times V_m$. Scalar multiplication in the product $V_1 \times \ldots \times V_m$ is defined by

$$\lambda(u_1, \ldots, u_m) = (\lambda u_1, \ldots, \lambda u_m), \tag{4.5}$$

where $\lambda \in \mathbb{F}$ and $(u_1, \ldots, u_m) \in V_1 \times \ldots \times V_m$.

Remark 4.19. *Note that the operations addition* (4.4) *and scalar multiplication* (4.5) *in a product of vector spaces are well defined. Really, let*

$(u_1,\ldots,u_m)$, $(v_1,\ldots,v_m)\in V_1\times\ldots\times V_m$ and $\lambda\in\mathbb{F}$. Then, using that V_j, $j\in\{1,\ldots,m\}$, are vector spaces, we have that

$$u_j+v_j\in V_j,\quad \lambda u_j\in V_j,\quad j\in\{1,\ldots,m\}.$$

Hence,

$$(u_1,\ldots,u_m)+(v_1,\ldots,v_m)\in V_1\times\ldots\times V_m$$

and

$$\lambda(u_1,\ldots,u_m)\in V_1\times\ldots\times V_m.$$

Example 4.20. *Consider $\mathscr{P}_2(\mathbb{R})\times\mathbb{R}^4$. Then*

$$(1+x+x^2,(0,1,-2,3))\in\mathscr{P}_2(\mathbb{R})\times\mathbb{R}^4.$$

Theorem 4.21. *Let V_1, …, V_m be vector spaces. Then $V_1\times\ldots\times V_m$ is a vector space.*

Proof. Let

$$u=(u_1,\ldots,u_m),\quad v=(v_1,\ldots,v_m),\quad w=(w_1,\ldots,w_m)\in V_1\times\ldots\times V_m$$

and $\lambda\in\mathbb{F}$ be arbitrarily chosen. Then, we have the following.

1.

$$\begin{aligned}
u+v &= (u_1,\ldots,u_m)+(v_1,\ldots,v_m)\\
&= (u_1+v_1,\ldots,u_m+v_m)\\
&= (v_1+u_1,\ldots,v_m+u_m)\\
&= (v_1,\ldots,v_m)+(u_1,\ldots,u_m)\\
&= v+u.
\end{aligned}$$

2.

$$\begin{aligned}
u+(v+w) &= (u_1,\ldots,u_m)+((v_1,\ldots,v_m)+(w_1,\ldots,w_m))\\
&= (u_1,\ldots,u_m)+(v_1+w_1,\ldots,v_m+w_m)
\end{aligned}$$

$$
\begin{aligned}
&= (u_1+(v_1+w_1),\ldots,u_m+(v_m+w_m)) \\
&= ((u_1+v_1)+w_1,\ldots,(u_m+v_m)+w_m) \\
&= (u_1+v_1,\ldots,u_m+v_m)+(w_1,\ldots,w_m) \\
&= ((u_1,\ldots,u_m)+(v_1,\ldots,v_m))+(w_1,\ldots,w_m) \\
&= (u+v)+w.
\end{aligned}
$$

3. Let

$$0=(0,\ldots,0).$$

Then

$$
\begin{aligned}
u+0 &= (u_1,\ldots,u_m)+(0,\ldots,0) \\
&= (u_1+0,\ldots,u_m+0) \\
&= (u_1,\ldots,u_m) \\
&= u.
\end{aligned}
$$

4. Let

$$-u=(-u_1,\ldots,-u_m).$$

Then

$$
\begin{aligned}
u+(-u) &= (u_1,\ldots,u_m)+(-u_1,\ldots,-u_m) \\
&= (u_1+(-u_1),\ldots,u_m+(-u_m)) \\
&= (0,\ldots,0) \\
&= 0.
\end{aligned}
$$

5.

$$
\begin{aligned}
\lambda(u+v) &= \lambda((u_1,\ldots,u_m)+(v_1,\ldots,v_m)) \\
&= \lambda(u_1+v_1,\ldots,u_m+v_m) \\
&= (\lambda(u_1+v_1),\ldots,\lambda(u_m+v_m)) \\
&= (\lambda u_1+\lambda v_1,\ldots,\lambda u_m+\lambda v_m) \\
&= (\lambda u_1,\ldots,\lambda u_m)+(\lambda v_1,\ldots,\lambda v_m) \\
&= \lambda(u_1,\ldots,u_m)+\lambda(v_1,\ldots,v_m) \\
&= \lambda u+\lambda v.
\end{aligned}
$$

6.

$$
\begin{aligned}
(\lambda+\mu)u &= (\lambda+\mu)(u_1,\ldots,u_m) \\
&= ((\lambda+\mu)u_1,\ldots,(\lambda+\mu)u_m) \\
&= (\lambda u_1+\mu u_1,\ldots,\lambda u_m+\mu u_m) \\
&= (\lambda u_1,\ldots,\lambda u_m)+(\mu u_1,\ldots,\mu u_m) \\
&= \lambda u+\mu u.
\end{aligned}
$$

This completes the proof.

□

Example 4.22. *Now, we will prove that $\mathbb{R}\times\mathbb{R}^2$ and $\mathbb{R}^3$ are isomorphic. For $(x_1,(x_2,x_3))\in\mathbb{R}\times\mathbb{R}^2$, $x_1\in\mathbb{R}$, $(x_2,x_3)\in\mathbb{R}^2$, define*

$$f(x_1,(x_2,x_3)) = (x_1,x_2,x_3).$$

We have $f : \mathbb{R} \times \mathbb{R}^2 \to \mathbb{R}^3$ is well defined. Let $(y_1, y_2, y_3) \in \mathbb{R}^3$ be arbitrarily chosen. Then $(y_1, (y_2, y_3)) \in \mathbb{R} \times \mathbb{R}^2$ and

$$f(y_1, (y_2, y_3)) = (y_1, y_2, y_3).$$

Therefore $f : \mathbb{R} \times \mathbb{R}^2 \to \mathbb{R}^3$ is surjective. Let

$$(z_1, (z_2, z_3)), \quad (v_1, (v_2, v_3)) \in \mathbb{R} \times \mathbb{R}^2$$

be arbitrarily chosen so that

$$f(z_1, (z_2, z_3)) = f(v_1, (v_2, v_3)).$$

Then

$$(z_1, z_2, z_3) = (v_1, v_2, v_3).$$

Hence,

$$\begin{aligned} z_1 &= v_1, \\ z_2 &= v_2, \\ z_3 &= v_3 \end{aligned}$$

and then

$$(z_2, z_3) = (v_2, v_3).$$

Therefore

$$(z_1, (z_2, z_3)) = (v_1, (v_2, v_3)).$$

Consequently $f : \mathbb{R} \times \mathbb{R}^2 \to \mathbb{R}^3$ is injective and $\mathbb{R} \times \mathbb{R}^2$ and $\mathbb{R}^3$ are isomorphic.

Exercise 4.23. *Prove that $\mathbb{R} \times \mathbb{R}^3$ and $\mathbb{R}^4$ are isomorphic.*

Example 4.24. *Now, we will show that*

$$\begin{aligned} u_1 &= (1, (0, 0, 0)), \\ u_2 &= (x, (0, 0, 0)), \\ u_3 &= (0, (1, 0, 0)), \end{aligned}$$

$$u_4 = (0,(0,1,0)),$$

$$u_5 = (0,(0,0,1))$$

form a basis of $\mathscr{P}_1(\mathbb{R}) \times \mathbb{R}^3$. *Take*

$$v = (b_1 + b_2 x, (b_3, b_4, b_5)) \in \mathscr{P}_1(\mathbb{R}) \times \mathbb{R}^3$$

arbitrarily. Let a_1, a_2, a_3, a_4, $a_5 \in \mathbb{F}$ *be such that*

$$a_1 u_1 + a_2 u_2 + a_3 u_3 + a_4 u_4 + a_5 u_5 = v.$$

Then, we have the following

$$\begin{aligned}(b_1 + b_2 x, (b_3, b_4, b_5)) &= a_1(1,(0,0,0)) + a_2(x,(0,0,0)) + a_3(0,(1,0,0)) \\ &\quad + a_4(0,(0,1,0)) + a_5(0,(0,0,1)) \\ &= (a_1,(0,0,0)) + (a_2 x,(0,0,0)) + (0,(a_3,0,0)) \\ &\quad + (0,(0,a_4,0)) + (0,(0,0,a_5)) \\ &= (a_1 + a_2 x, (a_3, a_4, a_5)).\end{aligned}$$

Hence, we get the system

$$\begin{aligned} b_1 + b_2 x &= a_1 + a_2 x \\ b_3 &= a_3 \\ b_4 &= a_4 \\ b_5 &= a_5. \end{aligned}$$

Therefore

$$b_1 = a_1$$

$$\begin{aligned} b_2 &= a_2 \\ b_3 &= a_3 \\ b_4 &= a_4 \\ b_5 &= a_5. \end{aligned}$$

In particular, if $v = (0,(0,0,0))$, *then*

$$\begin{aligned} a_1 &= 0 \\ a_2 &= 0 \\ a_3 &= 0 \\ a_4 &= 0 \\ a_5 &= 0. \end{aligned}$$

Thus, the list (u_1,u_2,u_3,u_4,u_5) *is linearly independent in* $\mathscr{P}_1(\mathbb{R}\times\mathbb{R}^3)$ *and spans* $\mathscr{P}_1(\mathbb{R})\times\mathbb{R}^3$. *Consequently the list* (u_1,u_2,u_3,u_4,u_5) *forms a basis of* $\mathscr{P}_1(\mathbb{R})\times\mathbb{R}^3$.

Exercise 4.25. *Find a basis of* $\mathscr{P}_2(\mathbb{R})\times\mathbb{R}^2$.

Theorem 4.26. . *Let* V_1, …, V_m *be finite-dimensional vector spaces. Then* $V_1\times\ldots\times V_m$ *is finite dimensional and*

$$\dim(V_1\times\ldots\times V_m) = \dim V_1 + \cdots + \dim V_m.$$

Proof. Take a basis of each V_j, $j\in\{1,\ldots,m\}$. For any basis vector v_{ij} of V_j, $i\in\{1,\ldots,\dim V_j\}$, $j\in\{1,\ldots,m\}$, consider the element

$$u_{ij} = (0,\ldots,0,v_{ij},0,\ldots,0), \quad i\in\{1,\ldots,\dim V_j\}, \quad j\in\{1,\ldots,m\},$$

of $V_1\times\ldots\times V_m$. The list of such vectors u_{ij}, $i\in\{1,\ldots,\dim V_j\}$, $j\in\{1,\ldots,m\}$, is linearly independent in $V_1\times\ldots\times V_m$ and spans $V_1\times\ldots\times V_m$. So, this list is a basis of $V_1 \times \ldots \times V_m$ and its length is

$$\dim V_1 + \cdots + \dim V_m.$$

This completes the proof. □

Theorem 4.27. *Let U_1, …, U_m be subspaces of U. Define the linear map $f : U_1 \times \ldots \times U_m \to U_1 + \cdots + U_m$ by*

$$f(u_1, \ldots, u_m) = u_1 + \cdots + u_m, \quad (u_1, \ldots, u_m) \in U_1 \times \ldots \times U_m.$$

Then $U_1 + \cdots + U_m$ is a direct sum if and only if f is injective.

Proof. Note that $f : U_1 \times \ldots \times U_m \to U_1 + \cdots + U_m$ is injective if and only if to write 0 as a sum $u_1 + \cdots + u_m$, where $u_j \in U_j$, $j \in \{1, \ldots, m\}$, is by taking each u_j equal to 0. Hence and Theorem 1.36, it follows that $f : U_1 \times \ldots \times U_m \to U_1 + \cdots + U_m$ is injective if and only if $U_1 + \cdots + U_m$ is a direct sum. This completes the proof. □

Theorem 4.28. *Let V be finite-dimensional and U_1, …, U_m be subspaces of V. Then $U_1 + \cdots + U_m$ is a direct sum if and only if*

$$\dim(U_1 + \cdots + U_m) = \dim U_1 + \cdots + \dim U_m.$$

Proof. Consider the map f, defined in Theorem 4.27. We have that $f : U_1 \times \ldots \times U_m \to U_1 + \cdots + U_m$ is injective if and only if $U_1 + \cdots + U_m$ is surjective because if $w_1 + \cdots + w_m \in U_1 + \cdots + U_m$ is arbitrarily chosen, then $(w_1, \ldots, w_m) \in U_1 \times \ldots \times U_m$ and

$$f(w_1, \ldots, w_m) = w_1 + \cdots + w_m.$$

Hence, applying Theorem 3.33, it follows that $f : U_1 \times \ldots \times U_m \to U_1 + \cdots + U_m$ is injective if and only if $U_1 + \cdots + U_m$ is injective if and only if $\ker f = \{0\}$. Now, applying the fundamental theorem of linear maps, Theorem 3.35, we have that f is injective if and only if

$$\begin{aligned} \dim(U_1 \times \ldots \times U_m) &= \dim \ker T + \dim \operatorname{range} T \\ \\ &= \dim(U_1 + \cdots + U_m). \end{aligned}$$

From here and from Theorem 4.27, it follows that $U_1 + \cdots + U_m$ is a direct sum if and only if

$$\dim(U_1 + \cdots + U_m) = \dim U_1 + \cdots + \dim U_m.$$

This completes the proof. □

4.6 Quotients of Vector Spaces

Definition 4.29. *Let $v \in V$ and U be a subspace of V. Then $v+U$ is the subset defined by*

$$v+U = \{v+u : u \in U\}.$$

The subset $v+U$ is said to be affine subset of V and parallel to U.

Example 4.30. *Consider $\mathbb{R}^2$ and*

$$U = \{(x, x+2) : x \in \mathbb{R}\}.$$

Then

$$(1,1)+U = \{(x+1, x+3) : x \in \mathbb{R}\}.$$

Example 4.31. *Consider $\mathbb{R}^3$ and*

$$U = \{(x, 3x, -x) : x \in \mathbb{R}\}.$$

Then

$$(-1,1,1)+U = \{(x-1, 3x+1, -x+1) : x \in \mathbb{R}\}.$$

Exercise 4.32. *Consider $\mathbb{R}^2$ and*

$$U = \{(x, x^2+x) : x \in \mathbb{R}\}.$$

Find $(-1,1)+U$.

Definition 4.33. *Suppose that U is a subspace of V. The quotient space V/U is the set of all affine subsets of V parallel to U, i.e.,*

$$V/U = \{v+U : v \in V\}.$$

Theorem 4.34. *Suppose that U is a subspace of V and $v, w \in V$. Then the following are equivalent.*

1. $v - w \in U$.

2. $v+U = w+U$.

3. $(v+U) \bigcap (w+U) \neq \emptyset$.

Proof. 1. Let $v-w \in U$. Let $u \in U$ be arbitrarily chosen. Then

$$\begin{aligned} v+u &= (w+(v-w))+u \\ &= w+((v-w)+u) \\ &\in w+U. \end{aligned}$$

Since $u \in U$ was arbitrarily chosen and we get that $v+u$ is an element of $w+U$, we conclude that

$$v+U \subseteq w+U. \tag{4.6}$$

Similarly,

$$w+U \subseteq v+U.$$

By the last inclusion and (4.6), we obtain

$$v+U = w+U.$$

So, $1 \Longrightarrow 2$.

2. Suppose that

$$v+U = w+U.$$

Then

$$(v+U)\bigcap(w+U) \neq \emptyset.$$

So, $2 \Longrightarrow 3$.

3. Let

$$(v+U)\bigcap(w+U) \neq \emptyset.$$

Then there are $u_1, u_2 \in U$ so that

$$v+u_1 = w+u_2.$$

Since U is a subspace of V, we find $u_2-u_1 \in U$. Moreover,

$$\begin{aligned} v-w &= u_2-u_1 \\ &\in U, \end{aligned}$$

i.e., $3 \Longrightarrow 1$. This completes the proof.

□

Remark 4.35. *By Theorem 4.34, it follows that two affine subsets parallel to U are equal or disjoint.*

Definition 4.36. *Let U be a subspace of V. Then, addition and scalar multiplication on V/U are defined by*

$$(v+U)+(w+U)=(v+w)+U$$

and

$$\lambda(v+U)=(\lambda v)+U,$$

respectively, for any $u,v\in V$ and any $\lambda\in\mathbb{F}$.

Remark 4.37. *Let U be a subspace of V. Suppose that $v,v_1,w,w_1\in U$ and $\lambda\in\mathbb{F}$ be such that*

$$\begin{aligned} v+U &= v_1+U, \\ w+U &= w_1+U. \end{aligned}$$

By Theorem 4.34, it follows that

$$v-v_1, \quad w-w_1\in U.$$

Since U is a subspace of V, we have

$$\begin{aligned} (v-v_1)+(w-w_1) &\in U, \\ (v-v_1)-(w-w_1) &\in U, \end{aligned}$$

whereupon

$$(v+w)-(v_1+w_1)\in U.$$

Again, applying Theorem 4.34, we find

$$(v+w)+U=(v_1+w_1)+U.$$

Therefore the definition of addition on U/V makes sense. Next, using that U is a subspace of V, we have

$$\begin{aligned} \lambda(v-v_1) &= (\lambda v)-(\lambda v_1) \\ &\in U. \end{aligned}$$

From here, applying Theorem 4.34, we get

$$(\lambda v) + U = (\lambda v_1) + U.$$

Thus, the definition of scalar multiplication on V/U makes sense.

Theorem 4.38. *Let U be a subspace of V. Then V/U is a vector space.*

Proof. Let $\lambda, \mu \in \mathbb{F}$ and $v, u, w \in V$ be arbitrarily chosen. Then we have the following.

1.

$$\begin{aligned} (v+U)+(u+U) &= (v+u)+U \\ &= (u+v)+U \\ &= (u+U)+(v+U). \end{aligned}$$

2.

$$\begin{aligned} (v+U)+((u+U)+(w+U)) &= (v+U)+((u+w)+U) \\ &= (v+(u+w))+U \\ &= ((v+u)+w)+U \\ &= ((v+u)+U)+(w+U) \\ &= ((v+U)+(u+U))+(w+U). \end{aligned}$$

3.

$$\begin{aligned} (v+U)+(0+U) &= (v+0)+U \\ &= v+U, \end{aligned}$$

i.e., $0+U$ is the additive identity of V/U.

4.

$$\begin{aligned}(v+U)+((-v)+U) &= (v+(-v))+U \\ &= 0+U.\end{aligned}$$

5.

$$\begin{aligned}\lambda((v+U)+(u+U)) &= \lambda((v+u)+U) \\ &= (\lambda(v+u))+U \\ &= (\lambda v+\lambda u)+U \\ &= ((\lambda v)+U)+((\lambda u)+U) \\ &= \lambda(v+U)+\lambda(u+U).\end{aligned}$$

6.

$$\begin{aligned}(\lambda+\mu)(v+U) &= ((\lambda+\mu)v)+U \\ &= (\lambda v+\mu v)+U \\ &= ((\lambda v)+U)+((\mu v)+U) \\ &= \lambda(v+U)+\mu(v+U).\end{aligned}$$

This completes the proof.

□

Definition 4.39. *Suppose that U is a subspace of V. The quotient map $\pi : V \to V/U$ is defined by*

$$\pi(v) = v+U, \quad v \in U.$$

Remark 4.40. *Note that π depends on U as well as V and these spaces are left out of the notation because they should be clear from the context.*

Theorem 4.41. *Let U be a subspace of V. Then the quotient map $\pi : V \to V/U$ is a linear map.*

Proof. Let $v_1, v_2 \in V$ and $\lambda \in \mathbb{F}$ be arbitrarily chosen. Then we have the following.

$$\begin{aligned} \pi(v_1+v_2) &= (v_1+v_2)+U \\ &= (v_1+U)+(v_2+U) \\ &= \pi(v_1)+\pi(v_2) \end{aligned}$$

and

$$\begin{aligned} \pi(\lambda v_1) &= (\lambda v_1)+U \\ &= \lambda(v_1+U) \\ &= \lambda\pi(v_1). \end{aligned}$$

Therefore $\pi : V \to V/U$ is a linear map. This completes the proof. □

Theorem 4.42. *Suppose that V is finite-dimensional and U is its subspace. Then*

$$\dim V/U = \dim V - \dim U.$$

Proof. Let $\pi : V \to V/U$ be the quotient map. Take $v \in \ker\pi$ arbitrarily. Then

$$\begin{aligned} \pi(v) &= 0+U \\ &= U. \end{aligned}$$

By the definition of the map π, we have

$$\pi(v) = v+U.$$

Hence,

$$v+U = U.$$

Therefore $v \in U$. Because $v \in \ker \pi$ was arbitrarily chosen and we get that it is an element of U, we get the inclusion

$$\ker \pi \subseteq U. \tag{4.7}$$

Let now, $w \in U$ be arbitrarily chosen. Then, using that U is a subspace of V, we have

$$\begin{aligned} w+U &= U \\ &= 0+U. \end{aligned}$$

Therefore $w \in \ker \pi$. Since $w \in U$ was arbitrarily chosen and we get that it is an element of $\ker \pi$, we find

$$U \subseteq \ker \pi.$$

By the last inclusion and (4.7), we obtain the equality.

$$\ker \pi = U.$$

Let now, $z \in V/U$ be arbitrarily chosen. Then there is a $v_1 \in V$ so that

$$z = v_1 + U.$$

Hence,

$$\begin{aligned} \pi(v_1) &= v_1 + U \\ &= z. \end{aligned}$$

Consequently

$$\mathrm{range}\pi = V/U.$$

Now, applying the fundamental theorem of linear maps, we obtain

$$\begin{aligned} \dim V &= \dim \ker \pi + \dim \mathrm{range}\pi \\ &= \dim U + \dim V/U, \end{aligned}$$

whereupon

$$\dim V/U = \dim V - \dim U.$$

This completes the proof.

□

Definition 4.43. *Suppose that $T \in \mathscr{L}(V,W)$. Define $\widetilde{T} : V/\ker T \to W$ by*

$$\widetilde{T}(v+\ker T) = Tv.$$

Remark 4.44. *We will show that the map $\widetilde{T} : V/\ker T \to W$ is well defined. Suppose that $u, v \in V$ be such that*

$$u+\ker T = v+\ker T.$$

Then, by Theorem 4.34, it follows that

$$u-v \in \ker T.$$

Hence,

$$\begin{aligned} 0 &= T(u-v) \\ &= Tu-Tv, \end{aligned}$$

i.e.,

$$Tu = Tv.$$

Thus, the definition of $\widetilde{T}$ makes sense.

Theorem 4.45. *Suppose that $T \in \mathscr{L}(V,W)$. Then $\widetilde{T} : V/\ker T \to W$ is a linear map.*

Proof. Let

$$v_1+\ker T, \quad v_2+\ker T \in V/\ker T$$

and $\lambda \in \mathbb{F}$ be arbitrarily chosen. Then

$$(v_1+\ker T)+(v_2+\ker T) = (v_1+v_2)+\ker T$$

and

$$\lambda(v_1+\ker T) = (\lambda v_1)+\ker T.$$

Hence,

$$\begin{aligned} \widetilde{T}((v_1+\ker T)+(v_2+\ker T)) &= \widetilde{T}((v_1+v_2)+\ker T) \\ &= T(v_1+v_2) \\ &= Tv_1+Tv_2 \end{aligned}$$

$$= \widetilde{T}(v_1 + \ker T) + \widetilde{T}(v_2 + \ker T).$$

Moreover,

$$\begin{aligned}\widetilde{T}(\lambda(v_1 + \ker T)) &= \widetilde{T}((\lambda v_1) + \ker T)\\ &= T(\lambda v_1)\\ &= \lambda T v_1\\ &= \lambda \widetilde{T}(v_1 + \ker T).\end{aligned}$$

Thus, $\widetilde{T} : V/\ker T \to W$ is a linear map. This completes the proof. □

Theorem 4.46. *Suppose that $T \in \mathscr{L}(V,W)$. Then $\widetilde{T} : V/\ker T \to W$ is injective.*

Proof. Let $v \in V$ be such that $v + \ker T \in \ker \widetilde{T}$. Then

$$\begin{aligned}0 &= \widetilde{T}(v + \ker T)\\ &= Tv,\end{aligned}$$

whereupon $v \in \ker T$. Because $v - 0 \in \ker T$, by Theorem 4.34, it follows that

$$v + \ker T = 0 + \ker T.$$

Therefore

$$\ker \widetilde{T} = \{0\}.$$

Now, applying Theorem 3.33, we conclude that $\widetilde{T} : V/\ker T \to W$ is injective. This completes the proof. □

Theorem 4.47. *Let $T \in \mathscr{L}(V,W)$. Then*

$$range T = range \widetilde{T}. \tag{4.8}$$

Proof. Let $v \in \text{range} T$ be arbitrarily chosen. Then there is a $v_1 \in V$ so that

$$T v_1 = v.$$

Hence,

$$\begin{aligned}\widetilde{T}(v_1 + \ker T) &= Tv_1 \\ &= v.\end{aligned}$$

Consequently $v \in \mathrm{range}\widetilde{T}$. Since $v \in \mathrm{range}T$ was arbitrarily chosen and we get that it is an element of $\mathrm{range}\widetilde{T}$, we obtain the inclusion

$$\mathrm{range}T \subseteq \mathrm{range}\widetilde{T}. \tag{4.9}$$

Let now, $w \in \mathrm{range}\widetilde{T}$ be arbitrarily chosen. Then there is a $w_1 \in V$ so that

$$\widetilde{T}(w_1 + \ker T) = w.$$

Hence, by the definition of $\widetilde{T}$, it follows that

$$Tw_1 = w.$$

So, $w \in \mathrm{range}T$. Because $w \in \mathrm{range}\widetilde{T}$ was arbitrarily chosen and we get that it is an element of $\mathrm{range}T$, we arrive at the inclusion

$$\mathrm{range}\widetilde{T} \subseteq \mathrm{range}T.$$

By the last inclusion and (4.9), we find (4.8). This completes the proof. □

Theorem 4.48. *Let $T \in \mathscr{L}(V,W)$. Then $V/\ker T$ and $\mathrm{range}T$ are isomorphic.*

Proof. By Theorem 4.47, we have (4.8). Hence, we have $\widetilde{T} : V/\ker T \to \mathrm{range}T$ and $\widetilde{T}$ is surjective. By Theorem 4.46, we have that $\widetilde{T} : V/\ker T \to \mathrm{range}T$ is injective. From here we conclude that $\widetilde{T} : V/\ker T \to \mathrm{range}T$ is invertible and then $V/\ker T$ and $\mathrm{range}T$ are isomorphic. This completes the proof. □

Definition 4.49. *Suppose that T is a function from V to W. The graph of T is the subset of $V \times W$ defined by*

$$graphT = \{(v, Tv) \in V \times W : v \in V\}.$$

Theorem 4.50. *Let T be a function from V to W. Then T is a linear map if and only if $graphT$ is a subspace of $V \times W$.*

Proof. 1. Let T be a linear map. Take $v_1, v_2, v_3 \in V$ and $\lambda, \mu \in \mathbb{F}$ arbitrarily. Then we have the following.

(a)

$$\begin{aligned}
(v_1, Tv_1) + (v_2, Tv_2) &= (v_1 + v_2, Tv_1 + Tv_2) \\
&= (v_1 + v_2, T(v_1 + v_2)) \\
&= (v_2 + v_1, T(v_2 + v_1)) \\
&= (v_2 + v_1, Tv_2 + Tv_1) \\
&= (v_2, Tv_2) + (v_1, Tv_1).
\end{aligned}$$

(b)

$$\begin{aligned}
&(v_1, Tv_1) + ((v_2, Tv_2) + (v_3, Tv_3)) \\
&= (v_1 + (v_2 + v_3), Tv_1 + (Tv_2 + Tv_3)) \\
&= ((v_1 + v_2) + v_3, (Tv_1 + Tv_2) + Tv_3) \\
&= (v_1 + v_2, TV_1 + Tv_2) + (v_3, Tv_3) \\
&= ((v_1, Tv_1) + (v_2, Tv_2)) + (v_3, Tv_3).
\end{aligned}$$

(c)

$$\begin{aligned}
(v_1, Tv_1) + (0, T0) &= (v_1 + 0, Tv_1 + T0) \\
&= (v_1, T(v_1 + 0)) \\
&= (v_1, Tv_1).
\end{aligned}$$

(d)

$$(v_1, Tv_1) + (-v_1, -Tv_1) = (v_1 + (-v_1), Tv_1 + (-Tv_1))$$

$$= \quad (0,0).$$

(e)

$$\begin{aligned}
\lambda((v_1, Tv_1) + (v_2, Tv_2)) &= \lambda(v_1 + v_2, Tv_1 + Tv_2) \\
&= (\lambda(v_1 + v_2), \lambda(Tv_1 + Tv_2)) \\
&= (\lambda v_1 + \lambda v_2, \lambda Tv_1 + \lambda Tv_2) \\
&= (\lambda v_1, \lambda Tv_1) + (\lambda v_2, \lambda Tv_2) \\
&= \lambda(v_1, Tv_1) + \lambda(v_2, Tv_2).
\end{aligned}$$

(f)

$$\begin{aligned}
(\lambda + \mu)(v_1, Tv_1) &= ((\lambda + \mu)v_1, (\lambda + \mu)Tv_1) \\
&= (\lambda v_1 + \mu v_1, \lambda Tv_1 + \mu Tv_1) \\
&= (\lambda v_1, \lambda Tv_1) + (\mu v_1, \mu Tv_1) \\
&= \lambda(v_1, Tv_1) + \mu(v_1, Tv_1).
\end{aligned}$$

Thus, graph$T$ is a subspace of $V \times W$.

2. Let graph$T$ be a subspace of $V \times W$. Take $v_1, v_2 \in V$ and $\lambda \in \mathbb{F}$ arbitrarily. Then

$$v_1 + v_2, \lambda v_1 \in V, \quad Tv_1, Tv_2, \lambda Tv_1, T(\lambda v_1) \in W$$

and

$$(v_1, Tv_1), (v_2, Tv_2), (v_1 + v_2, Tv_1 + Tv_2), (v_1 + v_2, T(v_1 + v_2)),$$
$$(\lambda v_1, T(\lambda v_1)), (\lambda v_1, \lambda Tv_1) \in W.$$

Hence,

$$(v_1 + v_2, Tv_1 + Tv_2) = (v_1 + v_2, T(v_1 + v_2))$$

and

$$(\lambda v_1, \lambda T v_1) = (\lambda v_1, T(\lambda v_1)).$$

Therefore

$$T(v_1 + v_2) = T v_1 + T v_2$$

and

$$T(\lambda v_1) = \lambda T v_1.$$

Thus, $T : V \to W$ is a linear map. This completes the proof.

□

4.7 Advanced Practical Problems

Problem 4.51. *Prove that v_1, v_2, v_3 is a basis of $\mathbb{R}^3$ and find $\mathscr{M}(w)$, where*

1.

$$v_1 = \begin{pmatrix} 3 \\ -1 \\ -2 \end{pmatrix}, \quad v_2 = \begin{pmatrix} -2 \\ -1 \\ 0 \end{pmatrix}, \quad v_3 = \begin{pmatrix} 1 \\ 3 \\ 3 \end{pmatrix}, \quad w = \begin{pmatrix} -2 \\ -2 \\ 0 \end{pmatrix}.$$

2.

$$v_1 = \begin{pmatrix} 1 \\ -1 \\ 1 \end{pmatrix}, \quad v_2 = \begin{pmatrix} -1 \\ -1 \\ 1 \end{pmatrix}, \quad v_3 = \begin{pmatrix} -1 \\ 0 \\ 2 \end{pmatrix}, \quad w = \begin{pmatrix} -3 \\ -4 \\ 1 \end{pmatrix}.$$

Problem 4.52. *Prove that $\mathbb{R}^2 \times \mathbb{R}^5$ and $\mathbb{R}^7$ are isomorphic.*

Problem 4.53. *Find a basis of $\mathscr{P}_3(\mathbb{R}) \times \mathbb{R}^4$.*

Problem 4.54. *Consider $\mathbb{R}^4$ and*

$$U = \{(x, x+1, y, 2y) : x, y \in \mathbb{R}\}.$$

Find $(-1, 2, 1, 3) + U$.

Problem 4.55. *Let V and W be finite-dimensional. Let also, $v \in V$ and*

$$E = \{T \in \mathscr{L}(V, W) : T v = 0\}.$$

Prove that E is a subspace of $\mathscr{L}(V, W)$.

Problem 4.56. *Let V be finite-dimensional and $S,T \in \mathscr{L}(V)$. Prove that $ST = I$ if and only if $TS = I$.*

Problem 4.57. *Prove that the following are equivalent.*

1. *The trivial solution $x = y = 0$ is the only solution to the homogeneous system*

$$\begin{aligned} a_{11}x + a_{12}y &= 0 \\ a_{21}x + a_{22}y &= 0. \end{aligned}$$

2. *For any $c_1, c_2 \in \mathbb{F}$, there exists a solution to the system*

$$\begin{aligned} a_{11}x + a_{12}y &= c_1 \\ a_{21}x + a_{22}y &= c_2. \end{aligned}$$

Problem 4.58. *Generalize Problem 4.57 for arbitrary $n \in \mathbb{N}$.*

Problem 4.59. *Suppose that A_1 and A_2 are affine subsets of V. Prove that $A_1 \bigcap A_2$ is either an affine subset of V or the empty set.*

Problem 4.60. *Suppose that $v_1, v_2 \in V$. Prove that*

$$A = \{\lambda_1 v_1 + \lambda_2 v_2 : \lambda_1, \lambda_2 \in \mathbb{F} : \lambda_1 + \lambda_2 = 1\}$$

is an affine subset of V.

Problem 4.61. *Generalize Problem 4.60 for arbitrary $n \in \mathbb{N}$.*

Problem 4.62. *Suppose that U is a subspace of V such that V/U is finite-dimensional. Prove that V is isomorphic to $U \times (V/U)$.*

Chapter 5

Invariant Subspaces, Eigenvalues and Eigenvectors

Suppose that V is a vector space over $\mathbb{F}$.

5.1 Invariant Subspaces

Definition 5.1. *Suppose that $T \in \mathscr{L}(V)$. A subspace U of V is called invariant under T if $u \in U$ implies that $Tu \in U$.*

Example 5.2. *Let $T \in \mathscr{L}(V)$.*

1. *Consider $U = \{0\}$. Then, $T0 = 0$ and U is invariant under T.*
2. *Consider V. Then, $v \in V$ implies that $Tv \in V$ because $T \in \mathscr{L}(V)$. Therefore V is invariant under T.*
3. *Let $U = \ker T$. Then, $u \in U$ implies that $Tu = 0 \in U$ and U is invariant under T.*
4. *Let $U = rangeT$. Then, $u \in U$ implies that $Tu \in U$ and U is invariant under T.*

Exercise 5.3. *Let $V = \mathbb{R}^2$,*

$$U = \{(x,y) \in \mathbb{R}^2 : |x| \le 2, |y| \le 2\}.$$

Check if U is invariant under T, where

1. $$T(x,y) = \left(\frac{1}{1+2x^2+y^2}, -1\right), \quad (x,y) \in \mathbb{R}^2.$$

2. $$T(x,y) = (3x+y, 4x^2), \quad (x,y) \in \mathbb{R}^2.$$

3. $$T(x,y) = \left(\frac{1}{1+x^2}, \frac{1}{1+y^2}\right), \quad (x,y) \in \mathbb{R}^2.$$

4. $$T(x,y) = \left(\frac{1}{1+(x+2y)^2}, -\frac{1}{1+(x-3y)^2}\right), \quad (x,y) \in \mathbb{R}^2.$$

5. $$T(x,y) = (x^2+4x+10, y), \quad (x,y) \in \mathbb{R}^2.$$

Answer 5.4. 1. *Yes.*

2. *No.*

3. *Yes.*

4. *Yes.*

5. *No.*

5.2 Eigenvalues and Eigenvectors

Definition 5.5. *Suppose that $T \in \mathscr{L}(V)$. A number $\lambda \in \mathbb{F}$ is called an eigenvalue of T if there is a $v \in V$ such that $v \neq 0$ and $Tv = \lambda v$.*

Remark 5.6. *In the definition above, we require $v \neq 0$ because for any scalar $\lambda \in \mathbb{F}$ we have*

$$T0 = \lambda 0.$$

Theorem 5.7. *Suppose that V is finite-dimensional, $T \in \mathscr{L}(V)$ and $\lambda \in \mathbb{F}$. Then the following are equivalent.*

1. *λ is an eigenvalue of T.*

2. *$T - \lambda I$ is not injective.*

3. *$T - \lambda I$ is not surjective.*

4. *$T - \lambda I$ is not invertible.*

Proof. 1. Let λ be an eigenvalue of T. Then there is a $v \in V$ so that $v \neq 0$ and

$$(T - \lambda I)v = 0.$$

Since

$$(T - \lambda I)0 = 0,$$

we conclude that $T - \lambda I$ is not injective. So, $1 \Longrightarrow 2$. Let now, $T - \lambda I$ is not injective. Then, applying Theorem 3.33, it follows that

$$\ker(T - \lambda I) \neq \{0\}.$$

Hence, there is a $v_1 \in V$, $v_1 \neq 0$, so that

$$(T - \lambda I)v_1 = 0.$$

Therefore λ is an eigenvalue of T and $2 \Longrightarrow 1$. Hence, $1 \Longleftrightarrow 2$.

2. Let $T - \lambda I$ is not injective. Assume that $T - \lambda I$ is surjective. Applying Theorem 4.17, we get that $T - \lambda I$ is injective. This is a contradiction. Thus, $T - \lambda I$ is not surjective and we get $2 \Longrightarrow 3$. Let now, $T - \lambda I$ is not surjective. Suppose that $T - \lambda I$ is injective. Hence, applying Theorem 4.17, we obtain that $T - \lambda I$ is surjective. This is a contradiction. Therefore $2 \Longleftrightarrow 3$.

3. Let $T - \lambda I$ is not surjective. Assume that $T - \lambda I$ is invertible. Then $T - \lambda I$ is surjective. This is a contradiction. Then $T - \lambda I$ is not invertible. Let now, $T - \lambda I$ is not invertible. Suppose that $T - \lambda I$ is surjective. From here, applying Theorem 4.17, it follows that $T - \lambda I$ is invertible. This is a contradiction. Thus, $3 \Longleftrightarrow 4$. This completes the proof.

□

Definition 5.8. *Suppose that $T \in \mathscr{L}(V)$ and $\lambda \in \mathbb{F}$ is an eigenvalue of T. A vector $v \in V$ is called an eigenvector corresponding to λ if $v \neq 0$ and*

$$Tv = \lambda v.$$

Because

$$Tv = \lambda v \quad \Longleftrightarrow (T - \lambda I)v = 0,$$

a vector $v \in V$, $v \neq 0$, is an eigenvector of T corresponding to λ if and only if

$$v \in \ker(T - \lambda I).$$

Example 5.9. *Consider $V = \mathbb{R}^2$ over $\mathbb{R}$. Let $T \in \mathscr{L}(V)$ be defined by*

$$T(x,y) = (-y,x), \quad (x,y) \in \mathbb{R}^2.$$

We will find the eigenvalues of T. Suppose that $(x,y) \in \mathbb{R}^2$, $(x,y) \neq (0,0)$, and $\lambda \in \mathbb{R}$ are such that

$$T(x,y) = \lambda(x,y).$$

Hence,

$$(-y,x) = \lambda(x,y),$$

or

$$(-y,x) = (\lambda x, \lambda y).$$

Thus, we get the system

$$\begin{aligned} -y &= \lambda x \\ x &= \lambda y. \end{aligned} \tag{5.1}$$

By the last system, we find

$$x = -\lambda^2 x.$$

If $x = 0$, then, using (5.1), we find $y = 0$, which is impossible. Therefore

$$\lambda^2 + 1 = 0.$$

The last equation has no real solutions and then the operator T has no eigenvalues.

Example 5.10. *Consider $V = \mathbb{C}^2$ over $F = \mathbb{C}$. Let $T \in \mathscr{L}(V)$ be defined by*

$$T(x,y) = ((1-2i)x - y, (-1+i)x + (1-i)y), \quad (x,y) \in V.$$

We will find the eigenvalues and the eigenvectors of T. Suppose that $(x,y) \in V$ and $\lambda \in \mathbb{F}$ are such that $(x,y) \neq (0,0)$ and

$$((1-2i)x - y, (-1+i)x + (1-i)y) = \lambda(x,y).$$

By the last equation, we get the following system

$$\begin{aligned}(1-2i)x-y &= \lambda x\\(-1+i)x+(1-i)y &= \lambda y,\end{aligned}$$

whereupon

$$\begin{aligned}y &= (1-2i)x-\lambda x\\(-1+i)x+(1-i)y &= \lambda y.\end{aligned}$$

By the last system, we obtain the equation

$$(-1+i)x+(1-i)((1-2i)x-\lambda x)=\lambda((1-2i)x-\lambda x),$$

or

$$(-1+i)x+(1-i)(1-2i)x-\lambda(1-i)x=\lambda(1-02i)x-\lambda^2x,$$

or

$$\lambda^2x-\lambda(1-i+1-2i)x+(-1+i)x+(1-2i-i-2)x=0,$$

or

$$\lambda^2x-\lambda(2-3i)x+(-1+i)x+(-1-3i)x=0,$$

or

$$\lambda^2x-\lambda(2-3i)x+(-2-2i)x=0,$$

or

$$(\lambda^2-\lambda(2-3i)-(2+2i))x=0.$$

If $x=0$, then $y=0$, which is impossible. Then

$$\lambda^2-\lambda(2-3i)-(2+2i)=0,$$

whereupon

$$\begin{aligned}\lambda_{1,2} &= \frac{2-3i\pm\sqrt{(2-3i)^2+4(2+2i)}}{2}\\&= \frac{2-3i\pm\sqrt{4-12i-9+8+8i}}{2}\end{aligned}$$

$$= \frac{2-3i\pm\sqrt{3-4i}}{2}.$$

Let

$$\sqrt{3-4i} = a+bi, \quad a,b \in \mathbb{R}.$$

Then

$$\begin{aligned} 3-4i &= (a+bi)^2 \\ &= a^2+2abi-b^2, \end{aligned}$$

from where we arrive at the system

$$\begin{aligned} a^2-b^2 &= 3 \\ 2ab &= -4, \end{aligned}$$

or

$$\begin{aligned} a^2-b^2 &= 3 \\ ab &= -2. \end{aligned}$$

Hence, $a \neq 0$ *and* $b \neq 0$*. Therefore*

$$\begin{aligned} a^2-b^2 &= 3 \\ b &= -\frac{2}{a}. \end{aligned}$$

Thus, we get the equation

$$a^2-\frac{4}{a^2}=3$$

or

$$a^4-3a^2-4=0.$$

So,

$$(a^2)_{1,2} = \frac{3\pm\sqrt{25}}{2}$$

$$= \frac{3 \pm 5}{2}.$$

Then

$$a^2 = 4$$

and

$$a_1 = 2, \quad a_2 = -2,$$

and

$$b_1 = -1, \quad b_2 = 1.$$

Therefore

$$\sqrt{3-4i} = \pm(2-i)$$

and

$$\lambda_{1,2} = \frac{2-3i \pm (2-i)}{2}.$$

Consequently

$$\begin{aligned} \lambda_1 &= \frac{2-3i+2-i}{2} \\ &= 2-2i, \\ \lambda_2 &= \frac{2-3i-2+i}{2} \\ &= -i. \end{aligned}$$

Hence, for $\lambda = \lambda_1$, we have

$$\begin{aligned} y &= (1-2i)x - \lambda_1 x \\ &= (1-2i)x - (2-2i)x \\ &= -x, \end{aligned}$$

and for $\lambda = \lambda_2$, we get

$$\begin{aligned} y &= (1-2i)x - \lambda_2 x \\ &= (1-2i)x + ix \end{aligned}$$

$$= \quad (1-i)x.$$

Thus, the eigenvalues of T are

$$\lambda_1 = 2-2i,$$

$$\lambda_2 = -i$$

and the corresponding eigenvectors are

$$v_1 = (x,-x),$$

$$v_2 = (x,(1-i)x), \quad x \in \mathbb{R}, \quad x \neq 0.$$

Example 5.11. *Consider $V = \mathbb{C}^3$ and $\mathbb{F} = \mathbb{C}$. Let $T \in \mathscr{L}(V)$ be defined by*

$$T(x,y,z) = (y,-4x+4y,-2x+y+2z), \quad (x,y,z) \in V.$$

We will find the eigenvalues and eigenvectors of T. Let $(x,y,z) \in V$ and $\lambda \in \mathbb{F}$ be such that $(x,y,z) \neq (0,0,0)$ and

$$T(x,y,z) = \lambda(x,y,z).$$

Then

$$(y,-4x+4y,-2x+y+2z) = \lambda(x,y,z).$$

Hence, we get the system

$$y = \lambda x$$

$$-4x+4y = \lambda y$$

$$-2x+y+2z = \lambda z,$$

whereupon

$$y = \lambda x$$

$$-4x+4\lambda x = \lambda^2 x$$

$$-2x+y+2z = \lambda z.$$

Consider the equation

$$-4x+4\lambda x = \lambda^2 c$$

or

$$(\lambda^2 - 4\lambda + 4)x = 0. \tag{5.2}$$

If $x = 0$, *then*

$$2z = \lambda z.$$

Hence, if $z = 0$, *we get* $(x,y,z) = (0,0,0)$, *which is impossible. Therefore* $\lambda = 2$ *and the corresponding eigenvector is* $(0,0,z)$, $z \in \mathbb{R}$, $z \neq 0$. *Let* $x \neq 0$. *Then*

$$\lambda^2 - 4\lambda + 4 = 0$$

or

$$(\lambda - 2)^2 = 0.$$

Therefore $\lambda = 2$ *and*

$$y = 2x, \quad x \in \mathbb{R}.$$

Hence, the eigenvalue of the operator T *is* $\lambda = 2$ *and the corresponding eigenvectors are*

$$v = x(1,2,0) + z(0,0,1), \quad (x,z) \in \mathbb{R}^2, \quad (x,z) \neq (0,0).$$

Exercise 5.12. *Let* $V = \mathbb{R}^2$ *and* $\mathbb{F} = \mathbb{R}$. *Find the eigenvalues and the eigenvectors of* $T \in \mathscr{L}(V)$, *where*

1. $$T(x,y) = (2x+y, -x), \quad (x,y) \in V.$$

2. $$T(x,y) = (x+2y, -x), \quad (x,y) \in V.$$

Answer 5.13. *1.*

$$\lambda = 1,$$

$$v = (x,-x), \quad x \in \mathbb{R}, \quad x \neq 0.$$

2. *No eigenvalues and eigenvectors.*

Exercise 5.14. *Let $V = \mathbb{C}^2$ and $\mathbb{F} = \mathbb{C}$. Find the eigenvalues and eigenvectors of $T \in \mathscr{L}(V)$, where*

$$T(x,y) = (-2x+y, -x-2y), \quad (x,y) \in V.$$

Answer 5.15.

$$\begin{aligned}
\lambda_1 &= -2+i, \\
\lambda_2 &= -2-i, \\
v_1 &= (x, ix), \\
v_2 &= (x, -ix), \quad x \in \mathbb{R}, \quad x \neq 0.
\end{aligned}$$

Exercise 5.16. *Let $V = \mathbb{C}^3$ and $\mathbb{F} = \mathbb{C}$. Find the eigenvalues and eigenvectors of the operator $T \in \mathscr{L}(V)$, where*

1.

$$T(x,y,z) = (2x-y+2z, 5x-3y+3z, -x-2z), \quad (x,y,z) \in V.$$

2.

$$T(x,y,z) = (4x-5y+2z, 5x-7y+3z, 6x-9y+4z), \quad (x,y,z) \in V.$$

Answer 5.17. 1.

$$\begin{aligned}
\lambda &= -1, \\
v &= (x, x, -x), \quad x \in \mathbb{R}, \quad x \neq 0.
\end{aligned}$$

2.

$$\begin{aligned}
\lambda_1 &= 1, \\
\lambda_2 &= 0, \\
v_1 &= (x, x, x), \\
v_2 &= (x, 2x, 3x), \quad x \in \mathbb{R}, \quad x \neq 0.
\end{aligned}$$

Theorem 5.18. *Let $T \in \mathscr{L}(V)$. Suppose that $\lambda_1, \ldots, \lambda_m$ are distinct eigenvalues of T and $v_1, \ldots, v_m$ are their corresponding eigenvectors. Then $v_1, \ldots, v_m$ are linearly independent.*

Proof. Suppose that $v_1, \ldots, v_m$ are linearly dependent. Let $k \in \{1, \ldots, m\}$ be the smallest integer so that

$$v_k \in \operatorname{span}(v_1, \ldots, v_{k-1}).$$

Then there exist scalars $a_1, \ldots, a_{k-1} \in \mathbb{F}$ so that

$$v_k = a_1 v_1 + \cdots + a_{k-1} v_{k-1}.$$

Then

$$\begin{aligned} \lambda_k v_k &= T v_k \\ &= T(a_1 v_1 + \cdots + a_{k-1} v_{k_1}) \\ &= T(a_1 v_1) + \cdots + T(a_{k-1} v_{k-1}) \\ &= a_1 T v_1 + \cdots + a_{k-1} T v_{k-1} \\ &= a_1 \lambda_1 v_1 + \cdots + a_{k-1} \lambda_{k-1} v_{k-1} \end{aligned}$$

and

$$\lambda_k v_k = a_1 \lambda_k v_1 + \cdots + a_{K-1} \lambda_k v_{k-1}.$$

We subtract from the last equation the previous equation and we find

$$0 = a_1(\lambda_k - \lambda_1) v_1 + \cdots + a_{k-1}(\lambda_k - \lambda_{k-1}) v_{k-1}.$$

By the definition of the integer k, it follows that $v_1, \ldots, v_{k-1}$ are linearly independent. Then, the last equation is possible if

$$\begin{aligned} a_1(\lambda_k - \lambda_1) &= 0 \\ &\vdots \\ a_{k-1}(\lambda_k - \lambda_{k-1}) &= 0. \end{aligned}$$

Because $\lambda_1,\ldots,\lambda_{k-1}$ are distinct, from the last system we find

$$\begin{aligned} a_1 &= 0 \\ &\vdots \\ a_{k-1} &= 0 \end{aligned}$$

and $v_k = 0$. This is impossible because v_k is an eigenvector of the operator T. Consequently $v_1,\ldots,v_m$ are linearly independent. This completes the proof. □

Theorem 5.19. *Let V be finite-dimensional. Then each operator on V has at most* $\dim V$ *distinct eigenvalues.*

Proof. Let $T \in \mathscr{L}(V)$ and $\lambda_1,\ldots,\lambda_m$ be distinct eigenvalues of T. Let also, $v_1,\ldots,v_m$ be their corresponding eigenvectors. Hence and Theorem 5.18, they are linearly independent. Now, applying Theorem 2.32, we obtain $m \leq \dim V$. This completes the proof. □

5.3 Restriction and Quotient Operators

Definition 5.20. *Suppose that $T \in \mathscr{L}(V)$ and U is a subspace of V invariant under T. The restriction operator $T|_U \in \mathscr{L}(U)$ is defined by*

$$T|_U(u) = u, \quad u \in U.$$

The quotient operator $T/U \in \mathscr{L}(V/U)$ is defined by

$$(T/U)(v+U) = Tv+U, \quad v \in V.$$

Let $v, w \in V$ be such that

$$v+U = w+U.$$

Then $v-w \in U$ and since U is invariant under T, we obtain

$$T(v-w) \quad = \quad Tv-Tw$$

$$\in \quad U.$$

Therefore

$$Tv + U = Tw + U.$$

Thus, the definition for quotient operator makes sense.

Example 5.21. *Let $T \in \mathscr{L}(\mathbb{F}^2)$ be defined by*

$$T(x,y) = (y,0), \quad (x,y) \in \mathbb{F}^2.$$

Let also,

$$U = \{(x,0) : x \in \mathbb{F}\}.$$

Then

$$\begin{aligned} T(x,0) &= (0,0) \\ &\in U. \end{aligned}$$

From here, we conclude that U is invariant under T. Next,

$$T|_U(x,0) = (0,0), \quad (x,0) \in U.$$

So, $T|_U$ is the zero operator. Moreover,

$$\begin{aligned} (T/U)((x,y)+U) &= T(x,y)+U \\ &= (y,0)+U \\ &= U \\ &= (0,0)+U. \end{aligned}$$

Therefore T/U is the zero operator.

Remark 5.22. *In the above example, we saw that $T|_U$ and T/U are the zero operator, but T is not the zero operator. Therefore $T|_U$ and T/U do not provide enough information for the operator T.*

Example 5.23. *Let $T \in \mathscr{L}(V)$ and U be a subspace of V such that $U \subset \ker T$. We will prove that U is invariant under T. Really, let $u \in U$ be arbitrarily chosen. Then $u \in \ker T$ and $Tu = 0$. Hence,*

$$Tu = 0$$

$$\in \quad U.$$

Because $u \in U$ was arbitrarily chosen, we conclude that U is invariant under T.

Example 5.24. *Let $S, T \in \mathscr{L}(V)$ be such that $ST = TS$. We will prove that* $\ker S$ *is invariant under T. Let $u \in \ker S$ be arbitrarily chosen. Then $Su = 0$ and hence,*

$$\begin{aligned} STu &= TSu \\ &= T0 \\ &= 0. \end{aligned}$$

Therefore $Tu \in U$. Because $u \in U$ was arbitrarily chosen, we get that U is invariant under T.

Exercise 5.25. *Let $S, T \in \mathscr{L}(V)$ be such that $ST = TS$. Prove that rangeS is invariant under T.*

5.4 Polynomials Applied to Operators

If $T \in \mathscr{L}(V)$, then TT makes sense and by Theorem 4.4, it follows that $TT \in \mathscr{L}(V)$. More generally, we have the following definition.

Definition 5.26. *Suppose that $T \in \mathscr{L}(V)$ and m is a positive integer. Then T^m is defined by*

$$T^m = \underbrace{T \dots T}_{m \quad times}.$$

T^0 is defined to be the identity operator. In addition, if T is invertible, then T^{-m} is defined to be the operator

$$T^{-m} = \left(T^{-1}\right)^m.$$

Note that

$$T^m T^n = T^{m+n},$$

$$(T^m)^n = T^{mn}$$

for any positive integers m and n.

Example 5.27. *Let $V = \mathbb{R}^2$ and $\mathbb{F} = \mathbb{R}$. Let also, $T \in \mathscr{L}(V)$ be defined by*

$$T(x,y) = (2x-3y,4x), \quad (x,y) \in V.$$

We will find

$$T^2(x,y), \quad T^3(x,y), \quad (x,y) \in V.$$

We have

$$\begin{aligned} T^2(x,y) &= TT(x,y) \\ &= T(2x-3y,4x) \\ &= (2(2x-3y)-3(4x),4(2x-3y)) \\ &= (4x-6y-12x,8x-12y) \\ &= (-8x-6y,8x-12y), \quad (x,y) \in V, \end{aligned}$$

and

$$\begin{aligned} T^3(x,y) &= TT^2(x,y) \\ &= T(-8x-6y,8x-12y) \\ &= (2(-8x-6y)-3(8x-12y),4(-8x-6y)) \\ &= (-16x-12y-24x+36y,-32x-24y) \\ &= (-40x+24y,-32x-24y), \quad (x,y) \in V. \end{aligned}$$

Example 5.28. *Let $V = \mathbb{F}^\infty$. Consider the shift operator*

$$T(x_1,x_2,x_3,\dots) = (x_2,x_3,x_4,\dots), \quad (x_1,x_2,x_3,\dots) \in \mathbb{F}^\infty.$$

We will find

$$T^2(x_1,x_2,x_3,\ldots),\quad T^4(x_1,x_2,x_3,\ldots),\quad (x_1,x_2,x_3,\ldots)\in V.$$

We have

$$\begin{aligned}T^2(x_1,x_2,x_3,\ldots) &= TT(x_1,x_2,x_3,\ldots)\\ &= T(x_2,x_3,x_4,\ldots)\\ &= (x_3,x_4,x_5,\ldots),\quad (x_1,x_2,x_3,\ldots)\in V,\end{aligned}$$

and

$$\begin{aligned}T^4(x_1,x_2,x_3,\ldots) &= T^2(x_3,x_4,x_5,\ldots)\\ &= (x_5,x_6,x_7,\ldots),\quad (x_1,x_2,x_3,\ldots)\in V.\end{aligned}$$

Example 5.29. *Let $V=\mathbb{R}$ and $F=\mathbb{R}$. Let also, $T\in\mathscr{L}(V)$ be defined by*

$$Tx = 2x+7,\quad x\in V.$$

We will find

$$T^5x,\quad x\in V.$$

We have

$$\begin{aligned}T^2x &= TTx\\ &= T(2x+7)\\ &= 2(2x+7)+7\\ &= 4x+14+7\\ &= 4x+21,\quad x\in V,\end{aligned}$$

and

$$T^4x = T^2T^2x$$

$$\begin{aligned}
&= T^2(4x+21) \\
&= 4(4x+21)+21 \\
&= 16x+84+21 \\
&= 16x+105, \quad x \in V,
\end{aligned}$$

and

$$\begin{aligned}
T^5x &= TT^4x \\
&= T(16x+105) \\
&= 2(16x+105)+7 \\
&= 32x+210+7 \\
&= 32x+217, \quad x \in V.
\end{aligned}$$

Exercise 5.30. *Let $V \in \mathbb{R}^3$, $\mathbb{F} = \mathbb{R}$ and $T \in \mathscr{L}(V)$ be defined by*

$$T(x,y,z) = (x+y, y+z, z+x), \quad (x,y,z) \in V.$$

Find

$$T^3(x,y,z), \quad (x,y,z) \in V.$$

Answer 5.31.

$$T^3(x,y,z) = (2x+3y+3z, 3x+2y+3z, 3x+3y+2z), \quad (x,y,z) \in V.$$

Definition 5.32. *Suppose that $T \in \mathscr{L}(V)$ and $p \in \mathscr{P}(\mathbb{F})$ is a polynomial given by*

$$p(z) = a_0 + a_1 z + a_2 z^2 + \cdots + a_m z^m, \quad z \in \mathbb{F}.$$

Then $p(T)$ is the operator defined by

$$p(T) = a_0 I + a_) 1T + a_2 T^2 + \cdots + a_m T^m.$$

Example 5.33. *Let $T \in \mathscr{L}(\mathbb{F})$ be defined by*

$$Tx = x - 1, \quad x \in \mathbb{F},$$

and

$$p(x) = 1 + 2x + 3x^2, \quad x \in \mathbb{F}.$$

We will find

$$(p(T))(x), \quad x \in \mathbb{F}.$$

We have

$$p(T) = I + 2T + 3T^2$$

and

$$\begin{aligned} T^2x &= TTx \\ &= T(x-1) \\ &= x - 1 - 1 \\ &= x - 2, \quad x \in \mathbb{F}. \end{aligned}$$

Hence,

$$\begin{aligned} (p(T))(x) &= (I + 2T + 3T^2)x \\ &= x + 2Tx + 3T^2x \\ &= x + 2(x-1) + 3(x-2)^2 \\ &= x + 2x - 2 + 3(x^2 - 4x + 4) \\ &= 3x - 2 + 3x^2 - 12x + 12 \\ &= 3x^2 - 9x + 10, \quad x \in \mathbb{F}. \end{aligned}$$

Exercise 5.34. *Let $T \in \mathscr{L}(\mathbb{F})$ be defined by*

$$Tx = 2x + 3, \quad x \in \mathbb{F},$$

and

$$p(x) = -1 + x + 2x^2, \quad x \in \mathbb{F}.$$

Find

$$(p(T))(x), \quad x \in \mathbb{F}.$$

Answer 5.35.

$$(p(T))(x) = 32x^2 + 145x + 165, \quad x \in \mathbb{F}.$$

Theorem 5.36. *Let $T \in \mathscr{L}(V)$. Then the function f from $\mathscr{P}(\mathbb{F})$ to $\mathscr{L}(V)$ defined by*

$$f(p) = p(T), \quad p \in \mathscr{P}(\mathbb{F}),$$

is a linear function.

Proof. Let

$$\begin{aligned} p(z) &= a_0 + a_1 z + a_2 z^2 + \cdots + a_{m_1} z^{m_1}, \\ q(z) &= b_0 + b_1 z + b_2 z^2 + \cdots + b_{m_2} z^{m_2}, \quad z \in \mathbb{F}, \end{aligned}$$

$a_1, \ldots, a_{m_1}$, $b_0, \ldots, b_{m_2} \in \mathbb{F}$, and $\lambda \in \mathbb{F}$ be arbitrarily chosen. Without loss of generality, suppose that $m_1 \geq m_2$. Then

$$\begin{aligned} (p+q)(z) &= p(z) + q(z) \\ &= a_0 + a_1 z + a_2 z^2 + \cdots + a_{m_2} z^{m_2} + a_{m_2+1} z^{m_2+1} + \cdots + a_{m_1} z^{m_1} \\ &\quad + b_0 + b_1 z + b_2 z^2 + \cdots + b_{m_2} z^{m_2} \\ &= (a_0 + b_0) + (a_1 + b_1) z + (a_2 + b_2) z^2 + \cdots + (a_{m_2} + b_{m_2}) z^{m_2} \\ &\quad + a_{m_2+1} z^{m_2+1} + \cdots + a_{m_1} z^{m_1}, \\ (\lambda p)(z) &= \lambda (a_0 + a_1 z + a_2 z^2 + \cdots + a_{m_1} z^{m_1}) \\ &= (\lambda a_0) + (\lambda_1) z + (\lambda a_2) z^2 + \cdots + (\lambda a_{m_1}) z^{m_1}, \quad z \in \mathbb{F}. \end{aligned}$$

Hence,

$$
\begin{aligned}
p(T) &= a_0 + a_1 T + a_2 T^2 + \cdots + a_{m_1} T^{m_1}, \\
q(T) &= b_0 + b_1 T + b_2 T^2 + \cdots + b_{m_2} T^{m_2}
\end{aligned}
$$

and

$$
\begin{aligned}
f(p+q) &= (p+q)(T)(a_0+b_0) + (a_1+b_1)T + (a_2+b_2)T^2 \\
&\quad + \cdots + (a_{m_2}+b_{m_2})T^{m_2} \\
&\quad + a_{m_2+1}T^{m_2+1} + \cdots + a_{m_1}T^{m_1} \\
&= (a_0 + a_1 T + a_2 T^2 + \cdots + a_{m_1} T^{m_1}) \\
&\quad + (b_0 + b_1 z + b_2 z^2 + \cdots + b_{m_1} T^{m_1}) \\
&= p(T) + q(T) \\
&= f(p) + f(q), \\
f(\lambda p) &= (\lambda p)(T) \\
&= (\lambda a_0) + (\lambda a_1)T + (\lambda a_2)T^2 + \cdots + (\lambda a_{m_1})T^{m_1} \\
&= \lambda \left(a_0 + a_1 T + a_2 T^2 + \cdots + a_{m_1} T^{m_1}\right) \\
&= \lambda p(T) \\
&= \lambda f(p).
\end{aligned}
$$

This completes the proof. □

Definition 5.37. *Let $p, q \in \mathscr{P}(\mathbb{F})$. Then $pq \in \mathscr{P}(\mathbb{F})$ is the polynomial defined by*

$$(pq)(z) = p(z)q(z), \quad z \in \mathbb{F}.$$

Theorem 5.38. *Suppose that $p, q \in \mathscr{P}(\mathbb{F})$ and $T \in \mathscr{L}(V)$. Then*

$$(pq)(T) = p(T)q(T)$$

and

$$p(T)q(T) = q(T)p(T).$$

Proof. Let

$$\begin{aligned} p(z) &= \sum_{j=0}^{m} a_j z^j, \\ q(z) &= \sum_{k=0}^{n} b_k z^k. \end{aligned}$$

Then

$$\begin{aligned} (pq)(z) &= p(z)q(z) \\ &= \left(\sum_{j=0}^{m} a_j z^j\right)\left(\sum_{k=0}^{n} b_k z^k\right) \\ &= \sum_{j=0}^{m}\sum_{k=0}^{n} a_j b_k z^{j+k}, \quad z \in \mathbb{F}. \end{aligned}$$

Therefore

$$\begin{aligned} (pq)(T) &= \sum_{j=0}^{m}\sum_{k=0}^{n} a_j b_k T^{j+k} \\ &= \sum_{j=0}^{m}\sum_{k=0}^{n} a_j b_k T^j T^k \\ &= \left(\sum_{j=0}^{m} a_j T^j\right)\left(\sum_{k=0}^{n} b_k T^k\right) \\ &= p(T)q(T). \end{aligned}$$

Hence,

$$p(T)q(T) = (pq)(T)$$

$$= (qp)(T)$$

$$= q(T)p(T).$$

This completes the proof. □

Theorem 5.39. *Every linear operator on a finite-dimensional, nonzero, complex vector space has an eigenvalue.*

Proof. Let V be a complex vector space with dimension $n > 0$ and $T \in \mathscr{L}(V)$. Take $v \in V$, $v \neq 0$. Such v exists because $n > 0$. Then the list

$$v, \quad Tv, \quad T^2v, \quad \ldots, T^n v$$

is linearly independent because it contains $n+1$ elements and $\dim V = n$. Hence, there are complex numbers a_0, a_1, ..., a_n, not all zero, such that

$$0 = a_0 v + a_1 Tv + a_2 T^2 v + \cdots + a_n T^n v.$$

If we assume that

$$a_1 = \cdots = a_n = 0,$$

then $a_0 \neq 0$ and $v = 0$, which is a contradiction. Therefore

$$(a_1, \ldots, a_n) \neq (0, \ldots, 0).$$

Consider

$$p(z) = a_0 + a_1 z + a_2 z^2 + \cdots + a_n z^n.$$

By the fundamental theorem of algebra, it follows that the polynomial p can be represented in the form

$$p(z) = c(z - \lambda_1) \cdots (z - \lambda_m),$$

where $c, \lambda_1, \ldots, \lambda_m \in \mathbb{C}$, $c \neq 0$. Hence,

$$\begin{aligned} 0 &= a_0 v + a_1 T^v + a_2 T^2 v + \cdots + a_n T^n v \\ &= (a_0 + a_1 T + a_2 T^2 + \cdots + a_n T^n) v \\ &= c(T - \lambda_1 I) \cdots (T - \lambda_m I) v. \end{aligned}$$

By the last equation, it follows that there is a $j \in \{1, \ldots, m\}$ such that $T - \lambda_j I$ is not injective. Hence, T has at least one eigenvalue. This completes the proof. □

5.5 Upper-Triangular Matrices

Definition 5.40. *The diagonal of a square matrix consists of the entries along the line from upper left corner to the bottom right corner.*

Example 5.41. *Let*

$$A = \begin{pmatrix} -3 & 1 & 2 \\ -2 & 0 & 1 \\ 1 & 2 & 3 \end{pmatrix}.$$

Then the diagonal of the matrix A consists of the entries -3, 0, 3.

Definition 5.42. *A matrix is called upper-triangular if all entries below the diagonal equal* 0.

Typically, we represent an upper-triangular matrix in the form

$$\begin{pmatrix} \lambda_1 & & \star \\ & \ddots & \\ 0 & & \lambda_n \end{pmatrix}.$$

Example 5.43. *The matrix*

$$A = \begin{pmatrix} -1 & 1 & 4 & 2 \\ 0 & 2 & 1 & 0 \\ 0 & 0 & 3 & -1 \\ 0 & 0 & 0 & 4 \end{pmatrix}$$

is an upper-triangular matrix.

Theorem 5.44. *Suppose that $T \in \mathscr{L}(V)$ and $v_1, \ldots, v_n$ is a basis in V. Then the following are equivalent.*

1. *The matrix $\mathscr{M}(T)$ of T is upper-triangular with respect to $v_1, \ldots, v_n$.*

2. *$Tv_j \in span(v_1, \ldots, v_j)$ for any $j \in \{1, \ldots, n\}$.*

3. *$span(v_1, \ldots, v_j)$ is invariant under T for any $j \in \{1, \ldots, n\}$.*

Proof. 1. Suppose that the matrix $\mathscr{M}(T)$ of T is upper-triangular with respect to $v_1, \ldots, v_n$. Then

$$\mathscr{M}(T) = \begin{pmatrix} a_{11} & a_{12} & \cdots & a_{1n} \\ 0 & a_{22} & \cdots & a_{2n} \\ \vdots & \vdots & \vdots & \vdots \\ 0 & 0 & \cdots & a_{nn} \end{pmatrix}. \tag{5.3}$$

Hence,

$$\begin{aligned} Tv_1 &= a_{11}v_1, \\ Tv_2 &= a_{12}v_1 + a_{22}v_2, \\ &\vdots \\ Tv_n &= a_{1n}v_1 + a_{2n}v_2 + \cdots + a_{nn}v_n, \end{aligned} \tag{5.4}$$

where $a_{ij} \in \mathbb{F}$, $i \geq j$, $i, j \in \{1, \ldots, n\}$. By the system (5.4), it follows that

$$Tv_j \in \operatorname{span}(v_1, \ldots, v_j), \quad j \in \{1, \ldots, n\}. \tag{5.5}$$

Thus, $1 \Longrightarrow 2$. Now, suppose that (5.5) holds. Then there are $a_{ij} \in \mathbb{F}$, $i, j \in \{1, \ldots, n\}$, so that $a_{ij} = 0$, $i < j$, $i, j \in \{1, \ldots, n\}$. Then (5.4) holds and from here, (5.3) holds. Thus, $\mathscr{M}(T)$ is upper-triangular and we get $1 \Longleftrightarrow 2$.

2. Suppose that (5.5) holds. Then, we have (5.4). Let $j \in \{1, \ldots, n\}$ be arbitrarily chosen and fixed. Take $w \in \operatorname{span}(v_1, \ldots, v_j)$. Then there are scalars $b_1, \ldots, b_j \in \mathbb{F}$ so that

$$w = b_1v_1 + \cdots + b_jv_j.$$

Hence,

$$\begin{aligned} Tw &= T(b_1v_1 + \cdots + b_jv_j) \\ &= T(b_1v_1) + \cdots + T(b_jv_j) \\ &= b_1Tv_1 + \cdots + b_jTv_j \\ &= b_1a_{11}v_1 + b_2(a_{12}v_1 + a_{22}v_2) + \cdots + b_j(a_{1j}v_1 + \cdots + a_{jj}v_j) \\ &= (b_1a_{11} + b_2a_{12} + \cdots + b_ja_{1j})v_j \\ &\quad +(b_2a_{22} + \cdots + b_ja_{2j})v_2 \\ &\quad +\cdots \end{aligned}$$

$$
\begin{aligned}
&+b_j a_{jj} v_j \\
&\in \quad \operatorname{span}(v_1, \ldots, v_j).
\end{aligned}
$$

Because $j \in \{1, \ldots, n\}$ was arbitrarily chosen, we conclude that span$(v_1, \ldots, v_j)$ is invariant under T for any $j \in \{1, \ldots, n\}$. So, $2 \Longrightarrow 3$. Let now, span$(v_1, \ldots, v_j)$ is invariant under T for any $j \in \{1, \ldots, n\}$. Then we have (5.5) for any $j \in \{1, \ldots, n\}$ and so, $2 \Longleftrightarrow 3$. This completes the proof.

□

Theorem 5.45. *Suppose that V is finite-dimensional complex vector space and $T \in \mathscr{L}(V)$. Then T has an upper-triangular matrix with respect to some basis of V.*

Proof. We will use induction on the dimension of V. Clearly the assertion is valid for $\dim V = 1$. Suppose that $\dim V > 1$ and the result holds for all complex vector spaces whose dimension is less than the dimension of V. Let λ be an eigenvalue of T. By Theorem 5.39, it follows that such eigenvalue exists. Set

$$U = \operatorname{range}(T - \lambda I).$$

By Theorem 5.7, it follows that $T - \lambda I$ is not surjective. Therefore $\dim U < \dim V$. Let $u \in U$. Then

$$Tu = (T - \lambda I)u + \lambda u.$$

Since $(T - \lambda I)u \in U$ and U is a subspace of V, we have $\lambda u \in U$ and then $Tu \in U$. Because $u \in U$ was arbitrarily chosen, we conclude that U is invariant under T. Thus, $T|_U$ is an operator on U. By the induction assumption, it follows that there is a basis $u_1, \ldots, u_m$ of U with respect to which $T|_U$ has an upper-triangular matrix. Hence and Theorem 5.44, we get

$$
\begin{aligned}
Tu_j &= T|_U(u_j) \\
&\in \operatorname{span}(u_1, \ldots, u_j).
\end{aligned}
\tag{5.6}
$$

We extend $u_1, \ldots, u_m$ to a basis $u_1, \ldots, u_m, v_1, \ldots, v_n$ of V. For each $k \in \{1, \ldots, n\}$, we have

$$Tv_k = (T - \lambda I)v_k + \lambda v_k.$$

By the definition of U, it follows that $(T-\lambda I)v_k \in U$. Hence,

$$Tv_k \in \operatorname{span}(u_1,\ldots,u_m,v_1,\ldots,v_k). \tag{5.7}$$

By (5.6) and (5.7), applying Theorem 5.44, we get that T has an upper-triangular matrix with respect to $u_1,\ldots,u_m,v_1,\ldots,v_n$. This completes the proof. □

Theorem 5.46. *Suppose that $T \in \mathscr{L}(V)$ has an upper-triangular matrix with respect to some basis of V. Then T is invertible if and only if all the entries on the diagonal of that upper-triangular matrix are nonzero.*

Proof. Suppose that $v_1,\ldots,v_n$ is a basis of V with respect to which T has an upper-triangular matrix

$$\mathscr{M}(T) = \begin{pmatrix} \lambda_1 & & & \star \\ & \lambda_2 & & \\ & & \ddots & \\ 0 & & & \lambda_n \end{pmatrix}.$$

1. Let $\lambda_1,\ldots,\lambda_n$ be all nonzero. Then

$$Tv_1 = \lambda_1 v_1. \tag{5.8}$$

   Hence,

$$T\left(\frac{v_1}{\lambda_1}\right) = v_1.$$

   Thus, $v_1 \in \operatorname{range} T$. Next,

$$Tv_2 = av_1 + \lambda_2 v_2,$$

   whereupon

$$T\left(\frac{v_2}{\lambda_2}\right) = v_2 + \frac{a}{\lambda_1}v_1,$$

   for some $a \in \mathbb{F}$. Hence, $v_2 \in \operatorname{range} T$, and so on, $v_n \in \operatorname{range} T$. Because $v_1,\ldots,v_n$ is a basis of V, we conclude that $\operatorname{range} T = V$. Thus, T is surjective. Hence and Theorem 4.17, we conclude that T is invertible.

2. Suppose that T is invertible. By (5.8), it follows that $\lambda_1 \neq 0$. Otherwise, we will have $v_1 \neq 0$ and $Tv_1 = 0$, which is impossible because

T is invertible. Assume that $\lambda_j = 0$ for some $j \in \{1,\ldots,n\}$. Then $T : \operatorname{span}(v_1,\ldots,v_j) \to \operatorname{span}(v_1,\ldots,v_{j-1})$. Since

$$\dim \operatorname{span}(v_1,\ldots,v_j) = j$$

and

$$\dim \operatorname{span}(v_1,\ldots,v_{j-1}) = j-1,$$

we conclude that $T|_{\operatorname{span}(v_1,\ldots,v_j)}$ is not injective. From here, it follows that there is a $v \in \operatorname{span}(v_1,\ldots,v_j)$ such that $v \neq 0$ and $Tv = 0$. Thus, T is not injective. This is a contradiction. Therefore $\lambda_j \neq 0$. This completes the proof.

□

Theorem 5.47. *Suppose that $T \in \mathscr{L}(V)$ has an upper-triangular matrix with respect to some basis of V. Then the eigenvalues of T are precisely the entries on the diagonal on that upper-triangular matrix.*

Proof. Suppose that $v_1,\ldots,v_n$ is a basis of V with respect to which T has an upper-triangular matrix

$$\mathscr{M}(T) = \begin{pmatrix} \lambda_1 & & & \star \\ & \lambda_2 & & \\ & & \ddots & \\ 0 & & & \lambda_n \end{pmatrix}.$$

Let $\lambda \in \mathbb{F}$. Then

$$\mathscr{M}(T-\lambda I) = \begin{pmatrix} \lambda_1 - \lambda & & & \star \\ & \lambda_2 - \lambda & & \\ & & \ddots & \\ 0 & & & \lambda_n - \lambda \end{pmatrix}.$$

Hence, $T - \lambda I$ is not invertible if and only if $\lambda = \lambda_j$ for some $j \in \{1,\ldots,n\}$. Thus, λ is an eigenvalue of T if and only if $\lambda = \lambda_j$ for some $j \in \{1,\ldots,n\}$. This completes the proof. □

5.6 Eigenspaces and Diagonal Matrices

Definition 5.48. *A diagonal matrix is a square matrix that is* 0 *everywhere except possibly along the diagonal.*

Example 5.49. *The matrix*

$$\begin{pmatrix} -1 & 0 & 0 \\ 0 & 0 & 0 \\ 0 & 0 & 2 \end{pmatrix}$$

is a diagonal matrix.

Definition 5.50. *Suppose that $T \in \mathscr{L}(V)$ and $\lambda \in \mathbb{F}$. The eigenspace of T corresponding to λ, denoted by $E(\lambda, T)$, is defined by*

$$E(\lambda, T) = \ker(T - \lambda I),$$

i.e., $E(\lambda, T)$ is the set of all eigenvectors of T corresponding to λ, along with the 0 vector.

Since the kernel of any linear map on V is a subspace of V, we have that $E(\lambda, T)$ is a subspace of V. By the above definition, it follows that λ is an eigenvalue of T if and only if $E(\lambda, T) \neq \{0\}$.

Example 5.51. *Suppose that the matrix of an operator $T \in \mathscr{L}(V)$ with respect to a basis v_1, v_2, v_3 of V is the matrix in Example 5.49. Then*

$$\begin{aligned} E(-1, T) &= span(v_1), \\ E(0, T) &= span(v_2), \\ E(2, T) &= span(v_3). \end{aligned}$$

Theorem 5.52. *Suppose that V is finite-dimensional and $T \in \mathscr{L}(V)$. Suppose also that $\lambda_1, \ldots, \lambda_m$ are distinct eigenvalues of T. Then*

$$E(\lambda_1, T) + \cdots + E(\lambda_m, T) \tag{5.9}$$

is a direct sum and

$$\dim E(\lambda_1, T) + \cdots + \dim E(\lambda_m, T) \leq \dim V. \tag{5.10}$$

Proof. Let $u_j \in E(\lambda_j, T)$, $j \in \{1, \ldots, m\}$, be such that

$$u_1 + \cdots + u_m = 0. \tag{5.11}$$

Note that u_j, $j \in \{1,\dots,m\}$, are eigenvectors of T corresponding to eigenvalues λ_j, $j \in \{1,\dots,m\}$. By Theorem 5.18, it follows that $u_1,\dots,u_m$ is linearly independent. Then the equation (5.11) is valid if

$$u_j = 0, \quad j \in \{1,\dots,m\}.$$

Hence, applying Theorem 1.36, we conclude that (5.9) is a direct sum and the inequality (5.10) holds. This completes the proof. □

Definition 5.53. *An operator $T \in \mathscr{L}(V)$ is called diagonalizable if the operator has a diagonal matrix with respect to some basis of V.*

Example 5.54. *Let $T \in \mathscr{L}(\mathbb{R}^2$ be defined by*

$$T(x,y) = (41x+7y, -20x+74y), \quad (x,y) \in \mathbb{R}^2.$$

Let

$$\begin{aligned} w_1 &= (1,0), \\ w_2 &= (0,1). \end{aligned}$$

Then

$$\mathscr{M}(T; w_1, w_2) = \begin{pmatrix} 41 & 7 \\ -20 & 74 \end{pmatrix},$$

which is not a diagonal matrix. Now, set

$$\begin{aligned} v_1 &= (1,4), \\ v_2 &= (7,5). \end{aligned}$$

Assume that there is a scalar $a \in \mathbb{F}$ so that

$$v_1 = a v_2.$$

Then

$$\begin{aligned} (1,4) &= a(7,5) \\ &= (7a, 5a). \end{aligned}$$

Hence,

$$\begin{aligned} 7a &= 1 \\ 5a &= 4. \end{aligned}$$

The last system is impossible and therefore the list v_1, v_2 is linearly independent and forms a basis of $\mathbb{R}^2$. We have

$$\begin{aligned} Tv_1 &= T(1,4) \\ &= (69,276), \\ Tv_2 &= T(7,5) \\ &= (322,230). \end{aligned}$$

Assume that $b_1, b_2, b_3, b_4 \in \mathbb{R}$ are such that

$$\begin{aligned} Tv_1 &= b_1v_1 + b_2v_2 \\ Tv_2 &= b_3v_1 + b_4v_2. \end{aligned}$$

Then

$$\begin{aligned} (69,276) &= b_1(1,4) + b_2(7,5) \\ &= (b_1, 4b_1) + (7b_2, 5b_2) \\ &= (b_1 + 7b_2, 4b_1 + 5b_2) \end{aligned}$$

and

$$\begin{aligned} (322,230) &= b_3(1,4) + b_4(7,5) \\ &= (b_3, 4b_3) + (7b_4, 5b_4) \\ &= (b_3 + 7b_4, 4b_3 + 5b_4). \end{aligned}$$

Thus, we get the systems

$$\begin{aligned} b_1+7b_2 &= 69 \\ 4b_1+5b_2 &= 276 \end{aligned} \tag{5.12}$$

and

$$\begin{aligned} b_3+7b_4 &= 322 \\ 4b_3+5b_4 &= 230. \end{aligned} \tag{5.13}$$

Consider the system (5.12). *We have*

$$\begin{aligned} b_1 &= 69-7b_2 \\ 4b_1+5b_2 &= 276. \end{aligned}$$

Hence,

$$\begin{aligned} 276 &= 4(69-7b_2)+5b_2 \\ &= 276-28b_2+5b_2 \\ &= 276-23b_2 \end{aligned}$$

and

$$\begin{aligned} b_1 &= 69 \\ b_2 &= 0. \end{aligned}$$

Now, we consider the system (5.13). *We have*

$$\begin{aligned} b_3 &= 322-7b_4 \\ 4b_3+5b_4 &= 230, \end{aligned}$$

whereupon

$$4(322-7b_4)+5b_4=230,$$

or

$$1288-28b_4+5b_4=230,$$

or

$$1058 = 23b_4.$$

Hence,

$$\begin{aligned} b_4 &= 46 \\ b_3 &= 322 - 7 \cdot 46, \end{aligned}$$

or

$$\begin{aligned} b_3 &= 322 - 322 \\ b_4 &= 46, \end{aligned}$$

or

$$\begin{aligned} b_3 &= 0 \\ b_4 &= 46. \end{aligned}$$

Therefore

$$\begin{aligned} Tv_1 &= 69v_1 \\ Tv_2 &= 46v_2. \end{aligned}$$

Consequently

$$\mathscr{M}(T; v_1, v_2) = \begin{pmatrix} 69 & 0 \\ 0 & 46 \end{pmatrix},$$

which is a diagonal matrix.

Theorem 5.55. *Suppose that V is a finite-dimensional vector space with $\dim V = n$ and $T \in \mathscr{L}(V)$. Let $\lambda_1, \ldots, \lambda_m$ be the distinct eigenvalues of T. Then the following are equivalent.*

1. *T is diagonalizable.*
2. *V has a basis of eigenvectors of T.*
3. *There exists one dimensional subspaces $U_1, \ldots, U_n$ of V that are invariant under T and*

$$V = U_1 \oplus \cdots \oplus U_n. \tag{5.14}$$

4.

$$V = E(\lambda_1, T) \oplus \cdots \oplus E(\lambda_m, T). \tag{5.15}$$

5.

$$\dim V = \dim E(\lambda_1, T) + \cdots + \dim E(\lambda_m, T). \tag{5.16}$$

Proof. 1. An operator $T \in \mathscr{L}(V)$ has a diagonal matrix

$$\begin{pmatrix} \lambda_1 & & 0 \\ & \ddots & \\ 0 & & \lambda_n \end{pmatrix}$$

with respect to a basis $v_1, \ldots, v_n$ of V if and only if

$$Tv_j = \lambda_j v_j, \quad j \in \{1, \ldots, n\}.$$

Thus, $1 \iff 2$.

2. Suppose that V has a basis $v_1, \ldots, v_n$ of eigenvectors of T. Define

$$U_j = \operatorname{span}(v_j), \quad j \in \{1, \ldots, n\}.$$

Then U_j, $j \in \{1, \ldots, n\}$, are one dimensional subspaces of V. If $u_j \in U_j$, $j \in \{1, \ldots, n\}$, then

$$u_j = a_j v_j, \quad a_j \in \mathbb{F}, \quad j \in \{1, \ldots, n\}.$$

Hence,

$$\begin{aligned} Tu_j &= T(a_j v_j) \\ &= a_j T v_j \\ &= a_j \lambda_j v_j \\ &\in U_j, \quad j \in \{1, \ldots, n\}. \end{aligned}$$

Therefore U_j, $j \in \{1, \ldots, n\}$, are invariant under T. Since $v_1, \ldots, v_n$ is a basis of V, then each element of V can be represented in a unique way as a linear combination of $v_1, \ldots, v_n$. Thus, we have (5.14) and we get $2 \Longrightarrow 3$. Now, suppose that U_j, $j \in \{1, \ldots, n\}$, are one dimensional subspaces of V so that (5.14) holds. Let $v_j \in U_j$, $v_j \neq 0$, $j \in \{1,\dots,n\}$. Then v_j, $j \in \{1,\dots,n\}$, are eigenvectors of T. Since each element of V can be written as a sum

$$u_{+}\cdots+u_n,$$

where $u_j \in U_j$, $j \in \{1,\dots,n\}$, we conclude that $v_1,\dots,v_n$ is a basis of V. So, $2 \iff 3$.

3. Suppose that V has a basis of eigenvectors of V. Then each element of V is a linear combination of eigenvectors of T and then we have (5.15). Thus, $2 \Longrightarrow 4$.

4. Suppose that (5.15) holds. Hence, we have that (5.16) holds. So, $4 \Longrightarrow 5$.

5. Suppose that (5.16) holds. Choose a basis of each $E(\lambda_j,T)$, $j \in \{1,\dots,m\}$, put all these bases together to form a list $v_1,\dots,v_n$ of eigenvectors of T. Let $a_j \in \mathbb{F}$, $j \in \{1,\dots,n\}$, be such that

$$a_1v_1+\cdots+a_nv_n=0.$$

Let u_j, $j \in \{1,\dots,m\}$, denote the sum of the terms a_kv_k such that $v_k \in E(\lambda_j,T)$, $j \in \{1,\dots,m\}$. Thus, $u_j \in E(\lambda_j,T)$, $j \in \{1,\dots,m\}$, and

$$u_1+\cdots+u_m=0.$$

Since u_j, $j \in \{1,\dots,m\}$, are eigenvectors of T corresponding to distinct eigenvalues of T, we get that the last sum is possible if

$$u_j=0, \quad j \in \{1,\dots,m\}.$$

Because u_j is a sum of the terms $a_kv_k \in E(\lambda_j,T)$, $j \in \{1,\dots,m\}$, we get

$$a_k=0, \quad k \in \{1,\dots,n\}.$$

Therefore the list $v_1,\dots,v_n$ is linearly independent and forms a basis of V. Thus, $5 \Longrightarrow 2$. This completes the proof.

□

Theorem 5.56. *Let $T \in \mathscr{L}(V)$ has $\dim V$ distinct eigenvectors. Then T is diagonalizable.*

Proof. Let $\dim V = n$ and $\lambda_1, \ldots, \lambda_n$ be distinct eigenvalues of T. Let also, $v_1, \ldots, v_n$ be eigenvectors of T corresponding to $\lambda_1, \ldots, \lambda_n$, respectively. Hence and Theorem 5.18, we get that the list $v_1, \ldots, v_n$ is linearly independent and then it is a basis of V. Now, applying Theorem 5.55, we conclude that T is diagonalizable. This completes the proof. □

Example 5.57. *Let $T \in \mathscr{L}(\mathbb{R}^3)$ be defined by*

$$T(x,y,z) = (2x+y, 5y+3z, 8z), \quad (x,y,z) \in \mathbb{R}^3.$$

We will find a basis of $\mathbb{R}^3$ with respect to which T has a diagonal matrix. Consider the standard basis of $\mathbb{R}^3$

$$\begin{aligned} w_1 &= (1,0,0), \\ w_2 &= (0,1,0), \\ w_3 &= (0,0,1). \end{aligned}$$

Then

$$\mathscr{M}(T; w_1, w_2, w_3) = \begin{pmatrix} 2 & 1 & 0 \\ 0 & 5 & 3 \\ 0 & 0 & 8 \end{pmatrix}.$$

This matrix is upper-triangular and then the eigenvalues of T are

$$\begin{aligned} \lambda_1 &= 2, \\ \lambda_2 &= 5, \\ \lambda_3 &= 8. \end{aligned}$$

Since $\dim \mathbb{R}^3 = 3$ and T has three distinct eigenvalues, we have that there is basis with respect to which T has a diagonal matrix. To find this basis, we will find an eigenvector corresponding to each eigenvalue of T. Consider the equation

$$T(x,y,z) = \lambda_1(x,y,z), \quad (x,y,z) \in \mathbb{R}^3.$$

We have

$$(2x+y, 5y+3z, 8z) = (2x, 2y, 2z),$$

whereupon

$$\begin{aligned} 2x+y &= 2x \\ 5y+3z &= 2y \\ 8z &= 2z, \end{aligned}$$

or

$$\begin{aligned} y &= 0 \\ z &= 0. \end{aligned}$$

Thus,

$$v_1 = (1,0,0)$$

is an eigenvector corresponding to λ_1. *Now, consider the equation*

$$T(x,y,z) = \lambda_2(x,y,z), \quad (x,y,z) \in \mathbb{R}^3.$$

We have

$$(2x+y, 5y+3z, 8z) = (5x, 5y, 5z),$$

whereupon

$$\begin{aligned} 2x+y &= 5x \\ 5y+3z &= 5y \\ 8z &= 5z, \end{aligned}$$

or

$$\begin{aligned} y &= 3x \\ z &= 0. \end{aligned}$$

Thus,

$$v_2 = (1,3,0)$$

is an eigenvector corresponding to λ_2. *Next, by the equation*

$$T(x,y,z) = \lambda_3(x,y,z), \quad (x,y,z) \in \mathbb{R}^3,$$

we find

$$(2x+y, 5y+3z, 8z) = (8x, 8y, 8z),$$

whereupon

$$\begin{aligned} 2x+y &= 8x \\ 5y+3z &= 8y \\ 8z &= 8z, \end{aligned}$$

or

$$\begin{aligned} y &= 6x \\ y &= z. \end{aligned}$$

So,

$$v_3 = (1,6,6)$$

is an eigenvector corresponding to λ_3. *Therefore*

$$\begin{aligned} v_1 &= (1,0,0), \\ v_2 &= (1,3,0), \\ v_3 &= (1,6,6) \end{aligned}$$

is a basis of $\mathbb{R}^3$ *with respect to which* T *has a diagonal matrix and this matrix is*

$$\mathscr{M}(T; v_1, v_2, v_3) = \begin{pmatrix} 2 & 0 & 0 \\ 0 & 5 & 0 \\ 0 & 0 & 8 \end{pmatrix}.$$

Example 5.58. *Let* $T \in \mathbb{L}(\mathbb{R}^3)$ *be defined by*

$$T(x,y,z) = (4x, 4y, 5z), \quad (x,y,z) \in \mathbb{R}^3.$$

The standard basis of $\mathbb{R}^3$ *is*

$$\begin{aligned} w_1 &= (1,0,0), \\ w_2 &= (0,1,0), \\ w_3 &= (0,0,1). \end{aligned}$$

Then

$$\mathscr{M}(T;w_1,w_2,w_3) = \begin{pmatrix} 4 & 0 & 0 \\ 0 & 4 & 0 \\ 0 & 0 & 5 \end{pmatrix}.$$

This matrix is a diagonal matrix, but T *has two distinct eigenvalues* 4 *and* 5*. Therefore the converse of Theorem 5.55 is not true.*

Exercise 5.59. *Find a basis of* $\mathbb{R}^3$ *with respect to which the operator* $T \in \mathscr{L}(\mathbb{R}^3)$ *has a diagonal matrix and find this matrix, where*

1. $T(x,y,z) = (-x+3y-z, -x+5y-z, -3x+3y+z)$, $(x,y,z) \in \mathbb{R}^3$.
2. $T(x,y,z) = (6x-2y-z, -x+5y-z, x-2y+4z)$, $(x,y,z) \in \mathbb{R}^3$.

Answer 5.60. *1.*

$$\begin{aligned} v_1 &= (3,1,3), \\ v_2 &= (0,1,3), \\ v_3 &= (1,2,1), \\ \mathscr{M}(T;v_1,v_2,v_3) &= \begin{pmatrix} -1 & 0 & 0 \\ 0 & 2 & 0 \\ 0 & 0 & 4 \end{pmatrix}. \end{aligned}$$

2.

$$\begin{aligned} v_1 &= (1,-1,1), \\ v_2 &= (1,1,-1), \end{aligned}$$

$$
\begin{aligned}
v_3 &= (1,1,1), \\
\mathscr{M}(T; v_1, v_2, v_3) &= \begin{pmatrix} 7 & 0 & 0 \\ 0 & 5 & 0 \\ 0 & 0 & 3 \end{pmatrix}.
\end{aligned}
$$

5.7 Advanced Practical Problems

Problem 5.61. *Let $V = \mathbb{R}^3$ and*

$$U = \{(x,y,z) \in \mathbb{R}^3 : |x| \le 1, |y| \le 2, |z| \le 3\}.$$

Check if U is invariant under T, where

1.
$$T(x,y,z) = (x, x+10, y+z), \quad (x,y,z) \in \mathbb{R}^3.$$

2.
$$T(x,y,z) = \left(\frac{1}{1+x^2+z^2}, \frac{1}{1+4y^2}, \frac{1}{1+5z^2} \right), \quad (x,y,z) \in \mathbb{R}^3.$$

3.
$$T(x,y,z) = \left(\frac{1}{1+(\sin x)^4}, \frac{1}{1+(\cos y)^6}, \cos z \right), \quad (x,y,z) \in \mathbb{R}^3.$$

4.
$$T(x,y,z) = \left(\frac{1}{1+x^2+y^2+z^2}, z+10, x+y-10 \right), \quad (x,y,z) \in \mathbb{R}^3.$$

5.
$$T(x,y,z) = \left(\frac{1}{2}(\cos(x+y+z))^2, \frac{1}{10}\sin(x-y+z), \quad 1+\frac{1}{4}\cos z \right),$$

$(x,y,z) \in \mathbb{R}^3$.

Answer 5.62. *1. No.*

2. Yes.

3. *Yes.*

4. *No.*

5. *Yes.*

Problem 5.63. *Let $V = \mathbb{R}^2$ and $\mathbb{F} = \mathbb{R}$. Find the eigenvalues and eigenvectors of the operator $T \in \mathscr{L}(V)$, where*

1.
$$T(x,y) = (x+y, x+y), \quad (x,y) \in V.$$

2.
$$T(x,y) = (2x+2y, -2x-2y), \quad (x,y) \in V.$$

Answer 5.64. 1.

$$\begin{aligned} \lambda_1 &= 0, \\ \lambda_2 &= 2, \\ v_1 &= (x, -x), \\ v_2 &= (x, x), \quad x \in \mathbb{R}, \quad x \neq 0. \end{aligned}$$

2.

$$\begin{aligned} \lambda &= 0, \\ v &= (x, -x), \quad x \in \mathbb{R}, \quad x \neq 0. \end{aligned}$$

Problem 5.65. *Let $V = \mathbb{C}^2$ and $\mathbb{F} = \mathbb{C}$. Find the eigenvalues and eigenvectors of the operator $T \in \mathscr{L}(V)$, where*

1.
$$T(x,y) = (-x+y, -2x+y), \quad (x,y) \in V.$$

2.
$$T(x,y) = (2y, -2x), \quad (x,y) \in V.$$

Answer 5.66. *1.*

$$\lambda_1 = i,$$

$$\lambda_2 = -i,$$

$$v_1 = (x,(1+i)x),$$

$$v_2 = (x,(1-i)x), \quad x \in \mathbb{R}, \quad x \neq 0.$$

2.

$$\lambda_1 = 2i,$$

$$\lambda_2 = -2i,$$

$$v_1 = (x, ix),$$

$$v_2 = (ix, x), \quad x \in \mathbb{R}, \quad x \neq 0.$$

Problem 5.67. *Let $V = \mathbb{C}^3$ and $\mathbb{F} = \mathbb{C}$. Find the eigenvalues and eigenvectors of the operator $T \in \mathscr{L}(V)$, where*

1.

$$T(x,y,z) = (x-3y+3z, -2x-6y+13z, -x-4y+8z), \quad (x,y,z) \in V.$$

2.

$$T(x,y,z) = (x-3y+4z, 4x-7y+8z, 6x-7y+7z), \quad (x,y,z) \in V.$$

3.

$$T(x,y,z) = (4x-2y-z, -x+3y-z, x-2y+2z), \quad (x,y,z) \in V.$$

Answer 5.68. *1.*

$$\lambda = 1,$$

$$v = (3x, x, x), \quad x \in \mathbb{R}, \quad x \neq 0.$$

2.

$$\begin{aligned}\lambda_1 &= 3,\\ \lambda_2 &= -1,\\ v_1 &= (x, 2x, 2x),\\ v_2 &= (x, 2x, x), \quad x \in \mathbb{R}, \quad x \neq 0.\end{aligned}$$

3.

$$\begin{aligned}\lambda_1 &= 5,\\ \lambda_2 &= 3,\\ \lambda_3 &= 1,\\ v_1 &= (x, -x, x),\\ v_2 &= (x, x, -x),\\ v_3 &= (x, x, x), \quad x \in \mathbb{R}, \quad x \neq 0.\end{aligned}$$

Problem 5.69. *Suppose that $T \in \mathscr{L}(V)$ and $U_1, \ldots, U_m$ are subspaces invariant under T. Prove that $U_1 + \cdots + U_m$ is invariant under T.*

Proof. Let $u_1 \in U_1$, ..., $u_m \in U_m$ be arbitrarily chosen. Since U_1, ..., U_m are invariant under T, we get $Tu_1 \in U_1$, ..., $Tu_m \in U_m$. From here, we find

$$\begin{aligned}T(u_1 + \cdots + u_m) &= Tu_1 + \cdots + Tu_m\\ &\in U_1 + \cdots + U_m.\end{aligned}$$

Because $u_1 + \cdots + u_m \in U_1 + \cdots + U_m$ was arbitrarily chosen, we conclude that $U_1 + \cdots + U_m$ is invariant under T. This completes the proof. □

Problem 5.70. *Let $T \in \mathscr{L}(V)$. Prove that the intersection of any collection of subspaces of V invariant under T is invariant under T.*

Proof. Let $\{U_\alpha : \alpha \in A\}$ be any collection of subspaces of V invariant under T. Here A is an index set. Let also, $u \in \bigcap_{\alpha \in A} U_\alpha$ be arbitrarily chosen. Then $u \in U_\alpha$ for any $\alpha \in A$. Since U_α is invariant under T for any $\alpha \in A$, we have $Tu \in U_\alpha$ for any $\alpha \in A$. Therefore $Tu \in \bigcap_{\alpha \in A} U_\alpha$ and $\bigcap_{\alpha \in A} U_\alpha$ is invariant under T. This completes the proof. □

Problem 5.71. *Suppose that $T \in \mathscr{L}(V)$ is invertible.*

1. *Suppose that $\lambda \in F$, $\lambda \neq 0$, is an eigenvalue of T. Prove that $\frac{1}{\lambda}$ is an eigenvalue of T^{-1}.*

2. *Prove that T and T^{-1} have the same eigenvectors.*

Proof. Since $\lambda \in \mathbb{F}$, $\lambda \neq 0$, is an eigenvalue of T, there is a $v \in V$, $v \neq 0$, so that

$$Tv = \lambda v.$$

Hence,

$$\begin{aligned} v &= T^{-1}Tv \\ &= T^{-1}(\lambda v) \\ &= \lambda T^{-1}v, \end{aligned}$$

whereupon

$$T^{-1}v = \frac{1}{\lambda}v.$$

Therefore $\frac{1}{\lambda}$ is an eigenvalue of T^{-1} and v is an eigenvector of T^{-1} corresponding to $\frac{1}{\lambda}$. This completes the proof. □

Problem 5.72. *Let $V = \mathbb{R}^3$, $\mathbb{F} = \mathbb{R}$ and $T \in \mathscr{L}(V)$ be defined by*

$$T(x,y,z) = (x-y, y-z, z-x), \quad (x,y,z) \in V.$$

Find

$$T^3(x,y,z), \quad (x,y,z) \in V.$$

Answer 5.73.

$$T^3(x,y,z) = (-3y+3z, 3x-3z, -3x+3y), \quad (x,y,z) \in V.$$

Problem 5.74. *Let $T \in \mathscr{L}(\mathbb{F})$ be defined by*

$$Tx = x-2, \quad x \in \mathbb{F},$$

and

$$p(x) = x + x^3, \quad x \in \mathbb{F}.$$

Find

$$(p(T))(x), \quad x \in \mathbb{F}.$$

Answer 5.75.

$$(p(T))(x) = x^3 - 18x^2 + 109x - 218, \quad x \in \mathbb{F}.$$

Problem 5.76. *Find a basis of $\mathbb{R}^3$ with respect to which the operator $T \in \mathscr{L}(\mathbb{R}^3)$ has a diagonal matrix and find this matrix, where*

1. $T(x,y,z) = (3x+y-z, 2x+2y-z, -2x+y+4z)$, $(x,y,z) \in \mathbb{R}^3$.
2. $T(x,y,z) = (5x+y-z, 2x+4y-z, -2x+y+6z)$, $(x,y,z) \in \mathbb{R}^3$.

Answer 5.77. *1.*

$$\begin{aligned} v_1 &= (1,1,-1), \\ v_2 &= (1,-1,1), \\ v_3 &= (1,1,1), \\ \mathscr{M}(T; v_1, v_2, v_3) &= \begin{pmatrix} 5 & 0 & 0 \\ 0 & 1 & 0 \\ 0 & 0 & 3 \end{pmatrix}. \end{aligned}$$

2.

$$\begin{aligned} v_1 &= (1,1,-1), \\ v_2 &= (1,-1,1), \end{aligned}$$

$$v_3 = (1,1,1),$$

$$\mathscr{M}(T; v_1, v_2, v_3) = \begin{pmatrix} 7 & 0 & 0 \\ 0 & 3 & 0 \\ 0 & 0 & 5 \end{pmatrix}.$$

Problem 5.78. *Suppose that V is finite-dimensional, $T \in \mathscr{L}(V)$ and there exists a positive integer n so that $T^n = 0$. Prove that $I - T$ is invertible and*

$$(I-T)^{-1} = I + T + \cdots + T^{n-1}.$$

Solution 5.79. *By Theorem 5.45, it follows that there is a basis $v_1, \ldots, v_m$ of V so that T has an upper-triangular matrix*

$$\mathscr{M}(T; v_1, \ldots, v_n) = \begin{pmatrix} a_{11} & a_{12} & a_{13} & \cdots & a_{1m} \\ 0 & a_{22} & a_{23} & \cdots & a_{2m} \\ 0 & 0 & a_{33} & \cdots & a_{3m} \\ \vdots & \vdots & \vdots & \vdots & \vdots \\ 0 & 0 & 0 & \cdots & a_{mm} \end{pmatrix}.$$

If $n = 1$, then $T = 0$ and the assertion follows because $I - T = I$. Let now, $n \geq 2$. Then

$$(\mathscr{M}(T; v_1, \ldots, v_m))^n = \begin{pmatrix} a_{11}^n & & & & \star \\ & a_{22}^n & & & \\ & & a_{33}^n & & \\ \vdots & \vdots & \vdots & \vdots & \vdots \\ 0 & & & & a_{mm}^n \end{pmatrix}$$

$$= \begin{pmatrix} 0 & & & & 0 \\ & 0 & & & \\ & & 0 & & \\ \vdots & \vdots & \vdots & \vdots & \vdots \\ 0 & & & & 0 \end{pmatrix}.$$

Therefore

$$a_{11} = 0$$

$$a_{22} = 0$$

$$\vdots$$

$$a_{mm} = 0.$$

Hence,

$$\mathcal{M}(T;v_1,\ldots,v_n) = \begin{pmatrix} 0 & a_{12} & a_{13} & \cdots & a_{1m} \\ 0 & 0 & a_{23} & \cdots & a_{2m} \\ 0 & 0 & 0 & \cdots & a_{3m} \\ \vdots & \vdots & \vdots & \vdots & \vdots \\ 0 & 0 & 0 & \cdots & 0 \end{pmatrix},$$

and

$$\mathcal{M}(I-T;v_1,\ldots,v_n) = \begin{pmatrix} 1 & a_{12} & a_{13} & \cdots & a_{1m} \\ 0 & 1 & a_{23} & \cdots & a_{2m} \\ 0 & 0 & 1 & \cdots & a_{3m} \\ \vdots & \vdots & \vdots & \vdots & \vdots \\ 0 & 0 & 0 & \cdots & 1 \end{pmatrix}.$$

Thus, all entries on the diagonal of $\mathcal{M}(I-T;v_1,\ldots,v_m)$ are nonzero. Hence, it follows that $I-T$ is invertible. Next,

$$\begin{aligned}(I-T)(I+T+T^2+\cdots+T^{n-1}) &= I+T+T^2+\cdots+T^{n-1} \\ &\quad -T-T^2-\cdots-T^{n-1}-T^n \\ &= I-T^n \\ &= I,\end{aligned}$$

whereupon

$$(I-T)^{-1} = I+T+\cdots+T^{n-1}.$$

Problem 5.80. *Let V be finite-dimensional, $T \in \mathcal{L}(V)$ and*

$$(T-3I)(T-5I)(T-7I) = 0. \tag{5.17}$$

Prove that 3 or 5, or 7 is an eigenvalue of T.

Solution 5.81. *By the equation* (5.17), *we have that*

$$T - 3I = 0$$

or

$$T - 5I = 0,$$

or

$$T - 7I = 0.$$

Hence, 3 *or* 5 *or* 7 *is an eigenvalue of* T.

Problem 5.82. *Let* V *be finite-dimensional,* $T \in \mathscr{L}(V)$, $T^2 = I$ *and* -1 *is not an eigenvalue of* T. *Prove that* $T^2 = I$.

Solution 5.83. *By Theorem 5.45, it follows that there is a basis* $v_1, \ldots, v_m$ *of* V *such that* T *has an upper-triangular matrix*

$$\mathscr{M}(T; v_1, \ldots, v_n) = \begin{pmatrix} a_{11} & a_{12} & a_{13} & \cdots & a_{1m} \\ 0 & a_{22} & a_{23} & \cdots & a_{2m} \\ 0 & 0 & a_{33} & \cdots & a_{3m} \\ \vdots & \vdots & \vdots & \vdots & \vdots \\ 0 & 0 & 0 & \cdots & a_{mm} \end{pmatrix}.$$

Then

$$\begin{aligned} (\mathscr{M}(T; v_1, \ldots, v_m))^2 &= \begin{pmatrix} a_{11}^2 & & & & \star \\ & a_{22}^2 & & & \\ & & a_{33}^2 & & \\ \vdots & \vdots & \vdots & \vdots & \vdots \\ 0 & & & & a_{mm}^2 \end{pmatrix} \\ &= \begin{pmatrix} 1 & & & & 0 \\ & 1 & & & \\ & & 1 & & \\ \vdots & \vdots & \vdots & \vdots & \vdots \\ 0 & & & & 1 \end{pmatrix}. \end{aligned}$$

Therefore

$$a_{11} = 1$$

$$a_{22} = 1$$

$$\vdots$$

$$a_{mm} = 1$$

because -1 *is not an eigenvalue of* T. *Moreover,*

$$\begin{aligned} 0 &= a_{11}a_{12}+a_{12}a_{22} \\ &= 2a_{12}, \end{aligned}$$

whereupon

$$a_{12}=0.$$

Next,

$$\begin{aligned} 0 &= a_{11}a_{13}+a_{12}a_{23}+a_{13}a_{33} \\ &= 2a_{13}, \end{aligned}$$

or

$$a_{13}=0,$$

and so on,

$$a_{mm-1}=0.$$

Therefore $T=I$.

Problem 5.84. *Let* V *be finite-dimensional,* $P \in \mathscr{L}(V)$, $P^2=P$. *Prove that*

$$V=\ker P \oplus rangeP. \tag{5.18}$$

Solution 5.85. *Let* $u \in \ker P \bigcap rangeP$ *be arbitrarily chosen. Then* $u \in \ker P$ *and*

$$Pu=0.$$

Hence,

$$\begin{aligned} 0 &= P0 \\ &= PPu \end{aligned}$$

$$= \quad u.$$

Because $u \in \ker P \bigcap rangeP$ *was arbitrarily chosen, we conclude that*

$$\ker P \bigcap rangeP = \{0\}.$$

Thus, (5.18) *holds.*

Problem 5.86. *Let* V *be finite-dimensional,* $S, T \in \mathscr{L}(V)$ *and* S *is invertible. Suppose that* $p \in \mathscr{P}(\mathbb{F})$ *be any polynomial. Prove that*

$$p(STS^{-1}) = Sp(T)S^{-1}.$$

Solution 5.87. *Let* $p \in \mathscr{P}(\mathbb{F})$ *be such that*

$$p(z) = a_0 + a_1 z + a_2 z^2 + \cdots + a_m z^m, \quad z \in \mathbb{F},$$

$a_0, a_1, a_2, \ldots, a_m \in \mathbb{F}$. *Prove that*

$$p(STS^{-1}) = Sp(T)S^{-1}.$$

Note that

$$\begin{aligned} (STS^{-1})^2 &= STS^{-1}STS^{-1} \\ &= ST^2S^{-1}. \end{aligned}$$

Assume that

$$(STS^{-1})^k = ST^kS^{-1} \tag{5.19}$$

for some $k \in \mathbb{N}$. *Then*

$$\begin{aligned} (STS^{-1})^{k+1} &= (STS^{-1})^k STS^{-1} \\ &= ST^kS^{-1}STS^{-1} \\ &= ST^{k+1}S^{-1}. \end{aligned}$$

Now, applying the principal of the mathematical induction, we conclude that (5.19) *holds for any* $k \in \mathbb{N}$. *Hence,*

$$p(STS^{-1}) \quad = \quad a_0 I + a_1 STS^{-1} + a_2 (STS^{-1})^2 + \cdots + a_m (STS^{-1})^m$$

$$\begin{aligned} &= a_0SS^{-1}+a_1STS^{-1}+a_2ST^2S^{-1}+\cdots+a_mST^mS^{-1} \\ &= S(a_0I+a_1T+a_2T^2+\cdots+a_mT^m)S^{-1} \\ &= Sp(T)S^{-1}. \end{aligned}$$

Problem 5.88. *Let V be finite-dimensional, $T \in \mathscr{L}(V)$ and U be a subspace of V invariant under T. Prove that U is invariant under $p(T)$ for any polynomial $p \in \mathscr{P}(\mathbb{F})$.*

Solution 5.89. *Since U is invariant under T, for any $u \in U$, we have*

$$\begin{aligned} Tu &\in U, \\ T^2u &\in U, \\ &\vdots \\ T^ku &\in U, \quad k \in \mathbb{N}. \end{aligned}$$

Let now,

$$p(z) = a_0+a_1z+a_2z^2+\cdots+a_mz^m, \quad z \in \mathbb{F},$$

$a_0, a_1, a_2, \ldots, a_m \in \mathbb{F}$. Then

$$p(T) = a_0I+a_1T+a_2T^2+\cdots+a_mT^m.$$

Now, using that U is a subspace of V, we get

$$\begin{aligned} a_0u &\in U, \\ a_1Tu &\in U, \\ a_2T^2u &\in U, \\ &\vdots \end{aligned}$$

$$a_m T^m u \in U, \quad u \in U.$$

Therefore

$$p(T)u \in U, \quad u \in U.$$

Problem 5.90. *Let V be finite-dimensional, $T \in \mathscr{L}(V)$, and $v \in V$ with $v \neq 0$. Let also, p be a nonzero polynomial of smallest degree such that*

$$p(T)v = 0.$$

Prove that any zero of p is an eigenvalue of T.

Solution 5.91. *Let*

$$p = a_0 + a_1 z + a_2 z^2 + \cdots + a_m z^m, \quad z \in \mathbb{F},$$

$a_0, a_1, a_2, \ldots, a_m \in \mathbb{F}$. Let also,

$$p(z) = c(z - z_1) \cdots (z - z_m),$$

where $c \in \mathbb{F}$ with $c \neq 0$ and $z_1, \ldots, z_m$ are the zeros of p. Assume that there is a $j \in \{1, \ldots, m\}$ such that z_j is not an eigenvalue of T. Then

$$(T - z_j I)v \neq 0.$$

Hence,

$$\begin{aligned} 0 &= p(T)v \\ &= c(T - z_1 I) \ldots (T - z_{j-1} I)(T - z_j I)(T - z_{j+1} I) \ldots (T - z_m I)v. \end{aligned}$$

From the last equation, we get

$$c(T - z_1 I) \ldots (T - z_{j-1} I)(T - z_{j+1} I) \ldots (T - z_m I)v = 0.$$

This is a contradiction because

$$\begin{aligned} \deg(c(z - z_1) \ldots (z - z_{j-1})(z - z_{j+1}) \ldots (z - z_m)) &= m - 1 \\ &< m. \end{aligned}$$

Thus, any zero of p is an eigenvalue of T.

Problem 5.92. *Suppose that $T \in \mathscr{L}(V)$ and v is an eigenvector of T with eigenvalue λ. Suppose that $p \in \mathscr{P}(\mathbb{F})$. Prove that*

$$p(T)v = p(\lambda)v.$$

Solution 5.93. *Since v is an eigenvector of T with eigenvalue λ, we have*

$$(T - \lambda I)v = 0.$$

Let

$$p(z) = a_0 + a_1 z + a_2 z^2 + \cdots + a_m z^m, \quad z \in \mathbb{F},$$

$a_0, a_1, a_2, \ldots, a_m \in \mathbb{F}$. *Then*

$$p(\lambda) = a_0 + a_1\lambda + a_2\lambda^2 + \cdots + a_m\lambda^m.$$

Hence,

$$\begin{aligned} p(z) - p(\lambda) &= a_0 + a_1 z + a_2 z^2 + \cdots + a_m z^m \\ &\quad -a_0 - a_1\lambda - a_2\lambda^2 - \cdots - a_m\lambda^m \\ &= a_1(z-\lambda) + a_2(z^2 - \lambda^2) + \cdots + a_m(z^m - \lambda^m) \\ &= a_1(z-\lambda) + a_2(z-\lambda)(z+\lambda) \\ &\quad + \cdots + a_m(z-\lambda)(z^{m-1} + z^{m-2}\lambda + \cdots + \lambda^{m-1}) \\ &= (a_1 + a_2(z+\lambda) \\ &\quad + \cdots + a_m(z^{m-1} + z^{m-2}\lambda + \cdots + \lambda^{m-1}))(z-\lambda). \end{aligned}$$

Hence,

$$\begin{aligned} p(T)v - p(\lambda)v &= (a_1 I + a_2(T + \lambda I) + \cdots + a_m(T^{m-1} + T^{m-2}\lambda \\ &\quad + \cdots + \lambda^{m-1} I))(T - \lambda I)v \\ &= 0, \end{aligned}$$

whereupon

$$p(T)v = p(\lambda)v.$$

Problem 5.94. *Suppose that V is finite-dimensional, $T \in \mathscr{L}(V)$ has* $\dim V$ *distinct eigenvalues, and $S \in \mathscr{L}(V)$ has the same eigenvectors as T. Prove that*

$$ST = TS. \tag{5.20}$$

Solution 5.95. *Let* $\dim V = n$*. Since V is finite-dimensional and T has n distinct eigenvalues, it follows that T is diagonalizable. Hence and Theorem 5.55, it follows that V has a basis $v_1, \ldots, v_n$ of eigenvectors of T. Let $\lambda_1, \ldots, \lambda_n, \nu_1, \ldots, \mu_n \in \mathbb{F}$ be such that*

$$\begin{aligned} Tv_j &= \lambda_j v_j, \\ Sv_j &= \mu_j v_j, \quad j \in \{1, \ldots, n\}. \end{aligned}$$

Let $u \in V$ be arbitrarily chosen. Then there are $a_1, \ldots, a_n \in \mathbb{F}$ so that

$$u = a_1 v_1 + \cdots + a_n v_n.$$

Hence,

$$\begin{aligned} STu &= ST(a_1 v_1 + \cdots + a_n v_n) \\ &= S(T(a_1 v_1 + \cdots + a_n v_n)) \\ &= S(T(a_1 v_1) + \cdots + T(a_n v_n)) \\ &= S(a_1 T v_1 + \cdots + a_n T v_n) \\ &= S(\lambda_1 a_1 v_1 + \cdots + \lambda_n a_n v_n) \\ &= S(\lambda_1 a_1 v_1) + \cdots + S(\lambda_n a_n v_n) \\ &= \lambda_1 a_1 S v_1 + \cdots + \lambda_n a_n S v_n \\ &= \lambda_1 \mu_1 a_1 v_1 + \cdots + \lambda_n \mu_n a_n v_n, \end{aligned}$$

i.e.,

$$STu = \lambda_1 \mu_1 a_1 v_1 + \cdots + \lambda_n \mu_n a_n v_n. \tag{5.21}$$

Next,

$$\begin{aligned} TSu &= TS(a_1v_1+\cdots+a_nv_n) \\ &= T(S(a_1v_1+\cdots+a_nv_n)) \\ &= T(S(a_1v_1)+\cdots+S(a_nv_n)) \\ &= T(a_1Sv_1+\cdots+a_nSv_n) \\ &= T(\mu_1a_1v_1+\cdots+\mu_na_nv_n) \\ &= T(\mu_1a_1v_1)+\cdots+T(\mu_na_nv_n) \\ &= \mu_1a_1Tv_1+\cdots+\mu_na_nTv_n \\ &= \lambda_1\mu_1a_1v_1+\cdots+\lambda_n\mu_na_nv_n. \end{aligned}$$

By the last equality and (5.21)*, we obtain* (5.20).

Problem 5.96. *Suppose that $T \in \mathscr{L}(V)$ is invertible. Prove that*

$$E(\lambda,T) = E\left(\frac{1}{\lambda},T^{-1}\right) \tag{5.22}$$

for any $\lambda \in \mathbb{F}$, $\lambda \neq 0$.

Solution 5.97. *Let $u \in E(\lambda,T)$ be arbitrarily chosen. Then*

$$Tu = \lambda u,$$

whereupon

$$\begin{aligned} u &= T^{-1}Tu \\ &= T^{-1}(\lambda u) \\ &= \lambda T^{-1}u \end{aligned}$$

and

$$T^{-1}u = \frac{1}{\lambda}u.$$

Because $u \in E(\lambda, T)$ *was arbitrarily chosen and we get that it is an element of* $E\left(\frac{1}{\lambda}, T^{-1}\right)$, *we obtain the inclusion*

$$E(\lambda, T) \subseteq E\left(\frac{1}{\lambda}, T^{-1}\right). \tag{5.23}$$

Let now, $v \in E\left(\frac{1}{\lambda}, T^{-1}\right)$ *be arbitrarily chosen. Then*

$$T^{-1}v = \frac{1}{\lambda}v.$$

From here,

$$\begin{aligned} v &= TT^{-1}v \\ &= T\left(\frac{1}{\lambda}v\right) \\ &= \frac{1}{\lambda}Tv \end{aligned}$$

and

$$Tv = \lambda v.$$

Because $v \in E\left(\frac{1}{\lambda}, T^{-1}\right)$ *was arbitrarily chosen and we get that it is an element of* $E(\lambda, T)$, *we arrive at the inclusion*

$$E\left(\frac{1}{\lambda}, T^{-1}\right) \subseteq E(\lambda, T).$$

By the last inclusion and (5.23), we find (5.22).

Chapter 6

Jordan Canonical Form

Suppose that V is a finite-dimensional nonzero vector space over $\mathbb{F}$.

6.1 Kernels of Powers of an Operator

Theorem 6.1. *Suppose that $T \in \mathscr{L}(V)$. Then*

$$\{0\} = \ker T^0 \subseteq \ker T^1 \subseteq \ker T^2 \ldots \subseteq \ker T^k \subseteq \ker T^{k+1} \subseteq \ldots. \tag{6.1}$$

Proof. We have $T^0 = I$ and then

$$\begin{aligned} \ker T^0 &= \{0\} \\ &\subseteq \ker T^1. \end{aligned}$$

Assume that

$$\ker T^k \subseteq \ker T^{k+1}$$

for some $k \in \mathbb{N}$. Let now, $v \in \ker T^{k+1}$ be arbitrarily chosen. Then

$$T^{k+1}v = 0.$$

Hence,

$$\begin{aligned} T^{k+2}v &= TT^{k+1}v \\ &= T(T^{k+1}v) \end{aligned}$$

$$= \quad T0$$

$$= \quad 0,$$

whereupon $v \in \ker T^{k+2}$. Because $v \in \ker T^{k+1}$ was arbitrarily chosen and we get that it is an element of $\ker T^{k+2}$, we obtain the inclusion

$$\ker T^{k+1} \subseteq \ker T^{k+2}.$$

Now, applying the principal of the mathematical induction, we conclude that (6.1) holds for any $k \in \mathbb{N}$. This completes the proof. □

Theorem 6.2. *Suppose that $T \in \mathscr{L}(V)$ and $m \in \mathbb{N}_0$ be such that*

$$\ker T^m = \ker T^{m+1}.$$

Then

$$\ker T^{m+k} = \ker T^{m+k+1} \tag{6.2}$$

for any $k \in \mathbb{N}_0$.

Proof. Take $k \in \mathbb{N}_0$ arbitrarily. By Theorem 6.1, we have that

$$\ker T^{m+k} \subseteq \ker T^{m+k+1}. \tag{6.3}$$

Let $u \in \ker T^{m+k+1}$ be arbitrarily chosen. Then

$$T^{m+k+1}u = 0.$$

Hence,

$$\begin{aligned} 0 &= T^{m+k+1}u \\ &= T^{m+1}T^k u \\ &= T^{m+1}(T^k u). \end{aligned}$$

Therefore $T^k u \in \ker T^{m+1}$. Since $\ker T^m = \ker T^{m+1}$, we get $T^k u \in \ker T^m$. Then

$$0 \quad = \quad T^m(T^k u)$$

$$= \quad T^m T^k u$$

$$= \quad T^{m+k} u.$$

Thus, $u \in \ker T^{m+k}$. Because $u \in \ker T^{m+k+1}$ was arbitrarily chosen and we get that it is an element of $\ker T^{m+k}$, we obtain the inclusion

$$\ker T^{m+k+1} \subseteq \ker T^{m+k}.$$

By the last inclusion and (6.3), we find (6.2). This completes the proof. □

Theorem 6.3. *Suppose* $T \in \mathscr{L}(V)$, $n = \dim V$. *Then*

$$\ker T^n = \ker T^{n+k} \tag{6.4}$$

for any $k \in \mathbb{N}$.

Proof. If there is an $l \in \{1, \ldots, n\}$ so that

$$\ker T^l = \ker T^{l+1}, \tag{6.5}$$

we get that

$$\ker T^{l+k} = \ker T^{l+k+1}, \quad k \in \mathbb{N}.$$

In particular, we get (6.5). Assume that

$$\{0\} \subsetneq \ker T^1 \subsetneq \ker T^2 \subsetneq \ldots \subsetneq \ker T^n \subsetneq \ker T^{n+1}.$$

Then

$$0 < \dim \ker T^1 < \dim \ker T^2 < \ldots < \dim \ker T^{n+1}.$$

Hence,

$$\dim \ker T^{n+1} \geq n+1,$$

which is a contradiction. Therefore there is an $l \in \{1, \ldots, n\}$ so that (6.5) holds. This completes the proof. □

Theorem 6.4. *Let* $T \in \mathscr{L}(V)$ *and* $n = \dim V$. *Then*

$$V = \ker T^n \oplus rangeT^n. \tag{6.6}$$

Proof. Let $v \in \ker T^n \bigcap \text{range} T^n$. Then $v \in \ker T^n$ and $v \in \text{range} T^n$. Hence,

$$T^n v = 0.$$

Because $v \in \text{range} T^n$, it follows that there is an $u \in V$ so that

$$T^n u = v.$$

Hence,

$$\begin{aligned} 0 &= T^n v \\ &= T^n(T^n u) \\ &= T^{2n} u. \end{aligned}$$

Therefore $u \in \ker T^{2n}$. Now, applying (6.4), we get $u \in \ker T^n$ and

$$T^n u = 0.$$

Therefore $v = 0$ and

$$\ker T^n \bigcap \text{range} T^n = \{0\}.$$

From here, we find (6.6). This completes the proof. □

6.2 Generalized Eigenvectors

Definition 6.5. *Suppose that $T \in \mathscr{L}(V)$ and $\lambda \in \mathbb{F}$ is its eigenvalue. A vector $v \in V$ is called a generalized eigenvector of T corresponding to λ if $v \neq 0$ and*

$$(T - \lambda I)^j v = 0$$

for some $j \in \mathbb{N}$. The set of all generalized eigenvectors corresponding to λ, along the 0 vector, is called generalized eigenspace of T and it is denoted by $G(\lambda, T)$.

We have

$$E(\lambda, T) \subseteq G(\lambda, T).$$

Theorem 6.6. *Suppose that $T \in \mathscr{L}(V)$, $n = \dim V$, and $\lambda \in \mathbb{F}$ is an eigenvalue of T. Then*

$$G(\lambda, T) = \ker(T - \lambda I)^n. \tag{6.7}$$

Proof. Let $v \in G(\lambda, T)$ be arbitrarily chosen. Then there is a $j \in \mathbb{N}$ so that

$$v \in \ker(T - \lambda I)^j.$$

If $j \in \{1, \ldots, n\}$, then it follows that

$$v \in \ker(T - \lambda I)^n. \tag{6.8}$$

If $j \geq n$, then (6.8) holds. Because $v \in G(\lambda, T)$ was arbitrarily chosen and we get that it is an element of $\ker(T - \lambda I)^n$, we obtain the inclusion

$$G(\lambda, T) \subseteq \ker(T - \lambda I)^n. \tag{6.9}$$

Let now, $u \in \ker(T - \lambda I)^n$ be arbitrarily chosen. Then

$$(T - \lambda I)^n u = 0.$$

Hence,

$$u \in G(\lambda, T).$$

Because $u \in \ker(T - \lambda I)^n$ was arbitrarily chosen and we obtain that it is an element of $G(\lambda, T)$, we conclude that

$$\ker(T - \lambda I)^n \subseteq G(\lambda, T).$$

By the last inclusion and (6.9), we get the equality (6.7). This completes the proof. □

Example 6.7. *Let $T \in \mathscr{L}(\mathbb{C}^2)$ be defined by*

$$T(x, y) = (x + y, y), \quad (x, y) \in \mathbb{C}^2.$$

We will find the eigenvalues of T, the corresponding eigenspaces and generalized eigenspaces. Let $\lambda \in \mathbb{C}$ and $(x, y) \in \mathbb{C}^2$, $(x, y) \neq (0, 0)$ be such that

$$T(x, y) = \lambda(x, y), \quad (x, y) \in \mathbb{C}^2.$$

Then

$$(x + y, y) = \lambda(x, y)$$

and

$$\begin{aligned} x + y &= \lambda x \\ y &= \lambda y. \end{aligned}$$

We have the following cases.

1. *Let* $y = 0$. *Then*
$$x = \lambda x.$$
   (a) *Let* $x = 0$. *Then*
   $$(x,y) = (0,0).$$
   (b) *Let* $x \neq 0$. *Then* $\lambda = 1$.
2. *Let* $y \neq 0$. *Then* $\lambda = 1$.

Therefore $\lambda = 1$ *is the unique eigenvalue of* T *and*
$$E(1,T) = \{(x,0) : x \in \mathbb{C}\}.$$
By Theorem 6.6, it follows that
$$G(1,T) = \ker(T-I)^2.$$
For $(x,y) \in \mathbb{C}^2$, *we have*
$$\begin{aligned}
(T-I)(x,y) &= T(x,y) - I(x,y) \\
&= (x+y,y) - (x,y) \\
&= (y,0), \\
(T-I)^2(x,y) &= (T-I)(T-I)(x,y) \\
&= (T-I)((T-I)(x,y)) \\
&= (T-I)(y,0) \\
&= T(y,0) - I(y,0) \\
&= (y,0) - (y,0) \\
&= (0,0).
\end{aligned}$$
Therefore
$$(G(1,T) = \mathbb{C}^2.$$

Example 6.8. *Let $T \in \mathscr{L}(\mathbb{C}^3)$ be defined by*

$$T(x,y,z) = (x+y, y+2z, 3z), \quad (x,y,z) \in \mathbb{C}^3.$$

We will find the eigenvalues of T, the corresponding eigenspaces and generalized eigenspaces. Let $\lambda \in \mathbb{C}$ and $(x,y,z) \in \mathbb{C}^3$, $(x,y,z) \neq (0,0,0)$, be such that

$$T(x,y,z) = \lambda(x,y,z).$$

We have

$$(x+y, y+2z, 3z) = (\lambda x, \lambda y, \lambda z).$$

Hence, we get the system

$$\begin{aligned} x+y &= \lambda x \\ y+2z &= \lambda y \\ 3z &= \lambda z, \end{aligned}$$

whereupon

$$\begin{aligned} y &= (\lambda-1)x \\ 2z &= (\lambda-1)y \\ 3z &= \lambda z. \end{aligned}$$

We have the following cases.

1. *Let $z = 0$. Then*

$$\begin{aligned} y &= (\lambda-1)x \\ 0 &= (\lambda-1)y. \end{aligned}$$

   (a) *Let $\lambda \neq 1$. Then*

$$\begin{aligned} y &= 0 \\ z &= 0. \end{aligned}$$

Hence,

$$(x,y,z) = (0,0,0).$$

This is impossible.

(b) Let $\lambda = 1$. *Then*

$$y = 0.$$

Moreover,

$$E(1,T) = \{(x,0,0) : x \in \mathbb{C}\}.$$

2. *Let* $z \neq 0$. *Then*

$$\lambda = 3$$

and

$$\begin{aligned} y &= 2x \\ 2z &= 2y, \end{aligned}$$

or

$$\begin{aligned} y &= 2x \\ y &= z. \end{aligned}$$

Thus,

$$E(3,T) = \{(x,2x,2x) : x \in \mathbb{C}\}.$$

Now, we will find $G(1,T)$ *and* $G(3,T)$. *For* $(x,y,z) \in \mathbb{C}^3$, *we have*

$$\begin{aligned} (T-I)^3(x,y,z) &= (T-I)^2(T-I)(x,y,z) \\ &= (T-I)^2((T-I)(x,y,z)) \\ &= (T-I)^2(T(x,y,z)-I(x,y,z)) \\ &= (T-I)^2((x+y,y+2z,3z)-(x,y,z)) \\ &= (T-I)^2(y,2z,2z) \end{aligned}$$

$$
\begin{aligned}
&= (T-I)(T-I)(y,2z,2z) \\
&= (T-I)((T-I)(y,2z,2z)) \\
&= (T-I)(T(y,2z,2z)-I(y,2z,2z)) \\
&= (T-I)(y+2z,2z+4z,6z)-(y,2z,2z)) \\
&= (T-I)((y+2z,6z,6z)-(y,2z,2z)) \\
&= (T-I)(2z,4z,4z) \\
&= T(2z,4z,4z)-I(2z,4z,4z) \\
&= (2z+4z,4z+8z,12z)-(2z,4z,4z) \\
&= (6z,12z,12z)-(2z,4z,4z) \\
&= (4z,8z,8z) \\
&= (0,0,0),
\end{aligned}
$$

whereupon $z=0$ *and*

$$G(1,T)=\{(x,y,0): x,y\in\mathbb{C}\}.$$

Next, for $(x,y,z)\in\mathbb{C}^3$, *we have*

$$
\begin{aligned}
(T-3I)^3(x,y,z) &= (T-3I)^2(T-3I)(x,y,z) \\
&= (T-3I)^2((T-3I)(x,y,z)) \\
&= (T-3I)^2(T(x,y,z)-3I(x,y,z)) \\
&= (T-3I)^2((x+y,y+2z,3z)-(3x,3y,3z))
\end{aligned}
$$

$$
\begin{aligned}
&= (T-3I)^2(y-2x,2z-2y,0)\\
&= (T-3I)(T-3I)(y-2x,2z-2y,0)\\
&= (T-3I)((T-3I)(y-2x,2z-2y,0))\\
&= (T-3I)(T(y-2x,2z-2y,0)-3I((y-2x,2z-2y,0))\\
&= (T-3I)\\
&\quad \times((y-2x+2z-2y,2z-2y,0)-(3y-6x,6z-6y,0))\\
&= (T-3I)\\
&\quad \times((-2x-y+2z,2z-2y,0)-(3y-6x,6z-6y,0))\\
&= (T-3I)(-2x-y+2z-3y+6x,2z-2y-6z+6y,0)\\
&= (T-3I)(4x-y+2z,4y-4z,0)\\
&= T(4x-y+2z,4y-4z,0)-3I(4x-y+2z,4y-4z,0)\\
&= (4x-y+2z+4y-4z,4y-4z,0)\\
&\quad -(12x-3y+6z,12y-12z,0)\\
&= (4x+3y-2z,4y-4z,0)\\
&\quad -(12x-3y+6z,12y-12z,0)\\
&= (4x+3y-2z-12x+3y-6z,4y-4z-12y+12z,0)
\end{aligned}
$$

$$
\begin{aligned}
&= (-8x+6y-8z,-8y+8z,0)\\
&= (0,0,0).
\end{aligned}
$$

Hence,

$$
\begin{aligned}
-8x+6y-8z &= 0\\
-8y+8z &= 0,
\end{aligned}
$$

or

$$
\begin{aligned}
y &= z\\
-8x-2y &= 0,
\end{aligned}
$$

or

$$
\begin{aligned}
y &= z\\
y &= -4x.
\end{aligned}
$$

Thus,

$$G(3,T) = \{(x,-4x,-4x) : x \in \mathbb{C}\}.$$

Exercise 6.9. *Let $T \in \mathscr{L}(\mathbb{C}^3)$ be defined by*

$$T(x,y,z) = (4y,0,5z), \quad (x,y,z) \in \mathbb{C}^3.$$

Find the eigenvalues of T and the corresponding eigenspaces and generalized eigenspaces.

Answer 6.10.

$$
\begin{aligned}
\lambda_1 &= 0,\\
\lambda_2 &= 5,\\
E(0,T) &= \{(x,0,0) : x \in \mathbb{C}\},
\end{aligned}
$$

$$
\begin{aligned}
E(5,T) &= \{(0,0,z) : z \in \mathbb{C}\}, \\
G(0,T) &= \{(x,y,0) : x,y \in \mathbb{C}\}, \\
G(5,T) &= \{(0,0,z) : z \in \mathbb{C}\}.
\end{aligned}
$$

Theorem 6.11. *Let $T \in \mathscr{L}(V)$. Suppose that $\lambda_1, \ldots, \lambda_m$ are distinct eigenvalues of T and $v_1, \ldots, v_m$ are corresponding generalized eigenvectors. Then the list $v_1, \ldots, v_m$ is linearly independent.*

Proof. Suppose that $a_1, \ldots, a_m \in \mathbb{C}$ are such that

$$a_1 v_1 + \cdots + a_m v_m = 0.$$

By Theorem 6.6, it follows that

$$(T - \lambda_j I)^n v_j = 0, \quad j \in \{1, \ldots, m\}.$$

Let k be the largest nonnegative integer such that

$$(T - \lambda_1 I)^k v_1 \neq 0.$$

Set

$$w = (T - \lambda_1 I)^k v_1.$$

We have $w \neq 0$ and

$$
\begin{aligned}
(T - \lambda_1 I) w &= (T - \lambda_1 I)(T - \lambda_1 I)^k v \\
&= (T - \lambda_1 I)^{k+1} v \\
&= 0.
\end{aligned}
$$

Hence,

$$Tw = \lambda_1 w$$

and

$$
(T - \lambda I) w = Tw - \lambda I w
$$

$$= \lambda_1 w - \lambda w$$

$$= (\lambda_1 - \lambda)w, \quad \lambda \in \mathbb{F}.$$

Assume that

$$(T - \lambda I)^l w = (\lambda_1 - \lambda)^l w, \quad \lambda \in \mathbb{F}. \tag{6.10}$$

Then

$$\begin{aligned}
(T - \lambda I)^{l+1} w &= (T - \lambda I)(T - \lambda I)^l w \\
&= (T - \lambda I)((T - \lambda I)^l w) \\
&= (T - \lambda I)((\lambda_1 - \lambda)^l w) \\
&= (\lambda_1 - \lambda)^l (T - \lambda I) w \\
&= (\lambda_1 - \lambda)^l (\lambda_1 - \lambda) w \\
&= (\lambda_1 - \lambda)^{l+1} w, \quad \lambda \in \mathbb{F}.
\end{aligned}$$

Now, applying the principal of the mathematical induction, we conclude that (6.10) holds for any $l \in \mathbb{N}$. From here,

$$\begin{aligned}
0 &= (T - \lambda_1 I)^k (T - \lambda_2 I)^n \ldots (T - \lambda_m I)^n (a_1 v_1 + \cdots + a_m v_m) \\
&= (T - \lambda_1 I)^k (T - \lambda_2 I)^n \ldots (T - \lambda_m I)^n (a_1 v_1) \\
&\quad + (T - \lambda_1 I)^k (T - \lambda_2 I)^n \ldots (T - \lambda_m I)^n (a_2 v_2) \\
&\quad + \cdots \\
&\quad + (T - \lambda_1 I)^k (T - \lambda_2 I)^n \ldots (T - \lambda_m I)^n (a_m v_m) \\
&= a_1 (T - \lambda_1 I)^k (T - \lambda_2 I)^n \ldots (T - \lambda_m I)^n v_1
\end{aligned}$$

$$
\begin{aligned}
&+a_2(T-\lambda_1 I)^k(T-\lambda_2 I)^n\dots(T-\lambda_m I)^n v_2\\
&+\cdots\\
&+a_m(T-\lambda_1 I)^k(T-\lambda_2 I)^n\dots(T-\lambda_m I)^n v_m\\
= \;& a_1(T-\lambda_2 I)^n\dots(T-\lambda_m I)^n(T-\lambda_1 I)^k v_1\\
&+a_2(T-\lambda_1 I)^k(T-\lambda_3 I)^n\dots(T-\lambda_m I)^n(T-\lambda_2 I)^n v_2\\
&+\cdots\\
&+a_m(T-\lambda_1 I)^k(T-\lambda_2 I)^n\dots(T-\lambda_m I)^n v_m\\
= \;& a_1(T-\lambda_2 I)^n\dots(T-\lambda_m I)^n w\\
= \;& a_1(T-\lambda_2 I)^n\dots(T-\lambda_{m-1} I)^n(\lambda_1-\lambda_m)^n w\\
= \;& a_1(\lambda_1-\lambda_m)^n(T-\lambda_2 I)^n\dots(T-\lambda_{m-1} I)^n w\\
\vdots\;&\\
= \;& a_1(\lambda_1-\lambda_m)^n\dots(\lambda_1-\lambda_3)^n(T-\lambda_2 I)^n w\\
= \;& a_1(\lambda_1-\lambda_m)^n\dots(\lambda_1-\lambda_3)^n(\lambda_1-\lambda_2)^n w,
\end{aligned}
$$

i.e.,

$$a_1(\lambda_1-\lambda_m)^n\dots(\lambda_1-\lambda_3)^n(\lambda_1-\lambda_2)^n w=0.$$

Because $\lambda_1,\dots,\lambda_m$ are distinct and $w\neq 0$, by the last equality, we get $a_1=0$ and then

$$a_2v_2+\cdots+a_mv_m=0.$$

Continuing as above, we get

$$a_j = 0, \quad j \in \{1, \ldots, m\},$$

and $v_1, \ldots, v_m$ are linearly independent. This completes the proof. □

6.3 Nilpotent Operators

Definition 6.12. *An operator is called nilpotent if some power of it equals 0.*

Example 6.13. *Let $T \in \mathscr{L}(\mathbb{C}^2)$ be defined by*

$$T(x,y) = (2x-y, 4x-2y), \quad (x,y) \in \mathbb{C}^2.$$

We have

$$\begin{aligned} T^2(x,y) &= TT(x,y) \\ &= T(T(x,y)) \\ &= T(2x-y, 4x-2y) \\ &= (2(2x-y)-(4x-2y), 4(2x-y)-2(4x-2y)) \\ &= (4x-2y-4x+2y, 8x-4y-8x+4y) \\ &= (0,0), \quad (x,y) \in \mathbb{C}^2. \end{aligned}$$

Thus, T is nilpotent.

Example 6.14. *Let $T \in \mathscr{L}(\mathbb{C}^4)$ be defined by*

$$A(x,y,z,t) = (y+2z+4t, 2z+t, 5t, 0), \quad (x,y,z,t)\mathbb{C}^4.$$

We have

$$T^2(x,y,z,t) = TT(x,y,z,t)$$

$$
\begin{aligned}
&= T(T(x,y,z,t))\\
&= T(y+2z+4t, 2z+t, 5t, 0)\\
&= (2z+t+2\cdot 5t+4\cdot 0, 2\cdot 5t+0, 5\cdot 0, 0)\\
&= (2z+11t, 10t, 0, 0),\\
T^3(x,y,z,t) &= TT^2(x,y,z,t)\\
&= T(T^2(x,y,z,t))\\
&= T(2z+11t, 10t, 0, 0)\\
&= (10t+2\cdot 0+4\cdot 0, 2\cdot 0+0, 5\cdot 0, 0)\\
&= (10t, 0, 0, 0),\\
T^4(x,y,z,t) &= TT^3(x,y,z,t)\\
&= T(T^3(x,y,z,t))\\
&= T(10t, 0, 0, 0)\\
&= (0,0,0,0), \quad (x,y,z,t)\in\mathbb{C}^4.
\end{aligned}
$$

Thus, T is nilpotent.

Example 6.15. *Let $T \in \mathscr{L}(\mathbb{C}^3)$ be defined by*

$$T(x,y,z) = (5x-3y+2z, 15x-9y+6z, 10x-6y+4z), \quad (x,y,z)\in\mathbb{C}^3.$$

We have

$$T^2(x,y,z) = TT(x,y,z)$$

$$\begin{aligned}
&= T(T(x,y,z)) \\
&= T(5x-3y+2z, 15x-9y+6z, 10x-6y+4z) \\
&= (5(5x-3y+2z)-3(15x-9y+6z)+2(10x-6y+4z), \\
&\quad 15(5x-3y+2z)-9(15x-9y+6z)+6(10x-6y+4z), \\
&\quad 10(5x-3y+2z)-6(15x-9y+6z)+4(10x-6y=4z)) \\
&= (25x-15y+10z-45x+27y-18z+20x-12y+8z, \\
&\quad 75x-45y+30z-135x+81y-54z+60x-36y+24z, \\
&\quad 50x-30y+20z-90x+54y-36z+40x-24y+16z) \\
&= (0,0,0), \quad (x,y,z) \in \mathbb{C}^3.
\end{aligned}$$

Thus, T is nilpotent.

Exercise 6.16. *Let $T \in \mathscr{L}(\mathbb{C}^3)$ be defined by*

$$T(x,y,z) = (y,z,0), \quad (x,y,z) \in \mathbb{C}^3.$$

Prove that T is nilpotent.

Hint 6.17. *Prove that $T^3 = 0$.*

Theorem 6.18. *Let $T \in \mathscr{L}(V)$ be nilpotent and $n = \dim V$. Then $T^n = 0$.*

Proof. Since $T \in \mathscr{L}(V)$ is nilpotent, there is a $k \in \mathbb{N}$ such that

$$T^k = 0.$$

Hence,

$$G(0,T) = V.$$

Now, applying Theorem 6.6, we get

$$\ker T^n = G(0,T)$$

$$= \quad V.$$

Hence,

$$\text{range} T^n \{0\}$$

and $T^n = 0$. This completes the proof. □

Theorem 6.19. *Suppose that $T \in \mathscr{L}(V)$ is nilpotent. Then there is a basis of V with respect to which the matrix of T has the form*

$$\begin{pmatrix} 0 & & \star \\ & \ddots & \\ 0 & & 0 \end{pmatrix}.$$

Proof. Choose a basis of $\ker T$. Then extend this basis to a basis of $\ker T^2$. We then extend the new basis obtained from this to $\ker T^3$. Continue in this fashion, we get a basis of V. The first column, and perhaps additional columns at the beginning, consists of all 0's, because the corresponding basis vectors are in $\ker T$. The next set of columns comes from basis vectors in $\ker T^2$. Applying T to any such vector, we get a vector in $\ker T$, i.e., we get a vector that is a linear combination of the previous basis vectors. Thus, all nonzero entries in these columns lie above the diagonal. The next set of columns comes from basis vectors in $\ker T^3$. Applying T to any such vector, we get a vector in $\ker T^2$, i.e., we get a vector that is a linear combination of the previous basis vectors. Thus, all nonzero entries in these columns lie above the diagonal. Continue in this fashion, we get the desired result. This completes the proof. □

6.4 Decomposition of Operators on Complex Vector Spaces

Theorem 6.20. *Suppose that $T \in \mathscr{L}(V)$ and $p \in \mathscr{P}(\mathbb{F})$. Then* $\ker p(T)$ *and* range$p(T)$ *are invariant under T.*

Proof. Suppose that $v \in \ker p(T)$ is arbitrarily chosen. Then

$$p(T)v = 0.$$

Hence,

$$p(T)(Tv) \quad = \quad p(T)Tv$$

$$
\begin{aligned}
&= Tp(T)v \\
&= T(p(T)v) \\
&= T0 \\
&= 0.
\end{aligned}
$$

Thus, $Tv \in \ker p(T)$. Because $v \in \ker p(T)$ was arbitrarily chosen, we conclude that $\ker p(T)$ is invariant under T. Let now, $u \in \mathrm{range}\, p(T)$ be arbitrarily chosen. Then there is a $w \in V$ such that

$$u = p(T)w.$$

Hence,

$$
\begin{aligned}
Tu &= T(p(T)w) \\
&= Tp(T)w \\
&= p(T)Tw \\
&= p(T)(Tw)
\end{aligned}
$$

and thus, $Tu \in \mathrm{range}\, p(T)$. Because $u \in \mathrm{range}\, p(T)$ was arbitrarily chosen, we conclude that $\mathrm{range}\, p(T)$ is invariant under T. This completes the proof. □

Theorem 6.21. *Suppose that V is a complex vector space and $T \in \mathscr{L}(V)$. Let $\lambda_1, \ldots, \lambda_m$ be the distinct eigenvalues of T. Then*

1. *each $G(\lambda_j, T)$ is invariant under T.*
2. *each $(T - \lambda_j I)|_{G(\lambda_j, T)}$ is nilpotent.*
3. *$V = G(\lambda_1, T) \oplus \cdots \oplus G(\lambda_j, T)$.*

Proof. 1. Let

$$p_j(z) = (z - \lambda_j I)^n, \quad j \in \{1, \ldots, m\}.$$

We have that $\ker p_j(T)$ and $\mathrm{range} p_j(T)$, $j \in \{1, \ldots, m\}$, are invariant under T.

2. Because for any $j \in \{1, \ldots, m\}$, there is an $l \in \mathbb{N}$ such that

$$(T - \lambda_j I)^l|_{G(\lambda_j, T)} v = 0$$

for any $v \in G(\lambda_j, T)$, $j \in \{1, \ldots, m\}$, we conclude that

$$(T - \lambda_j I)^l|_{G(\lambda_j, T)}$$

is nilpotent.

3. We will use the principal of the mathematical induction. If $n = 1$, then

$$V = G(\lambda_1, T).$$

We assume that $n > 1$ and that the desired result holds on all complex vector spaces of smaller dimension. By Theorem 5.52, it follows that T has at least one eigenvalue. Therefore $m \geq 1$. Note that $T - \lambda_1 I \in \mathscr{L}(V)$. Now, applying Theorem 6.4, we find

$$V = \ker(T - \lambda_1 I)^n \oplus \mathrm{range}(T - \lambda_1 I)^n.$$

Set

$$U = \mathrm{range}(T - \lambda_1 I)^n.$$

By Theorem 6.6, we have

$$G(\lambda_1, T) = \ker(T - \lambda_1 I)^n.$$

Therefore

$$V = G(\lambda_1, T) \oplus U. \tag{6.11}$$

Let

$$p(z) = (z - \lambda_1)^n, \quad z \in \mathbb{C}.$$

Then, by Theorem 6.21, we get that U is invariant under T. Thus, $T|_U$ is well defined. Because $G(\lambda_1, T) \neq \{0\}$, we have that

$$\dim U < n.$$

Therefore we can apply our induction hypothesis to $T|_U$. Note that the eigenvalues of $T|_U$ lies in the set $\{\lambda_2, \ldots, \lambda_m\}$. Then, by our induction hypothesis, we get

$$U = G(\lambda_2, T|_U) \oplus \cdots \oplus G(\lambda_m, T|_U).$$

Hence and (6.11), we obtain

$$V = G(\lambda_1, T) \oplus G(\lambda_2, T|_U) \oplus \cdots \oplus G(\lambda_m, T|_U). \tag{6.12}$$

Fix $k \in \{2, \ldots, m\}$ arbitrarily. We have

$$G(\lambda_k, T|_U) \subseteq G(\lambda_k, T). \tag{6.13}$$

Let now, $v \in G(\lambda_k, T)$ be arbitrarily chosen. Since $v \in V$, by (6.12), it follows that

$$v = v_1 + u,$$

where $v_1 \in G(\lambda_1, T)$ and $u \in U$. By the induction hypothesis, we find

$$u = v_2 + \cdot + v_m,$$

where each $v_j \in G(\lambda_j, T|_U)$, $j \in \{2, \ldots, m\}$. Because $G(\lambda_j, T|_U) \subseteq G(\lambda_j, T)$, we have $v_j \in G(\lambda_j, T)$, $j \in \{2, \ldots, m\}$. Thus,

$$v = v_1 + v_2 + \cdots + v_m. \tag{6.14}$$

By Theorem 6.6, we have that the generalized eigenvectors corresponding to the distinct eigenvalues are linearly independent. Thus, by (6.14), it follows that each v_j equals 0 except possibly when $j = k$. In particular, we have $v_1 = 0$. Then

$$v = u \in U.$$

Hence, $v \in G(\lambda_k, U|_T)$. Since $v \in G(\lambda_k, T)$ was arbitrarily chosen and we get that it is an element of $G(\lambda_k, T|_U)$, we obtain the inclusion

$$G(\lambda_k, T) \subseteq G(\lambda_k, T|_U).$$

By the last inclusion and the inclusion (6.13), we find

$$G(\lambda_k, T|_U) = G(\lambda_k, T).$$

Because $k \in \{1, \ldots, m\}$ was arbitrarily chosen, by (6.12), we obtain

$$V = G(\lambda_1, T) \oplus G(\lambda_2, T) \oplus \cdots \oplus G(\lambda_m, T).$$

This completes the proof.

□

Theorem 6.22. *Suppose that V is a complex vector space and $T \in \mathscr{L}(V)$. Then there is a basis of V consisting of generalized eigenvectors of T.*

Proof. Let $\lambda_1, \ldots, \lambda_m$ be the distinct eigenvalues of T. Choose a basis of each $G(\lambda_j, T)$, $j \in \{1, \ldots, m\}$. Put all these bases together to form a basis of V consisting of generalized eigenvectors of T. This completes the proof.

□

6.5 Multiplicity of an Eigenvalue

Definition 6.23. *Suppose that $T \in \mathscr{L}(V)$. The multiplicity of an eigenvalue λ of T is defined to be the dimension of the corresponding generalized eigenspace $G(\lambda, T)$, i.e.,*

$$\text{the multiplicity of} = \dim\ker(T - \lambda I)^n,$$

where $n = \dim V$.

Example 6.24. *Let $T \in \mathscr{L}(\mathbb{C}^3)$ be defined by*

$$T(z_1, z_2, z_3) = (6z_1 + 3z_2 + 4z_3, 6z_2 + 2z_3, 7z_3), \quad (z_1, z_2, z_3) \in \mathbb{C}^3.$$

Let $\lambda \in \mathbb{C}$ and $(z_1, z_2, z_3) \in \mathbb{C}^3$, $(z_1, z_2, z_3) \neq (0,0,0)$, be such that

$$T(z_1, z_2, z_3) = \lambda(z_1, z_2, z_3).$$

Then

$$(6z_1 + 3z_2 + 4z_3, 6z_2 + 2z_3, 7z_3) = (\lambda z_1, \lambda z_2, \lambda z_3),$$

whereupon we get the system

$$\begin{aligned} 6z_1 + 3z_2 + 4z_3 &= \lambda z_1 \\ 6z_2 + 2z_3 &= \lambda z_2 \\ 7z_3 &= \lambda z_3. \end{aligned}$$

1. *Let $z_3 = 0$. Then*

$$\begin{aligned} 6z_1 + 3z_2 &= \lambda z_1 \\ 6z_2 &= \lambda z_2. \end{aligned}$$

   (a) *Let $z_2 = 0$. Then*

$$6z_1 = \lambda z_1.$$

      i. *Let $z_1 = 0$. Then*

$$(z_1, z_2, z_3) = (0,0,0).$$

      This is impossible.

ii. *Let* $z_1 \neq 0$. *Then* $\lambda = 6$.

(b) *Let* $z_2 \neq 0$. *Then* $\lambda = 6$.

2. *Let* $z_3 \neq 0$. *Then* $\lambda = 7$.

Therefore the eigenvalues of the operator T *are*

$$\lambda_1 = 6 \quad \textit{and} \quad \lambda_2 = 7.$$

By Theorem 6.6, it follows that

$$\begin{aligned} G(\lambda_1, T) &= (T - \lambda_1 I)^3, \\ G(\lambda_2, T) &= (T - \lambda_2 I)^3. \end{aligned}$$

For $(z_1, z_2, z_3) \in \mathbb{C}^3$, *we have*

$$\begin{aligned} (T - \lambda_1 I)(z_1, z_2, z_3) &= (T - 6I)(z_1, z_2, z_3) \\ &= T(z_1, z_2, z_3) - 6(z_1, z_2, z_3) \\ &= (6z_1 + 3z_2 + 4z_3, 6z_2 + 2z_3, 7z_3) - (6z_1, 6z_2, 6z_3) \\ &= (3z_2 + 4z_3, 2z_3, z_3) \end{aligned}$$

and

$$\begin{aligned} (T - \lambda_1 I)^2(z_1, z_2, z_3) &= (T - 6I)^2(z_1, z_2, z_3) \\ &= (T - 6I)(T - 6I)(z_1, z_2, z_3) \\ &= (T - 6I)((T - 6I)(z_1, z_2, z_3)) \\ &= (T - 6I)(3z_2 + 4z_3, 2z_3, z_3) \\ &= (3 \cdot (2z_3) + 4z_3, 2z_3, z_3) \\ &= (10z_3, 2z_3, z_3), \end{aligned}$$

$$\begin{aligned}(T-\lambda_1 I)^3(z_1,z_2,z_3) &= (T-6I)^3(z_1,z_2,z_3)\\ &= (T-6I)(T-6I)^2(z_1,z_2,z_3)\\ &= (T-6I)((T-6I)^2(z_1,z_2,z_3))\\ &= (T-6I)(10z_3,2z_3,z_3)\\ &= (3\cdot(2z_3)+4z_3,2z_3,z_3)\\ &= (10z_3,2z_3,z_3).\end{aligned}$$

Then, for $(z_1,z_2,z_3)\in\mathbb{C}^3$, *we have*

$$(T-6I)^3(z_1,z_2,z_3)=(0,0,0)$$

if and only if $z_3=0$. *Consequently*

$$\begin{aligned}G(6,T) &= \{(z_1,z_2,0): z_1,z_2\in\mathbb{C}\}\\ &= span((1,0,0),(0,1,0))\end{aligned}$$

and

$$\dim G(6,T)=2,$$

i.e., the multiplicity of the eigenvalue $\lambda_1=6$ *equals* 2. *Next, for* $(z_1,z_2,z_3)\in\mathbb{C}$, *we get*

$$\begin{aligned}(T-7I)(z_1,z_2,z_3) &= T(z_1,z_z,z_3)-7(z_1,z_2,z_3)\\ &= (6z_1+3z_2+4z_3,6z_2+2z_3,7z_3)-(7z_1,7z_2,7z_3)\\ &= (3z_1+4z_3-z_1,2z_3-z_2,0)\end{aligned}$$

and

$$(T-7I)^2(z_1,z_2,z_3) = (T-7I)(T-7I)(z_1,z_2,z_3)$$

$$
\begin{aligned}
&= (T-7I)((T-7I)(z_1,z_2,z_3)) \\
&= (T-7I)(3z_3+4z_3-z_1, 2z_3-z_2, 0) \\
&= (3(2z_3-z_2)+4\cdot 0-(3z_2+4z_3-z_1), \\
&\quad 2\cdot 0-(2z_3-z_2), 0) \\
&= (6z_3-3z_2-3z_2-4z_3+z_1, z_2-2z_3, 0) \\
&= (2z_3-6z_2+z_1, z_2-2z_3, 0), \\
(T-7I)^3(z_1,z_2,z_3) &= (T-7I)(T-7I)^2(z_1,z_2,z_3) \\
&= (T-7I)((T-7I)^2(z_1,z_2,z_3)) \\
&= (T-7I)((2z_3-6z_2+z_1, z_2-2z_3, 0) \\
&= (3(z_2-2z_3)+4\cdot 0-(2z_3-6z_2+z_1), \\
&\quad 2\cdot 0-(z_2-2z_3), 0) \\
&= (3z_2-6z_3-2z_3+6z_2-z_1, 2z_3-z_2, 0) \\
&= (-z_1+9z_2-8z_3, 2z_3-z_2, 0).
\end{aligned}
$$

Then

$$(T-7I)^3(z_1,z_2,z_3) = 0$$

if and only if

$$(-z_1+9z_2-8z_3, 2z_3-z_2, 0) = (0,0,0)$$

if and only if

$$-z_1+9z_2-8z_3 = 0$$

$$2z_3 - z_2 = 0$$

if and only if

$$-z_1 + 18z_3 - 8z_3 = 0$$

$$z_2 = 2z_3$$

if and only if

$$z_1 = 10z_3$$

$$z_2 = 2z_3.$$

Consequently

$$G(7,T) = \{(10z_3, 2z_3, z_3) : z_3 \in \mathbb{C}\}$$

$$= span((10,2,1))$$

and

$$\dim G(7,T) = 1,$$

i.e., the multiplicity of the eigenvalue $\lambda_2 = 7$ *equals* 1.

Example 6.25. *Let* $T \in \mathscr{L}(\mathbb{C}^2)$ *be defined as in Example 5.10. Then, by the computations in Example 5.10, we have that the eigenvalues of* T *are*

$$\lambda_1 = 2 - 2i,$$

$$\lambda_2 = -i.$$

We will find their multiplicities. For $(z_1, z_2) \in \mathbb{C}^2$*, we have*

$$(T - \lambda_1 I)(z_1, z_2) = (T - (2-2i)I)(z_1, z_2)$$

$$= T(z_1, z_2) - (2-2i)(z_1, z_2)$$

$$
\begin{aligned}
&= ((1-2i)z_1 - z_2, (-1+i)z_1 \\
&= (1-i)z_2) - ((2-2i)z_1, (2-2i)z_2) \\
&= (-z_1 - z_2, (-1+i)z_1 + (-1+i)z_2), \\
(T-\lambda_1 I)^2(z_1,z_2) &= (T-(2-2i)I)^2(z_1,z_2) \\
&= (T-(2-2i)I)(T-(2-2i)I)(z_1,z_2) \\
&= (T-(2-2i)I)((T-(2-2i)I)(z_1,z_2)) \\
&= (T-(2-2i)I)(-z_1 - z_2, \\
&\quad (-1+i)z_1 + (-1+i)z_2) \\
&= (z_1 + z_2 - (-1+i)z_1 - (-1+i)z_2, \\
&\quad (-1+i)(-z_1 - z_2 + (-1+i)z_1 + (-1+i)z_2)) \\
&= ((2-i)z_1 + (2-i)z_2, (-1+i)(-2+i)(z_1+z_2)) \\
&= ((2-i)(z_1+z_2), (-1+i)(-2+i)(z_1+z_2)).
\end{aligned}
$$

Hence,

$$(T-\lambda_1 I)(z_1,z_2) = 0, \quad (z_1,z_2) \in \mathbb{C}^2,$$

if and only if

$$((2-i)(z_1+z_2), (-1+i)(-2+i)(z_1+z_2)) = 0, \quad (z_1,z_2) \in \mathbb{C}^2,$$

if and only if

$$z_2 = -z_1, \quad z_1, z_2 \in \mathbb{C}.$$

Consequently

$$G(2-2i,T) = \{(z_1,-z_1) : z_1 \in \mathbb{C}\}$$

$$= \quad span((1,-1)),$$

$$\dim F(2-2i,T) \quad = \quad 1,$$

i.e., the multiplicity of the eigenvalue $\lambda_1 = 2-2i$ *equals* 1. *Next, for* $(z_1,z_2) \in \mathbb{C}$, *we have*

$$\begin{aligned}
(T-\lambda_2 I)(z_1,z_2) &= (T+iI)(z_1,z_2) \\
&= T(z_1,z_2)+i(z_1,z_2) \\
&= ((1-2i)z_1-z_2,(-1+i)z_1+(1-i)z_2)+(iz_1,iz_2) \\
&= ((1-i)z_1-z_2,(-1+i)z_1+z_2), \\
(T-\lambda_2 I)^2(z_1,z_2) &= (T+iI)^2(z_1,z_2) \\
&= (T+iI)(T+iI)(z_1,z_2) \\
&= (T+iI)((T+iI)(z_1,z_2)) \\
&= (T+iI)((1-i)z_1-z_2,-((1-i)z_1-z_2)) \\
&= ((1-i)((1-i)z_1-z_2)+((1-i)z_1-z_2), \\
&\quad -((1-i)((1-i)z_1-z_2)+(1-i)z_1-z_2)) \\
&= ((2-i)((1-i)z_1-z_2),-(2-i)((1-i)z_1-z_2)).
\end{aligned}$$

Therefore

$$(T-\lambda_2 I)^2(z_1,z_2) = (0,0), \quad (z_1,z_2) \in \mathbb{C}^2,$$

if and only if

$$((2-i)((1-i)z_1-z_2),-(2-i)((1-i)z_1-z_2)) = (0,0), \quad (z_1,z_2) \in \mathbb{C}^2,$$

if and only if

$$z_2 = (1-i)z_1, \quad z_1, z_2 \in \mathbb{C}.$$

Consequently

$$\begin{aligned} G(-i,T) &= \{(z_1,(1-i)z_1) : z_1 \in \mathbb{C}\} \\ &= span((1,1-i)), \\ \dim G(-i,T) &= 1, \end{aligned}$$

i.e., the multiplicity of the eigenvalue $\lambda_2 = -i$ *equals* 1.

Example 6.26. *Let* $T \in \mathscr{L}(\mathbb{C}^3)$ *be defined as in Example 5.11. By the computations in Example 5.11, it follows that* $\lambda = 2$ *is the eigenvalue of* T*. For* $(z_1,z_2,z_3) \in \mathbb{C}^3$*, we have*

$$\begin{aligned} (T-2I)(z_1,z_2,z_3) &= T(z_1,z_2,z_3) - 2(z_1,z_2,z_3) \\ &= (z_2, -4z_1+4z_2, -2z_1+z_2+2z_3) \\ &\quad +(-2z_1,-2z_2,-2z_3) \\ &= (-2z_1+z_2, -4z_1+2z_2, -2z_1+z_2), \\ (T-2I)^2(z_1,z_2,z_3) &= (T-2I)(T-2I)(z_1,z_2,z_3) \\ &= (T-2I)((T-2I)(z_1,z_2,z_3)) \\ &= (T-2I)(-2z_1+z_2, -4z_1+2z_2, -2z_1+z_2) \\ &= (-2(-2z_1+z_2) - 4z_1 + 2z_2, \\ &\quad 2(-2(-2z_1+z_2) - 4z_1 + 2z_2), \\ &\quad -2(-2z_1+z_2) - 4z_1 + 2z_2) \end{aligned}$$

$$= \quad (0,0,0),$$

$$(T-2I)^3(z_1,z_2,z_3) \quad = \quad (0,0,0).$$

Consequently

$$G(2,T) \quad = \quad \mathbb{C}^3,$$

$$\dim G(2,T) \quad = \quad 3,$$

i.e., the multiplicity of $\lambda = 2$ *equals* 3.

Exercise 6.27. *Let* $T \in \mathscr{L}(\mathbb{C}^2)$ *be defined as in Exercise 5.14. Find the multiplicities of the eigenvalues of* T.

Answer 6.28. 1.

Exercise 6.29. *Let* $T \in \mathscr{L}(\mathbb{C}^3)$ *be defined as in Exercise 5.16. Find the multiplicities of the eigenvalues of* T.

Answer 6.30. *1.* 3.

2. *the multiplicity of* $\lambda = 1$ *equals* 1. *The multiplicity of* $\lambda = 0$ *equals* 2.

Theorem 6.31. *Suppose that* V *is a complex vector space and* $T \in \mathscr{L}(V)$. *Then the sum of the multiplicities of the distinct eigenvalues of* T *equals* $\dim V$.

Proof. Let $\lambda_1, \ldots, \lambda_m$ be the distinct eigenvalues of T. By Theorem 6.21, it follows that

$$V = G(\lambda_1,T) \oplus \cdots \oplus G(\lambda_m,T).$$

Hence,

$$\begin{aligned} \dim V &= \dim G(\lambda_1,T) + \cdots + \dim G(\lambda_m T) \\ &= \text{multiplicity of } \lambda_1 + \cdots + \text{multiplicity of } \lambda_m. \end{aligned}$$

This completes the proof. □

Definition 6.32. *Let V be a complex vector space, $T \in \mathscr{L}(V)$ and λ be an eigenvalue of T. Then*

$$\begin{aligned} \textit{algebraic multiplicity of } \lambda &= \dim G(\lambda, T), \\ \textit{geometric multiplicity of } \lambda &= \dim E(\lambda, T). \end{aligned}$$

Example 6.33. *Let $T \in \mathscr{L}(\mathbb{C}^2)$ be defined as in Example 5.10. Then*

$$\textit{geometric multiplicity of } 2-2i = 1,$$

$$\textit{geometric multiplicity of } -i = 1.$$

Example 6.34. *Let $T \in \mathscr{L}(\mathbb{C}^2)$ be defined as in Example 5.11. Then*

$$\textit{geometric multiplicity of } 2 = 1.$$

Exercise 6.35. *Let $T \in \mathscr{L}(\mathbb{C}^2)$ be defined as in Exercise 5.14. Find the geometric multiplicities of the eigenvalues of T.*

Answer 6.36. 1.

Exercise 6.37. *Let $T \in \mathscr{L}(\mathbb{C}^3)$ be defined as in Exercise 5.16. Find the geometric multiplicities of the eigenvalues of T.*

Answer 6.38. 1.

6.6 Block Diagonal Matrices

Definition 6.39. *A block diagonal matrix is a square matrix of the form*

$$\begin{pmatrix} A_1 & & 0 \\ & \ddots & \\ 0 & & A_m \end{pmatrix},$$

where $A_1, \ldots, A_m$ are square matrices lying along the diagonal and all the other entries of the matrix equal 0.

Example 6.40. *The* 5×5 *matrix*

$$A = \begin{pmatrix} 4 & 0 & 0 & 0 & 0 \\ 0 & 2 & -3 & 0 & 0 \\ 0 & 0 & 2 & 0 & 0 \\ 0 & 0 & 0 & 1 & 7 \\ 0 & 0 & 0 & 0 & 1 \end{pmatrix}$$

is a block diagonal matrix with

$$\begin{pmatrix} A_1 & & 0 \\ & A_2 & \\ 0 & & A_3 \end{pmatrix},$$

where

$$\begin{aligned} A_1 &= (4), \\ A_2 &= \begin{pmatrix} 2 & -3 \\ 0 & 2 \end{pmatrix}, \\ A_3 &= \begin{pmatrix} 1 & 7 \\ 0 & 1 \end{pmatrix}. \end{aligned}$$

Theorem 6.41. *Suppose that* V *is a complex vector space and* $T \in \mathscr{L}(V)$. *Let* $\lambda_1, \ldots, \lambda_m$ *be distinct eigenvalues of* T *with multiplicities* $\lambda_1, \ldots, \lambda_m$, *respectively. Then there is a basis of* V *with respect to which* T *has a block diagonal matrix of the form*

$$\begin{pmatrix} A_1 & & 0 \\ & \ddots & \\ 0 & & A_m \end{pmatrix},$$

where each A_j *is a* $d_j \times d_j$ *upper triangular matrix of the form*

$$A_j = \begin{pmatrix} \lambda_j & & \star \\ & \lambda_j & \\ 0 & & \lambda_j \end{pmatrix}.$$

Proof. By Theorem 6.21, it follows that $(T - \lambda_j I)|_{G(\lambda_j, T)}$ is nilpotent for each $j \in \{1, \ldots, m\}$. For each $j \in \{1, \ldots, m\}$ we choose a basis of $G(\lambda_j, T)$

such that $(T-\lambda_j I)|_{G(\lambda_j,T)}$ with respect to this basis has the form

$$\begin{pmatrix} 0 & & \star \\ & \ddots & \\ 0 & & 0 \end{pmatrix}.$$

This is possible because we can apply Theorem 6.19. Then

$$\mathscr{M}\left(T|_{G(\lambda_j,T)}\right) = \mathscr{M}\left((T-\lambda_j I)|_{G(\lambda_j,T)}\right) + \mathscr{M}\left((\lambda_j I)|_{G(\lambda_j,T)}\right).$$

This completes the proof. □

6.7 Square Roots

Definition 6.42. *A square root of an operator $T \in \mathscr{L}(V)$ is an operator $R \in \mathscr{L}(V)$ such that $R^2 = T$.*

Theorem 6.43. *Suppose that $N \in \mathscr{L}(V)$ is nilpotent. Then $I+N$ has a square root.*

Proof. Consider the Taylor series of the function $\sqrt{1+x}$. We have that

$$\sqrt{1+x} = 1 + a_1 x + a_2 x^2 + \cdots .$$

Since N is nilpotent, there is an integer $m \in \mathbb{N}$ so that

$$N^m = 0.$$

Hence,

$$I + a_1 N + a_2 N^2 + \cdots = I + a_1 N + a_2 N^2 + \cdots + a_{m-1} N^{m-1}.$$

Note that we have that

$$\begin{aligned} I+N &= (I + a_1 N + a_2 N^2 + \cdots + a_{m-1} N^{m-1})^2 \\ &= I + 2a_1 N + (2a_2 + a_1^2) N^2 + (2a_3 + 2a_1 a_2) N^3 \\ &\quad + \cdots \end{aligned}$$

$$+(2a_{m-1} + \text{terms} \quad \text{involving} \quad a_1, \ldots, a_{m-2})N^{m-1}$$

if and only if

$$\begin{aligned} 2a_1 &= 1 \\ 2a_2 + a_1^2 &= 0 \\ 2a_3 + 2a_1a_2 &= 0 \\ &\vdots \\ 2a_{m-1} + \text{terms} \quad \text{involving} \quad a_1, \ldots, a_{m-2} &= 0 \end{aligned}$$

if and only if

$$\begin{aligned} a_1 &= \frac{1}{2} \\ a_2 &= -\frac{1}{8} \\ a_3 &= \frac{1}{16} \\ &\vdots \end{aligned}$$

This completes the proof. □

Theorem 6.44. *Suppose that V is a complex vector space and $T \in \mathscr{L}(V)$ is invertible. Then T has a square root.*

Proof. Let $\lambda_1, \ldots, \lambda_m$ be the distinct eigenvalues of the operator T. By Theorem 6.21, it follows that

$$(T - \lambda_j I)|_{G(\lambda_j, T)}, \quad j \in \{1, \ldots, m\},$$

is nilpotent. Hence, there exist nilpotent operators $N_j \in \mathscr{L}(V)$, $j \in \{1, \ldots, m\}$, such that

$$T|_{G(\lambda_j, T)} = \lambda_j I + N_j, \quad j \in \{1, \ldots, m\}.$$

Since T is invertible, we have that $\lambda_j \neq 0$, $j \in \{1,\ldots,m\}$. Then $\frac{N_j}{\lambda_j}$, $j \in \{1,\ldots,m\}$, is nilpotent. By Theorem 6.43, it follows that

$$I+\frac{N_j}{\lambda_j}, \quad j \in \{1,\ldots,m\},$$

has a square root. Because

$$T|_{G(\lambda_j,T)} = \lambda_j\left(I+\frac{N_j}{\lambda_j}\right), \quad j \in \{1,\ldots,m\},$$

we conclude that there is a square root R_j of $T|_{G(\lambda_j,T)}$, $j \in \{1,\ldots,m\}$. Let $v \in V$. Then it can be represented in the form

$$v = u_1+\cdots+u_m,$$

where $u_j \in G(\lambda_j,T)$, $j \in \{1,\ldots,m\}$. Define $R \in \mathscr{L}(V)$ such that

$$Rv = R_1u_1+\cdots+R_mu_m.$$

Hence,

$$\begin{aligned}
R^2v &= RRv \\
&= R(Rv) \\
&= R(R_1u_1+\cdots+R_mu_m) \\
&= R_1^2u_1+\cdots+R-m^2u_m \\
&= T|_{G(\lambda_1,T)}u_1+\cdots+T|_{G(\lambda_m,T)}u_m \\
&= T(u_1+\cdots+u_m) \\
&= Tv.
\end{aligned}$$

Thus, R is a square root of T. This completes the proof. □

Example 6.45. *Let $T \in \mathscr{L}(\mathbb{C}^2)$ be defined as in Example 6.13. By the computations in Example 6.13, it follows that*

$$T^2 = 0.$$

Then, using the proof of Theorem 6.43, we get

$$\begin{aligned}\sqrt{I+T}(z_1,z_2) &= \left(I+\frac{1}{2}T\right)(z_1,z_2)\\ &= (z_1,z_2)+\frac{1}{2}T(z_1,z_2)\\ &= (z_1,z_2)+\frac{1}{2}(2z_1-z_2,4z_1-2z_2)\\ &= (z_1,z_2)+\left(z_1-\frac{1}{2}z_2,2z_1-z_2\right)\\ &= \left(2z_1-\frac{1}{2}z_2,2z_1\right),(z_1,z_2)\in\mathbb{C}^2.\end{aligned}$$

Example 6.46. *Let $T \in \mathscr{L}(\mathbb{C}^4)$ be the operator defined as in Example 6.14. By the computations in Example 6.14, we have that*

$$T^4 = 0.$$

Then, using the proof of Theorem 6.43, we get

$$\begin{aligned}\sqrt{I+T}(z_1,z_2,z_3,z_4) &= \left(I+\frac{1}{2}T-\frac{1}{8}T^2+\frac{1}{16}T^3\right)(z_1,z_2,z_3,z_4)\\ &= (z_1,z_2,z_3,z_4)+\frac{1}{2}T(z_1,z_2,z_3,z_4)-\frac{1}{8}T^2(z_1,z_2,z_3,z_4)\\ &\quad+\frac{1}{16}T^3(z_1,z_2,z_3,z_4)\\ &= (z_1,z_2,z_3,z_4)+\frac{1}{2}(z_2+2z_3+4z_4,2z_3+z_4,5z_4,0)\\ &\quad-\frac{1}{8}(2z_3+11z_4,10z_4,0,0)+\frac{1}{16}(10z_4,0,0,0)\end{aligned}$$

$$
\begin{aligned}
&= (z_1,z_2,z_3,z_4)+\left(\frac{1}{2}z_2+z3+2z_4,z_3+\frac{1}{2}z_4,\frac{5}{2}z_4,0\right)\\
&\quad+\left(-\frac{1}{4}z_3-\frac{11}{8}z_4,-\frac{5}{4}z_4,0,0\right)+\left(\frac{5}{8}z_4,0,0,0\right)\\
&= \left(z_1+\frac{1}{2}z_2+z_3+2z_4-\frac{1}{4}z_3-\frac{11}{8}z_4,z_2+z_3+\frac{1}{2}z_4-\frac{5}{4}z_4,\right.\\
&\quad\left.z_3+\frac{5}{8}z_4-\frac{5}{4}z_4,z_4\right)\\
&= \left(z_1+\frac{1}{2}z_2-\frac{3}{4}z_3+\frac{5}{8}z_4,z_2+z_3-\frac{3}{4}z_4,z_3+\frac{5}{4}z_4,z_4\right),
\end{aligned}
$$

$(z_1,z_2,z_3,z_4)\in\mathbb{C}^4$.

Example 6.47. *Let $T\in\mathscr{L}(\mathbb{C}^3)$ be the operator defined as in Example 6.15. By the computations in Example 6.15, we find*

$$
\begin{aligned}
&\sqrt{I+T}(z_1,z_2,z_3)=\left(I+\frac{1}{2}T\right)(z_1,z_2,z_3)\\
&= (z_1,z_2,z_3)+\frac{1}{2}T(z_1,z_2,z_3)\\
&= (z_1,z_2,z_3)+\frac{1}{2}(5z_1-3z_2+2z_3,15z_1-9z_2+6z_3,10z_1-6z_2+4z_3)\\
&= (z_1,z,z_3)+\left(\frac{5}{2}z_1-\frac{3}{2}z_2+z_3,\frac{15}{2}z_1-\frac{9}{2}z_2+3z_3,5z_1-3z_2+2z_3\right)\\
&= \left(\frac{7}{2}z_1-\frac{3}{2}z_2+z_3,\frac{15}{2}z_1-\frac{7}{2}z_2+3z_3,5z_1-3z_2+3z_3\right),
\end{aligned}
$$

$(z_1,z_2,z_3)\in\mathbb{C}^3$.

Exercise 6.48. *Let $T\in\mathscr{L}(\mathbb{C}^3)$ be defined as in Exercise 6.16. Find*

$$\sqrt{I+T}(z_1,z_2,z_3),\quad (z_1,z_2,z_3)\in\mathbb{C}^3.$$

Answer 6.49.

$$\left(z_1+\frac{1}{2}z_2,z_2+\frac{1}{2}z_3,z_3\right),\quad (z_1,z_2,z_3)\in\mathbb{C}^3.$$

6.8 The Cayley-Hamilton Theorem

Definition 6.50. *Suppose that V is a complex vector space and $T \in \mathscr{L}(V)$. Let $\lambda_1, \ldots, \lambda_m$ be the distinct eigenvalues of T with multiplicities $d_1, \ldots, d_m$, respectively. The polynomial*

$$(z-\lambda_1)^{d_1} \ldots (z-\lambda_m)^{d_m}$$

is called the characteristic polynomial.

Example 6.51. *Let $T \in \mathscr{L}(\mathbb{C}^3)$ be defined as in Example 6.24. Then its characteristic polynomial is*

$$(z-6)^2(z-7).$$

Example 6.52. *Let $T \in \mathscr{L}(\mathbb{C}^2)$ be defined as in Example 5.10. Then its characteristic polynomial is*

$$(z-2+2i)(z+i).$$

Example 6.53. *Let $T \in \mathscr{L}(\mathbb{C}^3)$ be defined as in Example 5.11. Then its characteristic polynomial is*

$$(z-2)^3.$$

Exercise 6.54. *Let $T \in \mathscr{L}(\mathbb{C}^2)$ be defined as in Exercise 5.14. Find its characteristic polynomial.*

Answer 6.55.

$$(z+2-i)(z+2+i).$$

Exercise 6.56. *Let $T \in \mathscr{L}(\mathbb{C}^3)$ be defined as in Exercise 5.16. Find its characteristic polynomial.*

Answer 6.57.

1. $(z+1)^3$.
2. $(z-1)z^2$.

Remark 6.58. *Note that the degree of the characteristic polynomial of any operator $T \in \mathscr{L}(V)$ equals* $\dim V$. *Moreover, by the definition of the characteristic polynomial, it follows that its zeros are the eigenvalues of T.*

Theorem 6.59 (Cayley-Hamilton Theorem). *Suppose that V is a complex vector space and $T \in \mathscr{L}(V)$. Let q denote its characteristic polynomial. Then*

$$q(T) = 0. \tag{6.15}$$

Proof. Let $\lambda_1, \ldots, \lambda_m$ be the distinct eigenvalues of T and $d_1, \ldots, d_m$ be their multiplicities, respectively. By Theorem 6.21, it follows that

$$(T-\lambda_j I)|_{G(\lambda_j,T)}, \quad j \in \{1,\ldots,m\},$$

are nilpotent and

$$V = G(\lambda_1,T)\oplus\cdots\oplus G(\lambda_m,T),$$

where $G(\lambda_j,T)$, $j \in \{1,\ldots,m\}$, are the corresponding generalized eigenspaces corresponding to λ_j, $j \in \{1,\ldots,m\}$. Fix $j \in \{1,\ldots,m\}$. We have

$$q(T) = (T-\lambda_1 I)^{d_1}\ldots(T-\lambda_m I)^{d_m}.$$

The operators on the right hand side of the above equation commute and then we can move $(T-\lambda_j I)^{d_j}$ to be the last term in the expression above. Since

$$(T-\lambda_j I)^{d_j}|_{G(\lambda_j,T)} = 0,$$

we get

$$q(T)|_{G(\lambda_j,T)} = 0.$$

Therefore we have (6.15). This completes the proof. □

6.9 The Minimal Polynomial

Definition 6.60. *A monic polynomial is a polynomial whose highest degree coefficient equals* 1.

Example 6.61. *The polynomial*

$$z^6+1$$

is a monic polynomial of degree 6.

Example 6.62. *The polynomial*

$$3z^5-z^4+z^3$$

is not a monic polynomial. It is a polynomial of degree 5.

Example 6.63. *The polynomial*

$$z^9-2z^8+3z^7+4z^5-z^4+z^3+2z^2-7z+10$$

is a monic polynomial of degree 9.

Exercise 6.64. *Determine which polynomials are monic.*

1. $z^4 - 3z^2 + 1$.
2. $3z^4 + 2z - 10$.
3. $z^7 - z^6 + z$.
4. $z^5 + 3z^4 - z^2 + z$.
5. $z^6 - z^7 + 2$.

Answer 6.65.

1. *Yes.*
2. *No.*
3. *Yes.*
4. *Yes.*
5. *No.*

Theorem 6.66. *Suppose that $T \in \mathscr{L}(V)$. Then there is a unique monic polynomial p of smallest degree such that*

$$p(T) = 0.$$

Proof. Let $\dim V = n$. Then, by Theorem 4.11, we have

$$\begin{aligned} \dim \mathscr{L}(V) &= (\dim V)^2 \\ &= n^2. \end{aligned}$$

Then the list

$$I, T, T^2, \ldots, T^{n^2}$$

is linearly independent because it contains $n^2 + 1$ elements. Let m be the smallest integer such that the list

$$I, T, T^2, \ldots, T^m$$

is linearly dependent. Then one of its element is a linear combination of the previous ones. By the choice of the integer m, it follows that T^m is a linear combination of $I, T, T^2, \ldots, T^{m-1}$. Hence, there exist scalars $a_0, a_1, \ldots, a_{m-1}$ such that

$$a_0 I + a_1 T + a_2 T^2 + \cdots + a_{m-1} T^{m-1} + T^m = 0. \tag{6.16}$$

Define a monic polynomial $p \in \mathscr{P}(\mathbb{F})$ by

$$p(z) = a_0 + a_1 z + a_2 z^2 + \cdots + a_{m-1} z^{m-1} + z^m.$$

Then, by (6.16), we get

$$p(T) = 0.$$

Suppose that $q \in \mathscr{P}(\mathbb{F})$ is a monic polynomial such that

$$q(T) = 0.$$

Then

$$(p - q)(T) = 0$$

and

$$\deg(p - q) < m.$$

Thus, the choice of the integer m implies $p = q$. This completes the proof. □

Definition 6.67. *Suppose that $T \in \mathscr{L}(V)$. Then the minimal polynomial of T is the unique monic polynomial of smallest degree such that*

$$p(T) = 0.$$

Remark 6.68. *By the Cayley-Hamilton theorem, it follows that if V is a complex vector space, then the minimal polynomial of each operator on V has degree at most* $\dim V$.

Remark 6.69. *Suppose that we have a matrix of an operator $T \in \mathscr{L}(V)$. To find the minimal polynomial of T, we consider*

$$a_0 \mathscr{M}(I) + a_1 \mathscr{M}(T) + \cdots + a_{m-1} \mathscr{M}(T^{m-1}) = -\mathscr{M}(T^m)$$

for successive values $m = 1, 2, \ldots$ until this system has a solution

$$a_0, a_1, a_2, \ldots, a_{m-1}.$$

The scalars $a_0, a_1, a_2, \ldots, a_{m-1}, 1$ will then be the coefficients of the minimal polynomial of T.

Example 6.70. *Let the matrix of the operator $T \in \mathscr{L}(\mathbb{C}^2)$ be given by*

$$\mathscr{M}(T) = \begin{pmatrix} 1 & 1 \\ 1 & 1 \end{pmatrix}.$$

We will find the minimal polynomial of the operator T. Consider the equation

$$-\mathscr{M}(T) = a_0 \mathscr{M}(I),$$

where $a_0 \in \mathbb{F}$ is unknown. We have

$$-\begin{pmatrix} 1 & 1 \\ 1 & 1 \end{pmatrix} = a_0 \begin{pmatrix} 1 & 0 \\ 0 & 1 \end{pmatrix},$$

which is impossible. Note that

$$\begin{aligned} (\mathscr{M}(T))^2 &= \mathscr{M}(T)\mathscr{M}(T) \\ &= \begin{pmatrix} 1 & 1 \\ 1 & 1 \end{pmatrix}\begin{pmatrix} 1 & 1 \\ 1 & 1 \end{pmatrix} \\ &= \begin{pmatrix} 2 & 2 \\ 2 & 2 \end{pmatrix}. \end{aligned}$$

Consider the equation

$$-(\mathscr{M}(T))^2 = a_0 \mathscr{M}(T) + a_1 \mathscr{M}(T),$$

where $a_0, a_1 \in \mathbb{F}$ are unknown. We have

$$\begin{aligned} -\begin{pmatrix} 2 & 2 \\ 2 & 2 \end{pmatrix} &= a_0 \begin{pmatrix} 1 & 0 \\ 0 & 1 \end{pmatrix} + a_1 \begin{pmatrix} 1 & 1 \\ 1 & 1 \end{pmatrix} \\ &= \begin{pmatrix} a_0 & 0 \\ 0 & a_0 \end{pmatrix} + \begin{pmatrix} a_1 & a_1 \\ a_1 & a_1 \end{pmatrix} \\ &= \begin{pmatrix} a_0 + a_1 & a_1 \\ a_1 & a_0 + a_1 \end{pmatrix}, \end{aligned}$$

or

$$\begin{pmatrix} -2 & -2 \\ -2 & -2 \end{pmatrix} = \begin{pmatrix} a_0 + a_1 & a_1 \\ a_1 & a_0 + a_1 \end{pmatrix}.$$

Hence, we get the system

$$a_0 + a_1 = -2$$

$$a_1 = -2,$$

whereupon

$$a_0 = 0$$

$$a_1 = -2.$$

Thus, the minimal polynomial

$$p(z) = z^2 - 2z.$$

Example 6.71. *Let the matrix of the operator T be given by*

$$\mathscr{M}(T) = \begin{pmatrix} 2 & -1 & 2 \\ 5 & -3 & 3 \\ -1 & 0 & -2 \end{pmatrix}.$$

We will find the minimal polynomial of T. Consider the equation

$$a_0\mathscr{M}(I) = -\mathscr{M}(T),$$

where $a_0 \in \mathbb{F}$ is unknown. We have

$$-\begin{pmatrix} 2 & -1 & 2 \\ 5 & -3 & 3 \\ -1 & 0 & -2 \end{pmatrix} = a_0 \begin{pmatrix} 1 & 0 & 0 \\ 0 & 1 & 0 \\ 0 & 0 & 1 \end{pmatrix}$$

or

$$\begin{pmatrix} -2 & 1 & -2 \\ -5 & 3 & -3 \\ 1 & 0 & 2 \end{pmatrix} = \begin{pmatrix} a_0 & 0 & 0 \\ 0 & a_0 & 0 \\ 0 & 0 & a_0 \end{pmatrix},$$

which is impossible. Note that

$$\begin{aligned} (\mathscr{M}(T))^2 &= \begin{pmatrix} 2 & -1 & 2 \\ 5 & -3 & 3 \\ -1 & 0 & -2 \end{pmatrix} \begin{pmatrix} 2 & -1 & 2 \\ 5 & -3 & 3 \\ -1 & 0 & -2 \end{pmatrix} \\ &= \begin{pmatrix} -3 & 1 & -3 \\ -8 & 4 & -7 \\ 0 & 1 & 2 \end{pmatrix}. \end{aligned}$$

Consider the equation

$$a_0\mathcal{M}(T)+a_1\mathcal{M}(T)=-(\mathcal{M}(T))^2,$$

where $a_0, a_1 \in \mathbb{F}$ *are unknown. We have*

$$-\begin{pmatrix} -3 & 1 & -3 \\ -8 & 4 & -7 \\ 0 & 1 & 2 \end{pmatrix} = a_0 \begin{pmatrix} 1 & 0 & 0 \\ 0 & 1 & 0 \\ 0 & 0 & 1 \end{pmatrix} + a_1 \begin{pmatrix} 2 & -1 & 2 \\ 5 & -3 & 3 \\ -1 & 0 & -2 \end{pmatrix}$$

$$= \begin{pmatrix} a_0 & 0 & 0 \\ 0 & a_0 & 0 \\ 0 & 0 & a_0 \end{pmatrix} + \begin{pmatrix} 2a_1 & -a_1 & 2a_1 \\ 5a_1 & -3a_1 & 3a_1 \\ -a_1 & 0 & -2a_1 \end{pmatrix}$$

$$= \begin{pmatrix} a_0+2a_1 & -a_1 & 2a_1 \\ 5a_1 & a_0-3a_1 & 3a_1 \\ -a_1 & 0 & a_0-2a_1 \end{pmatrix}$$

or

$$\begin{pmatrix} 3 & -1 & 3 \\ 8 & -4 & 7 \\ 0 & -1 & -2 \end{pmatrix} = \begin{pmatrix} a_0+2a_1 & -a_1 & 2a_1 \\ 5a_1 & a_0-3a_1 & 3a_1 \\ -a_1 & 0 & a_0-2a_1. \end{pmatrix}$$

Hence, we get

$$\begin{aligned} a_0+2a_1 &= 3 \\ -a_1 &= -1 \\ 2a_1 &= 3, \end{aligned}$$

which is impossible. Next,

$$\begin{aligned} (\mathcal{M}(T))^3 &= \mathcal{M}(T)(\mathcal{M}(T))^2 \\ &= \begin{pmatrix} 2 & -1 & 2 \\ 5 & -3 & 3 \\ -1 & 0 & -2 \end{pmatrix} \begin{pmatrix} -3 & 1 & -3 \\ -8 & 4 & -7 \\ 0 & 1 & 2 \end{pmatrix} \\ &= \begin{pmatrix} 2 & 0 & 3 \\ -9 & -4 & 12 \\ 3 & -3 & -1 \end{pmatrix}. \end{aligned}$$

Consider the equation

$$a_0\mathscr{M}(I)+a_1\mathscr{M}(T)+a_2(\mathscr{M}(T))^2=-(\mathscr{M}(T))^3,$$

where $a_0,a_1,a_2\in\mathbb{F}$ are unknown. We have

$$-\begin{pmatrix} 2 & 0 & 3 \\ -9 & -4 & 12 \\ 3 & -3 & -1 \end{pmatrix} = a_0\begin{pmatrix} 1 & 0 & 0 \\ 0 & 1 & 0 \\ 0 & 0 & 1 \end{pmatrix} + a_1\begin{pmatrix} 2 & -1 & 2 \\ 5 & -3 & 3 \\ -1 & 0 & -2 \end{pmatrix}$$

$$+a_2\begin{pmatrix} -3 & 1 & -3 \\ -8 & 4 & -7 \\ 0 & 1 & 2 \end{pmatrix}$$

$$= \begin{pmatrix} a_0 & 0 & 0 \\ 0 & a_0 & 0 \\ 0 & 0 & a_0 \end{pmatrix} + \begin{pmatrix} 2a_1 & -a_1 & 2a_1 \\ 5a_1 & -3a_1 & 3a_1 \\ -a_1 & 0 & -2a_1 \end{pmatrix}$$

$$+\begin{pmatrix} -3a_2 & a_2 & -3a_2 \\ -8a_2 & 4a_2 & -7a_2 \\ 0 & a_2 & 2a_2 \end{pmatrix}$$

$$= \begin{pmatrix} a_0+2a_1 & -a_1 & 2a_1 \\ 5a_1 & a_0-3a_1 & 3a_1 \\ -a_1 & 0 & a_0-2a_1 \end{pmatrix} + \begin{pmatrix} -3a_2 & a_2 & -3a_2 \\ -8a_2 & 4a_2 & -7a_2 \\ 0 & a_2 & 2a_2 \end{pmatrix}$$

$$= \begin{pmatrix} a_0+2a_1-3a_2 & -a_1+a_2 & 2a_1-3a_2 \\ 5a_1-8a_2 & a_0-3a_1+4a_2 & 3a_1-7a_2 \\ -a_1 & a_2 & a_0-2a_1+2a_2 \end{pmatrix}$$

or

$$\begin{pmatrix} -2 & 0 & -3 \\ 9 & 4 & -12 \\ -3 & 3 & 1 \end{pmatrix} = \begin{pmatrix} a_0+2a_1-3a_2 & -a_1+a_2 & 2a_1-3a_2 \\ 5a_1-8a_2 & a_0-3a_1+4a_2 & 3a_1-7a_2 \\ -a_1 & a_2 & a_0-2a_1+2a_2 \end{pmatrix}.$$

Hence, we get the system

$$\begin{aligned} a_0+2a_1-3a_2 &= -2 \\ -a_1+a_2 &= 0 \\ 2a_1-3a_2 &= -3 \end{aligned}$$

$$\begin{aligned} 5a_1 - 8a_2 &= 9 \\ a_0 - 3a_1 + 4a_2 &= 4 \\ 3a_1 - 7a_2 &= -12 \\ -a_1 &= -3 \\ a_2 &= 3. \end{aligned}$$

Therefore

$$\begin{aligned} a_0 &= 1 \\ a_1 &= 3 \\ a_2 &= 3. \end{aligned}$$

The minimal polynomial of T is

$$p(z) = z^3 + 3z^2 + 3z + 1, \quad z \in \mathbb{F}.$$

Example 6.72. *Let the matrix of the operator $T \in \mathscr{L}(\mathbb{C}^3)$ with respect to the standard basis be given by*

$$\mathscr{M}(T) = \begin{pmatrix} 0 & 1 & 0 \\ -4 & 4 & 0 \\ -2 & 1 & 2 \end{pmatrix}.$$

We will find the minimal polynomial of T. Consider the equation

$$-\mathscr{M}(T) = a_0 \mathscr{I},$$

where $a_0 \in \mathbb{F}$ is unknown. We have

$$-\begin{pmatrix} 0 & 1 & 0 \\ -4 & 4 & 0 \\ -2 & 1 & 2 \end{pmatrix} = a_0 \begin{pmatrix} 1 & 0 & 0 \\ 0 & 1 & 0 \\ 0 & 0 & 1 \end{pmatrix}$$

or

$$\begin{pmatrix} 0 & -1 & 0 \\ 4 & -4 & 0 \\ 2 & -1 & -2 \end{pmatrix} = a_0 \begin{pmatrix} 1 & 0 & 0 \\ 0 & 1 & 0 \\ 0 & 0 & 1 \end{pmatrix},$$

which is impossible. Next,

$$\begin{aligned} (\mathscr{M}(T))^2 &= \mathscr{M}(T)\mathscr{M}(T) \\ &= \begin{pmatrix} 0 & 1 & 0 \\ -4 & 4 & 0 \\ -2 & 1 & 2 \end{pmatrix} \begin{pmatrix} 0 & 1 & 0 \\ -4 & 4 & 0 \\ -2 & 1 & 2 \end{pmatrix} \\ &= \begin{pmatrix} -4 & 4 & 0 \\ -16 & 12 & 0 \\ -8 & 4 & 4 \end{pmatrix}. \end{aligned}$$

Now, we consider the equation

$$-(\mathscr{M}(T))^2 = a_0\mathscr{M}(I) + a_1\mathscr{M}(T).$$

We have

$$\begin{aligned} -\begin{pmatrix} -4 & 4 & 0 \\ -16 & 12 & 0 \\ -8 & 4 & 4 \end{pmatrix} &= a_0 \begin{pmatrix} 1 & 0 & 0 \\ 0 & 1 & 0 \\ 0 & 0 & 1 \end{pmatrix} + a_1 \begin{pmatrix} 0 & 1 & 0 \\ -4 & 4 & 0 \\ -2 & 1 & 2 \end{pmatrix} \\ &= \begin{pmatrix} a_0 & 0 & 0 \\ 0 & a_0 & 0 \\ 0 & 0 & a_0 \end{pmatrix} + \begin{pmatrix} 0 & a_1 & 0 \\ -4a_1 & 4a_1 & 0 \\ -2a_1 & a_1 & 2a_1 \end{pmatrix} \\ &= \begin{pmatrix} a_0 & a_1 & 0 \\ -4a_1 & a_0+4a_1 & 0 \\ -2a_1 & a_1 & a_0+2a_1 \end{pmatrix}, \end{aligned}$$

or

$$\begin{pmatrix} 4 & -4 & 0 \\ 16 & -12 & 0 \\ 8 & -4 & -4 \end{pmatrix} = \begin{pmatrix} a_0 & a_1 & 0 \\ -4a_1 & a_0+4a_1 & 0 \\ -2a_1 & a_1 & a_0+2a_1 \end{pmatrix}.$$

Thus, we get the system

$$a_0 = 4$$

$$\begin{aligned} a_1 &= -4 \\ -4a_1 &= 16 \\ a_0+4a_1 &= -12 \\ -2a_1 &= 8 \\ a_1 &= -4 \\ a_0+2a_1 &= -4. \end{aligned}$$

Therefore the minimal polynomial of T is

$$p(z) = z^2 - 4z + 4, \quad z \in \mathbb{F}.$$

Example 6.73. *Let the matrix of the operator $T \in \mathscr{L}(\mathbb{C}^3)$ be given by*

$$\mathscr{M}(T) = \begin{pmatrix} -2 & 0 & 0 \\ 0 & -2 & 0 \\ 0 & 0 & -2 \end{pmatrix}.$$

We will find the minimal polynomial of T. We have

$$\begin{aligned} -\mathscr{M}(T) &= -\begin{pmatrix} -2 & 0 & 0 \\ 0 & -2 & 0 \\ 0 & 0 & -2 \end{pmatrix} \\ &= \begin{pmatrix} 2 & 0 & 0 \\ 0 & 2 & 0 \\ 0 & 0 & 2 \end{pmatrix} \\ &= 2\begin{pmatrix} 1 & 0 & 0 \\ 0 & 1 & 0 \\ 0 & 0 & 1 \end{pmatrix} \\ &= 2\mathscr{M}(I). \end{aligned}$$

Therefore the minimal polynomial of T is

$$p(z) = z + 2, \quad z \in \mathbb{F}.$$

Exercise 6.74. *Find the minimal polynomial of the operator T, where*

1.
$$\mathscr{M}(T) = \begin{pmatrix} 2 & 2 \\ -2 & -2 \end{pmatrix},$$

2.
$$\mathscr{M}(T) = \begin{pmatrix} 4 & -5 & 2 \\ 5 & -7 & 3 \\ 6 & -9 & 4 \end{pmatrix}.$$

Answer 6.75. 1. $p(z) = z^2$, $z \in \mathbb{F}$.

2. $p(z) = z^3 - z^2$, $z \in \mathbb{F}$.

Theorem 6.76. *Suppose that $T \in \mathscr{L}(V)$ and $q \in \mathscr{P}(\mathbb{F})$. Then $q(T) = 0$ if and only if q is a polynomial multiple of the minimal polynomial of T.*

Proof. Let p be the minimal polynomial of T.

1. Let $q \in \mathscr{P}(\mathbb{F})$ be a polynomial multiple of p. Then there is a polynomial $r \in \mathscr{P}(\mathbb{F})$ such that

$$q(z) = p(z)r(z), \quad z \in \mathbb{F}.$$

Hence,

$$\begin{aligned} q(T) &= p(T)r(T) \\ &= 0. \end{aligned}$$

2. Suppose that

$$q(T) = 0.$$

By the division algorithm for polynomials, it follows that there exist polynomials $r, s \in \mathscr{P}(\mathbb{F})$ such that

$$q = ps + r$$

and $\deg r < \deg p$. Then

$$\begin{aligned} 0 &= q(T) \\ &= p(T)s(T) + r(T) \end{aligned}$$

$$= \quad r(T).$$

Hence, using that $\deg r < \deg p$, we get that $r = 0$. Thus,

$$q = ps,$$

i.e., q is multiple of p. This completes the proof.

□

Theorem 6.77. *Suppose that $\mathbb{F} = \mathbb{C}$ and $T \in \mathscr{L}(V)$. Then the characteristic polynomial of T is a polynomial multiple of the minimal polynomial of T.*

Proof. Let p and q be the minimal polynomial and the characteristic polynomial of T, respectively. By the Cayley-Hamilton theorem, Theorem 6.59, it follows that

$$q(T) = 0.$$

Hence and Theorem 6.76, we conclude that q is multiple of p. This completes the proof. □

Theorem 6.78. *Let $T \in \mathscr{L}(V)$. Then the zeros of the minimal polynomial p of T are precisely the eigenvalues of T.*

Proof. Let

$$p(z) = z^m + a_{m-1}z^{m-1} + \cdots + a_z + a_0, \quad z \in \mathbb{F}.$$

Suppose that $\lambda \in \mathbb{F}$ is a zero of p. Then

$$p(z) = (z - \lambda)q(z), \quad z \in \mathbb{F},$$

where $q \in \mathscr{P}(\mathbb{F})$ so that $\deg q < \deg p$. Since

$$p(T) = 0,$$

we get

$$(T - 0\lambda I)q(T)v = 0$$

for any $v \in V$. Since $\deg q < \deg p$, it follows that there exists a $v_1 \in V$ so that

$$q(T)v_1 \neq 0.$$

Therefore

$$(T - \lambda I)v_1 = 0,$$

i.e., λ is an eigenvalue of T. Let $\mu \in \mathbb{F}$ be an eigenvalue of T. Then there is a $v_2 \in V$, $v_2 \neq 0$, so that

$$Tv_2 = \mu v_2.$$

Hence,

$$T^j v_2 = \mu^j v_2$$

for any $j \in \mathbb{N}$. Thus,

$$\begin{aligned}
0 &= p(T)v_2 \\
& (a_0 I + a_1 T + \cdots + a_{m-1}T^{m-1} + T^m)v_2 \\
&= a_0 v_2 + a_1 T v_2 + \cdots a_{m-1}T^{m-1}v_2 + T^m v_2 \\
&= a_0 v_2 + a_1 \mu v_2 + \cdots + a_{m-1}\mu^{m-1}v_2 + \mu_2^m \\
&= (a_0 + a_1\mu + \cdots + a_{m-1}\mu^{m-1} + \mu^m)v_2 \\
&= p(\mu)v_2.
\end{aligned}$$

Since, $v_2 \neq 0$, we obtain

$$p(\mu) = 0.$$

Thus, μ is a zero of p. This completes the proof. □

Example 6.79. *Let $T \in \mathscr{L}(\mathbb{C}^3)$ be as in Example 6.73. Then $\lambda = -2$ is the eigenvalue of T with multiplicity 3. The characteristic polynomial of T is the polynomial*

$$q(z) = (z+2)^3, \quad z \in \mathbb{F}.$$

The minimal polynomial of T is

$$p(z) = z + 2, \quad z \in \mathbb{F}.$$

Theorem 6.80. *Let $N \in \mathscr{L}(V)$ be nilpotent. Then there exist vectors $v_1, \ldots, v_n \in V$ and nonnegative integers $m_1, \ldots, m_n$ such that*

1.

$$N^{m_1}v_1, \ldots, Nv_1, v_1, \ldots, N^{m_n}v_n, \ldots, Nv_n, v_n \tag{6.17}$$

is a basis of V.

2.
$$N^{m_1+1}v_1 = \ldots = N^{m_n+1}v_n = 0. \tag{6.18}$$

Proof. We will prove the assertion by induction on $\dim V$.

1. Let $\dim V = 1$. Then the only nilpotent operator is the zero operator and we take v_1 to be any nonzero vector. Then $m_1 = 0$.

2. Assume that $\dim V > 1$ and that the desired result holds on all vector spaces of smaller dimension. Since N is nilpotent, it is not injective. Thus, N is not surjective. Therefore range$N$ is a subspace of V that has a smaller dimension than V. Then we can apply the induction hypothesis to the restriction operator $N|_{\mathrm{range}N} \in \mathscr{L}(\mathrm{range}N)$. Hence, there exist vectors $v_1, \ldots, v_n \in \mathrm{range}N$ and nonnegative integers $m_1, \ldots, m_n$ such that (6.17) and (6.18) hold. Since $v_j \in \mathrm{range}N$, $j \in \{1, \ldots, n\}$, there exists $u_j \in V$, $j \in \{1, \ldots, n\}$, such that
$$v_j = Nu_j, \quad j \in \{1, \ldots, n\}.$$
Thus,
$$N^{k+1}u_j = N^k v_j, \quad j \in \{1, \ldots, n\},$$
for each nonnegative integer k. Consider the list
$$N^{m_1+1}u_1, \ldots, Nu_1, u_1, \ldots, N^{m_n+1}u_n, \ldots, Nu_n, u_n. \tag{6.19}$$
Suppose that some linear combination of (6.19) equals zero. Applying N to this linear combination, we get a linear combination (6.17) equals 0. Since (6.17) is linearly independent, all coefficients of the original linear combination equal 0 except possibly the coefficients of the vectors
$$N^{m_1+1}u_1, \ldots, N^{m_n+1}u_n$$
which equal the vectors
$$N^{m_1}v_1, \ldots, N^{m_n}v_n.$$
Again using that the list (6.17) is linearly independent, we conclude that these coefficients also equal 0. Therefore (6.19) is linearly independent. We extend (6.19) to a basis
$$N^{m_1+1}u_1, \ldots, Nu_1, u_1, \ldots, N^{m_n+1}u_n, \ldots, Nu_n, u_n, w_1, \ldots, w_p \tag{6.20}$$

of V. Each $Nw_j \in \text{range}N$, $j \in \{1,\ldots,p\}$, and hence it belongs to the span of (6.17). Then there exists x_j, $j \in \{1,\ldots,p\}$, in the span of (6.19) such that

$$Nw_j = Nx_j, \quad j \in \{1,\ldots,p\}.$$

Let

$$u_{n+j} = w_j - x_j, \quad j \in \{1,\ldots,p\}.$$

Then

$$Nu_{n+j} = 0, \quad j \in \{1,\ldots,p\},$$

and

$$N^{m_1+1}u_1,\ldots,Nu_1,u_1,\ldots,N^{m_n+1}u_n,\ldots,Nu_n,u_n,u_{n+1},\ldots,u_{n+p} \tag{6.21}$$

spans V because it contains each x_j and each u_{n+j}, $j \in \{1,\ldots,p\}$, and hence, it contains each w_j and (6.20) spans V. Therefore (6.21) is a basis of V. This basis has the required form. This completes the proof.

□

Definition 6.81. *Suppose that $T \in \mathscr{L}(V)$. A basis of V is called a Jordan basis for T if with respect to this basis T has a block diagonal matrix*

$$\begin{pmatrix} A_1 & & 0 \\ & \ddots & \\ 0 & & A_p \end{pmatrix},$$

where each A_j, $j \in \{1,\ldots,p\}$, is an upper-triangular matrix of the form

$$A_j = \begin{pmatrix} \lambda_j & 1 & & 0 \\ & \ddots & \ddots & \\ & & \ddots & 1 \\ 0 & & & \lambda_j \end{pmatrix}, \quad j \in \{1,\ldots,p\}.$$

Theorem 6.82. *Suppose that V is a complex vector space. If $T \in \mathscr{L}(V)$, then there is a basis of V that is a Jordan basis for T.*

Proof. 1. Let $N \in \mathscr{L}(V)$ and let $v_1, \ldots, v_n \in V$ be given as in Theorem 6.80. For each $j \in \{1, \ldots, n\}$, N sends the first vector in the list

$$N^{m_j} v_j, \ldots, N v_j, v_j$$

to 0 and N sends each vector in the list other than the first vector to the previous vector. Thus, (6.17) gives a basis of V with respect to which N has a block diagonal matrix in the form

$$\begin{pmatrix} 0 & 1 & & 0 \\ & \ddots & \ddots & \\ & & \ddots & 1 \\ 0 & & & 0 \end{pmatrix}.$$

Thus, the result holds for N.

2. Let now, $T \in \mathscr{L}(V)$ and $\lambda_1, \ldots, \lambda_m$ be the distinct eigenvalues of T. We have that

$$V = G(\lambda_1, T) \oplus \cdots \oplus G(\lambda_m, T)$$

and $(T - \lambda_j I)|_{G(\lambda_j, T)}$ is nilpotent. Thus, some basis of each $G(\lambda_j, T)$, $j \in \{1, \ldots, m\}$, is a Jordan basis for $(T - \lambda_j I)|_{G(\lambda_j, T)}$. Put these bases together to get a basis of V that is a Jordan basis for T. This completes the proof.

□

6.10 An Algorithm for Jordan Canonical Form

Let V be a complex vector space with $\dim V = n$ and $e_1, \ldots, e_n$ be a basis of V. Let also, $T \in \mathscr{L}(V)$ and $\mathscr{M}(T, (e_1, \ldots, e_n), (e_1, \ldots, e_n))$ be the matrix of T with respect to the basis $e_1, \ldots, e_n$.

1. We find the distinct eigenvalues of T: $\lambda_1, \ldots, \lambda_m$.

2. We find

$$G(\lambda_1, T), \ldots, G(\lambda_m, T)$$

and their dimensions $d_1, \ldots, d_m$, respectively.

3. Take $v_j \in G(\lambda_j, T)$, $v_j \neq 0$, $j \in \{1, \ldots, m\}$, arbitrarily. Then we find nonnegative integers m_j, $j \in \{1, \ldots, m\}$, so that

$$(T - \lambda_j I) v_j, \ldots, (T - \lambda_j)^{m_j} v_j \neq 0, \quad j \in \{1, \ldots, m\}.$$

By Theorem 6.21, we have

$$(T-\lambda_j I)^{m_j+1}v_j = 0, \quad j \in \{1,\dots,m\}.$$

4. Form a Jordan basis for T

$$(T-\lambda_1 I)^{m_1}v_1,\dots,(T-\lambda_1 I)v_1, v_1,\dots,$$

$$(T-\lambda_m I)^{m_m}v_m,\dots,(T-\lambda_m I)v_m, v_m.$$

Set

$$\begin{aligned} u_1 &= v_1, \\ &\vdots \\ u_{d_1} &= (T-\lambda_1 I)^{m_1}, \\ &\vdots \\ u_{n-d_m} &= v_m \\ &\vdots \\ u_n &= (T-\lambda_m I)^{m_m}v_m. \end{aligned}$$

5. Find the matrices

$$\mathscr{M}(I,(e_1,\dots,e_n),(u_1,\dots,u_n))$$

and

$$\mathscr{M}(I,(u_1,\dots,u_n),(e_1,\dots,e_n)).$$

6. By Theorem 6.82, the matrix

$$\mathscr{M}(I,(u_1,\dots,u_n),(e_1,\dots,e_n))\mathscr{M}(T,(e_1,\dots,e_n),(e_1,\dots,e_n))$$

$$\times\mathscr{M}(I,(e_1,\dots,e_n),(u_1,\dots,u_n))$$

is the Jordan canonical form of the matrix of the operator T.

Example 6.83. *Let $T \in \mathscr{L}(\mathbb{C}^3)$ be defined as in Example 6.24. By the computations in Example 6.24, we have that $\lambda_1 = 6$ and $\lambda_2 = 7$ are the eigenvalues of T, and*

$$G(6,T) = \{(z_1,z_2,0) : z_1,z_2 \in \mathbb{C}\},$$

$$G(7,T) = \{(10z_3,2z_3,z_3) : z_3 \in \mathbb{C}\},$$

and

$$\dim G(6,T) = 2,$$

$$\dim G(7,T) = 1,$$

and

$$(T-6I)(z_1,z_2,z_3) = (3z_2+4z_3,2z_3,z_3),$$

$$(T-7I)(z_1,z_2,z_3) = (3z_2+4z_3-z_1,2z_3-z_2,0), \quad (z_1,z_2,z_3) \in \mathbb{C}^3.$$

With e_1,e_2,e_3 we will denote the standard basis in $\mathbb{C}^3$, i.e.,

$$e_1 = (1,0,0),$$

$$e_2 = (0,1,0),$$

$$e_3 = (0,0,1).$$

Then

$$\mathscr{M}(T,(e_1,e_2,e_3),(e_1,e_2,e_3)) = \begin{pmatrix} 6 & 0 & 0 \\ 3 & 6 & 0 \\ 4 & 2 & 7 \end{pmatrix}.$$

Take

$$v_1 = (1,1,0).$$

Then

$$(T-6I)v_1 = (T-6I)(1,1,0)$$

$$= (3,0,0),$$

and

$$\begin{aligned}(T-6I)^2v_1 &= (T-6I)(T-6I)v_1\\ &= (T-6I)((T-6I)(1,1,0))\\ &= (T-6I)(3,0,0)\\ &= (0,0,0).\end{aligned}$$

Next, take

$$v_2 = (10,2,1).$$

Then

$$\begin{aligned}(T-7I)v_2 &= (T-7I)(10,2,1)\\ &= (3\cdot 2+4\cdot 1-10, 2\cdot 1-2, 0)\\ &= (0,0,0).\end{aligned}$$

Set

$$\begin{aligned}u_1 &= (1,1,0),\\ u_2 &= (3,0,0),\\ u_3 &= (10,2,1).\end{aligned}$$

Let $a_{ij}\in\mathbb{C}$, $i,j\in\{1,2,3\}$, be such that

$$\begin{aligned}e_1 &= a_{11}u_1+a_{12}u_2+a_{13}u_3\\ e_2 &= a_{21}u_1+a_{22}u_2+a_{23}u_3\\ e_3 &= a_{31}u_1+a_{32}u_2+a_{33}u_3.\end{aligned}$$

Then

$$(1,0,0) = a_{11}(1,1,0)+a_{12}(3,0,0)+a_{13}(10,2,1)$$

$$\begin{aligned}(0,1,0) &= a_{21}(1,1,0)+a_{22}(3,0,0)+a_{23}(10,2,1)\\(0,0,1) &= a_{31}(1,1,0)+a_{32}(3,0,0)+a_{33}(10,2,1),\end{aligned}$$

or

$$\begin{aligned}(1,0,0) &= (a_{11},a_{11},0)+(3a_{12},0,0)+(10a_{13},2a_{13},a_{13})\\(0,1,0) &= (a_{21},a_{21},0)+(3a_{22},0,0)+(10a_{23},2a_{23},a_{23})\\(0,0,1) &= (a_{31},a_{31},0)+(3a_{32},0,0)+(10a_{33},2a_{33},a_{33}),\end{aligned}$$

or

$$\begin{aligned}(1,0,0) &= (a_{11}+3a_{12}+10a_{13},a_{11}+2a_{13},a_{13})\\(0,1,0) &= (a_{21}+3a_{22}+10a_{23},a_{21}+2a_{23},a_{23})\\(0,0,1) &= (a_{31}+3a_{32}+10a_{33},a_{31}+2a_{33},a_{33}).\end{aligned}$$

Therefore

$$\begin{aligned}a_{11}+3a_{12}+10a_{13} &= 1\\a_{11}+2a_{13} &= 0\\a_{13} &= 0,\end{aligned}$$

and

$$\begin{aligned}a_{21}+3a_{22}+10a_{23} &= 0\\a_{21}+2a_{23} &= 1\\a_{23} &= 0,\end{aligned}$$

and

$$a_{31}+3a_{32}+10a_{33} = 0$$

$$\begin{aligned} a_{31} + 2a_{33} &= 0 \\ a_{33} &= 1. \end{aligned}$$

Hence,

$$\begin{aligned} a_{11} &= 0 \\ a_{12} &= \frac{1}{3} \\ a_{13} &= 0, \end{aligned}$$

and

$$\begin{aligned} a_{21} &= 1 \\ a_{22} &= -\frac{1}{3} \\ a_{23} &= 0, \end{aligned}$$

and

$$\begin{aligned} a_{31} &= -2 \\ a_{32} &= -\frac{8}{3} \\ a_{33} &= 1. \end{aligned}$$

Consequently

$$\begin{aligned} e_1 &= \frac{1}{3}u_2 \\ e_2 &= u_1 - \frac{1}{3}u_2 \\ e_3 &= -2u_1 - \frac{8}{3}u_2 + u_3. \end{aligned}$$

Therefore

$$\mathscr{M}(I,(u_1,u_2,u_3),(e_1,e_2,e_3)) = \begin{pmatrix} 1 & 1 & 0 \\ 3 & 0 & 0 \\ 10 & 2 & 1 \end{pmatrix},$$

$$\mathscr{M}(I,(e_1,e_2,e_3),(u_1,u_2,u_3)) = \begin{pmatrix} 0 & \frac{1}{3} & 0 \\ 1 & -\frac{1}{3} & 0 \\ -2 & -\frac{8}{3} & 1 \end{pmatrix}.$$

The Jordan canonical form of the matrix of T is

$$\mathscr{M}(I,(u_1,u_2,u_3),(e_1,e_2,e_3))\mathscr{M}(T,(e_1,e_2,e_3),(e_1,e_2,e_3))$$
$$\times\mathscr{M}(I,(e_1,e_2,e_3),(u_1,u_2,u_3))$$
$$= \begin{pmatrix} 1 & 1 & 0 \\ 3 & 0 & 0 \\ 10 & 2 & 1 \end{pmatrix}\begin{pmatrix} 6 & 0 & 0 \\ 3 & 6 & 0 \\ 4 & 2 & 7 \end{pmatrix}\begin{pmatrix} 0 & \frac{1}{3} & 0 \\ 1 & -\frac{1}{3} & 0 \\ -2 & -\frac{8}{3} & 1 \end{pmatrix}$$
$$= \begin{pmatrix} 1 & 1 & 0 \\ 3 & 0 & 0 \\ 10 & 2 & 1 \end{pmatrix}\begin{pmatrix} 0 & 2 & 0 \\ 6 & -1 & 0 \\ -12 & -18 & 7 \end{pmatrix}$$
$$= \begin{pmatrix} 6 & 1 & 0 \\ 0 & 6 & 0 \\ 0 & 0 & 7 \end{pmatrix}.$$

Example 6.84. *Let $T \in \mathscr{L}(\mathbb{C}^3)$ be defined as in Example 6.25. Then $\lambda = 2$ is the unique eigenvalue of T and*

$$G(2,T) = \mathbb{C}^3,$$

$$\dim G(2,T) = 3.$$

The matrix of T with respect to the standard basis is

$$\mathscr{M}(T,(e_1,e_2,e_3),(e_1,e_2,e_3)) = \begin{pmatrix} 0 & -4 & -2 \\ 1 & 4 & 1 \\ 0 & 0 & 2 \end{pmatrix}.$$

Take

$$\begin{aligned} v_1 &= (1,0,0), \\ v_2 &= (0,0,1). \end{aligned}$$

Then

$$\begin{aligned} (T-2I)v_1 &= (-2,-4,-2), \\ (T-2I)^2 v_1 &= (T-2I)(T-2I)v_1 \\ &= (T-2I)((T-2I)v_1) \\ &= (0,0,0), \\ (T-2I)v_2 &= (0,0,0). \end{aligned}$$

Set

$$\begin{aligned} u_1 &= (1,0,0), \\ u_2 &= (-2,-4,-2), \\ u_3 &= (0,0,1). \end{aligned}$$

Then

$$\mathscr{M}(I,(u_1,u_2,u_3),(e_1,e_2,e_3)) = \begin{pmatrix} 1 & 0 & 0 \\ -2 & -4 & -2 \\ 0 & 0 & 1 \end{pmatrix}.$$

Let $a_{ij} \in \mathbb{C}$, $i,j \in \{1,2,3\}$, be such that

$$\begin{aligned} e_1 &= a_{11}u_1 + a_{12}u_2 + a_{13}u_3 \\ e_2 &= a_{21}u_1 + a_{22}u_2 + a_{23}u_3 \\ e_3 &= a_{31}u_1 + a_{32}u_2 + a_{33}u_3. \end{aligned}$$

Then

$$\begin{aligned}(1,0,0) &= a_{11}(1,0,0)+a_{12}(-2,-4,-2)+a_{13}(0,0,1)\\(0,1,0) &= a_{21}(1,0,0)+a_{22}(-2,-4,-2)+a_{23}(0,0,1)\\(0,0,1) &= a_{31}(1,0,0)+a_{32}(-2,-4,-2)+a_{33}(0,0,1),\end{aligned}$$

or

$$\begin{aligned}(1,0,0) &= (a_{11},0,0)+(-2a_{12},-4a_{12},-2a_{12})+(0,0,a_{13})\\(0,1,0) &= (a_{21},0,0)+(-2a_{22},-4a_{22},-2a_{22})+(0,0,a_{23})\\(0,0,1) &= (a_{31},0,0)+(-2a_{32}-4a_{32}-2a_{32})+(0,0,a_{33}),\end{aligned}$$

or

$$\begin{aligned}(1,0,0) &= (a_{11}-2a_{12},-4a_{12},-2a_{12}+a_{13})\\(0,1,0) &= (a_{21}-2a_{22},-4a_{22},-2a_{22}+a_{23})\\(0,0,1) &= (a_{31}-2a_{32},-4a_{32},-2a_{32}+a_{33}).\end{aligned}$$

Thus, we get the systems

$$\begin{aligned}a_{11}-2a_{12} &= 1\\-4a_{12} &= 0\\-2a_{12}+a_{13} &= 0,\end{aligned}$$

and

$$\begin{aligned}a_{21}-2a_{22} &= 0\\-4a_{22} &= 1\end{aligned}$$

$$-2a_{22}+a_{23} = 0,$$

and

$$a_{31}-2a_{32} = 0$$

$$-4a_{32} = 0$$

$$-2a_{32}+a_{33} = 1.$$

Therefore

$$a_{11} = 1$$

$$a_{12} = 0$$

$$a_{13} = 0,$$

and

$$a_{21} = -\frac{1}{2}$$

$$a_{22} = -\frac{1}{4}$$

$$a_{23} = -\frac{1}{2},$$

and

$$a_{31} = 0$$

$$a_{32} = 0$$

$$a_{33} = 1.$$

Hence,

$$e_1 = u_1$$

$$e_2 = -\frac{1}{2}u_1 - \frac{1}{4}u_2 - \frac{1}{2}u_3$$

$$e_3 = u_3$$

and

$$\mathscr{M}(I,(e_1,e_2,e_3),(u_1,u_2,u_3)) = \begin{pmatrix} 1 & 0 & 0 \\ -\frac{1}{2} & -\frac{1}{4} & -\frac{1}{2} \\ 0 & 0 & 1 \end{pmatrix}.$$

Consequently the Jordan canonical form of the matrix of T is

$$\mathscr{M}(I,(u_1,u_2,u_3),(e_1,e_2,e_3))\mathscr{M}(T,(e_1,e_2,e_3),(e_1,e_2,e_3))$$
$$\mathscr{M}(I,(e_1,e_2,e_3),(u_1,u_2,u_3))$$

$$= \begin{pmatrix} 1 & 0 & 0 \\ -2 & -4 & -2 \\ 0 & 0 & 1 \end{pmatrix}\begin{pmatrix} 0 & -4 & -2 \\ 1 & 4 & 1 \\ 0 & 0 & 2 \end{pmatrix}\begin{pmatrix} 1 & 0 & 0 \\ -\frac{1}{2} & -\frac{1}{4} & -\frac{1}{2} \\ 0 & 0 & 1 \end{pmatrix}$$

$$= \begin{pmatrix} 1 & 0 & 0 \\ -2 & -4 & -2 \\ 0 & 0 & 1 \end{pmatrix}\begin{pmatrix} 2 & 1 & 0 \\ -1 & -1 & -1 \\ 0 & 0 & 2 \end{pmatrix}$$

$$= \begin{pmatrix} 2 & 1 & 0 \\ 0 & 2 & 0 \\ 0 & 0 & 2 \end{pmatrix}.$$

Example 6.85. *Let the matrix of the operator $T \in \mathscr{L}(\mathbb{C}^4)$ be given by the matrix*

$$\begin{pmatrix} 0 & 1 & -1 & 1 \\ 0 & 0 & 1 & 0 \\ 0 & -1 & 2 & 0 \\ -2 & 1 & -2 & 3 \end{pmatrix}.$$

We will find its Jordan canonical form. We have

$$T(z_1,z_2,z_3,z_4) = (-2z_4, z_1 - z_3 + z_4, -z_1 + z_2 + 2z_3 - 2z_4, z_1 + 3z_4),$$

$(z_1,z_2,z_3,z_4) \in \mathbb{C}^4$. *Let* $(z_1,z_2,z_3,z_4) \in \mathbb{C}^4$, $(z_1,z_2,z_3,z_4) \neq (0,0,0,0)$, *and* $\lambda \in \mathbb{C}$ *be such that*

$$T(z_1,z_2,z_3,z_4) = \lambda(z_1,z_2,z_3,z_4).$$

Then

$$(-2z_4, z_1 - z_3 + z_4, -z_1 + z_2 + 2z_3 - 2z_4, z_1 + 3z_4) = (\lambda z_1, \lambda z_2, \lambda z_3, \lambda z_4).$$

We get the system

$$\begin{aligned} -2z_4 &= \lambda z_1 \\ z_1 - z_3 + z_4 &= \lambda z_2 \\ -z_1 + z_2 + 2z_3 - 2z_4 &= \lambda z_3 \\ z_1 + 3z_4 &= \lambda z_4, \end{aligned}$$

whereupon

$$\begin{aligned} z_4 &= -\frac{\lambda}{2} z_1 \\ z_3 &= \frac{2-\lambda}{2} z_1 - \lambda z_2 \\ -z_1 + z_2 + (2-\lambda)z_1 - 2\lambda z_2 + \lambda z_1 &= \frac{\lambda(2-\lambda)}{2} z_1 - \lambda^2 z_2 \\ z_1 - \frac{3\lambda}{2} z_1 &= -\frac{\lambda^2}{2} z_1, \end{aligned}$$

or

$$\begin{aligned} z_4 &= -\frac{\lambda}{2} z_1 \\ z_3 &= \frac{2-\lambda}{2} z_1 - \lambda z_2 \\ \frac{\lambda^2 - 2\lambda + 2}{2} z_1 &= -(\lambda - 1)^2 z_2 \\ (\lambda^2 - 3\lambda + 2) z_1 &= 0, \end{aligned}$$

or

$$z_4 = -\frac{\lambda}{2} z_1$$

$$\begin{aligned} z_3 &= \frac{2-\lambda}{2}z_1 - \lambda z_2 \\ (\lambda^2 - 2\lambda + 2)z_1 &= -(\lambda - 1)^2 z_2 \\ (\lambda - 1)(\lambda - 2)z_1 &= 0. \end{aligned}$$

We consider the following cases.

1. *Let $\lambda = 1$. Then*

$$\begin{aligned} z_4 &= -\frac{1}{2}z_1 \\ z_3 &= \frac{1}{2}z_1 - z_2 \\ z_1 &= 2z_2, \end{aligned}$$

or

$$\begin{aligned} z_4 &= -z_2 \\ z_3 &= 0 \\ z_1 &= 2z_2. \end{aligned}$$

2. *Let $\lambda = 2$. Then*

$$\begin{aligned} z_4 &= -z_1 \\ z_3 &= -2z_2 \\ z_1 &= 0, \end{aligned}$$

or

$$z_1 = 0$$

$$z_3 = -2z_2$$

$$z_4 = 0.$$

3. Let $\lambda \neq 1$ and $\lambda_2 \neq 0$. Then

$$z_1 = 0$$

$$z_2 = 0$$

$$z_3 = 0$$

$$z_4 = 0,$$

which is impossible.

Thus, the eigenvalues of the operator T are $\lambda_1 = 1$ and $\lambda_2 = 2$. Now, take $(z_1, z_2, z_3, z_4) \in \mathbb{C}^4$ arbitrarily. Then, we have the following.

$$(T-I)(z_1,z_2,z_3,z_4) = T(z_1,z_2,z_3,z_4) - I(z_1,z_2,z_3,z_4)$$

$$= (-2z_4, z_1 - z_3 + z_4, -z_1 + z_2 + 2z_3 - 2z_4, z_1 + 3z_4) - (z_1, z_2, z_3, z_4)$$

$$= (-z_1 - 2z_4, z_1 - z_2 - z_3 + z_4, -z_1 + z_2 + z_3 - 2z_4, z_1 + 2z_4)$$

and

$$(T-I)^2(z_1,z_2,z_3,z_4) = (T-I)(T-I)(z_1,z_2,z_3,z_4)$$

$$= (T-I)((T-I)(z_1,z_2,z_3,z_4))$$

$$= (T-I)(-z_1 - 2z_4, z_1 - z_2 - z_3 + z_4, -z_1 + z_2 + z_3 - 2z_4, z_1 + 2z_4)$$

$$= (z_1 + 2z_4 - 2z_1 - 4z_4,$$

$$-z_1 - 2z_4 - z_1 + z_2 + z_3 - z_4 + z_1 - z_2 - z_3 + 2z_4 + z_1 + 2z_4,$$

$$z_1+2z_4+z_1-z_2-z_3+z_4-z_1+z_2+z_3-2z_4-2z_1=4z_4,$$

$$-z_1-2z_44+2z_1+4z_4)$$

$$= \quad (-z_1-2z_4,z_4-z_1-3z_4,z_1+2z_4),$$

and

$$(T-I)^3(z_1,z_2,z_3,z_4)=(T-I)(T-I)^2(z_1,z_2,z_3,z_4)$$

$$= \quad (T-I)((T-I)^2(z_1,z_2,z_3,z_4))$$

$$= \quad (T-I)(-z_1-2z_4,z_4,-z_1-3z_4,z_1+2z_4)$$

$$= \quad (z_1+2z_4-2z_1-4z_4,-z_1-2z_4-z_4+z_1+3z_4+z_1+2z_4,$$

$$z_1+2z_4+z_4-z_1-3z_4-2z_1-4z_4,-z_1-2z_4+2z_1+4z_4)$$

$$= \quad (-z_1-2z_4,z_1+2z_4,-2z_1-4z_4,z_1+2z_4),$$

and

$$(T-I)^4(z_1,z_2,z_3,z_4)=(T-I)(T-I)^3(z_1,z_2,z_3,z_4)$$

$$= \quad (T-I)((T-I)^3(z_1,z_2,z_3,z_4))$$

$$= \quad (T-I)(-z_1-2z_4,z_1+2z_4,-2z_1-4z_4,z_1+2z_4)$$

$$= \quad (z_1+2z_4-2z_1-4z_4,-z_1-2z_4-z_1-2z_4+2z_1+4z_4+z_1+2z_4,$$

$$z_1+2z_4+z_1+2z_4-2z_1-4z_4-2z_1-4z_4,$$

$$-z_1-2z_4+2z_1+4z_4)$$

$$= \quad (-z_1-2z_4,z_1+2z_4,-2z_1-4z_4,z_1+2z_4).$$

Hence,

$$(T-I)^4(z_1,z_2,z_3,z_4)=(0,0,0,0)$$

if and only if

$$z_1=-2z_4.$$

Consequently

$$G(1,T)=\{(-2z_4,z_2,z_3,z_4):z_2,z_3,z_4\in\mathbb{C}\}.$$

Next, for $(z_1,z_2,z_3,z_4)\in\mathbb{C}^4$, *we have*

$$\begin{aligned}(T-2I)(z_1,z_2,z_3,z_4) &= T(z_1,z_2,z_3,z_4)-2I(z_1,z_2,z_3,z_4)\\ &= (-2z_4,z_1-z_3+z_4,-z_1+z_2+2z_3-2z_4,z_1+3z_4)\\ &\quad -(2z_1,2z_2,2z_3,2z_4)\\ &= (-2z_1-2z_4,z_1-2z_2-z_3+z_4,-z_1+z_2-2z_4,z_1+z_4),\end{aligned}$$

and

$$\begin{aligned}(T-2I)^2(z_1,z_2,z_3,z_4) &= (T-2I)(T-2I)(z_1,z_2,z_3,z_4)\\ &= (T-2I)((T-2I)(z_1,z_2,z_3,z_4))\\ &= (T-2I)(-2z_1-2z_4,z_1-2z_2-z_3+z_4,-z_1+z_2-2z_4,z_1+z_4)\\ &= (4z_1+4z_4-2z_1-2z_4,\\ &\quad -2z_1-2z_4-2z_1+4z_2+2z_3-2z_4+z_1-z_2+2z_4+z_1+z_4,\\ &\quad 2z_1+2z_4+z_1-2z_2-z_3+z_4-2z_1-2z_4,-2z_1-2z_4+z_1+z_4)\\ &= (2z_1+2z_4,-2z_1+3z_2+2z_3-z_4,z_1-2z_2-z_3+z_4,-z_1-z_4),\end{aligned}$$

and

$$(T-2I)^3(z_1,z_2,z_3,z_4)=(T-2I)(T-2I)^2(z_1,z_2,z_3,z_4)$$

$$
\begin{aligned}
&= (T-2I)((T-2I)^2(z_1,z_2,z_3,z_4)) \\
&= (T-2I)(2z_1+2z_4,-2z_1+3z_2+2z_3-z_4,z_1-2z_2-z_3+z_4,-z_1-z_4) \\
&= (-4z_1-4z_4+2z_1+2z_4, \\
&\quad 2z_1+2z_4+4z_1-6z_2-4z_3+2z_4-z_1+2z_2+z_3-z_4-z_1-z_4, \\
&\quad -2z_1-2z_4-2z_1+3z_2+2z_3-z_4+2z_1+2z_4, \\
&\quad 2z_1+2z_4-z_1-z_4) \\
&= (-2z_1-2z_4,4z_1-4z_2-3z_3+2z_4,-2z_1+3z_2+2z_3-z_4,z_1+z_4),
\end{aligned}
$$

and

$$
\begin{aligned}
&(T-2I)^4(z_1,z_2,z_3,z_4) = (T-2I)(T-2I)^3(z_1,z_2,z_3,z_4) \\
&= (T-2I)((T-2I)^3(z_1,z_2,z_3,z_4)) \\
&= (T-2I)(-2z_1-2z_4,4z_1-4z_2-3z_3+2z_4,-2z_1+3z_2+2z_3-z_4,z_1+z_4) \\
&= (4z_1+4z_4-2z_1-2z_4, \\
&\quad -2z_12-2z_4-8z_1+8z_2+6z_3-4z_4+2z_1-3z_2-2z_3+z_4+z_1+z_4, \\
&\quad 2z_1+2z_4+4z_1-4z_2-3z_3+2z_4-2z_1-2z_4, \\
&\quad -2z_1-2z_4+z_1+z_4) \\
&= (2z_1+2z_4,-7z_1+5z_2+4z_3-4z_4,4z_1-4z_2-3z_3+2z_4,-z_1-z_4).
\end{aligned}
$$

Hence,

$$(T-2I)^4(z_1,z_2,z_3,z_4) = 0$$

if and only if

$$(2z_1+2z_4, -7z_1+5z_2+4z_3-4z_4, 4z_1-4z_2-3z_3+2z_4, -z_1-z_4) = (0,0,0,0)$$

if and only if

$$\begin{aligned} z_4 &= -z_1 \\ -7z_1+5z_2+4z_3-4z_4 &= 0 \\ 4z_1-4z_2-3z_3+2z_4 &= 0, \end{aligned}$$

if and only if

$$\begin{aligned} z_4 &= -z_1 \\ -3z_1+5z_2+4z_3 &= 0 \\ 2z_1-4z_2-3z_3 &= 0, \end{aligned}$$

if and only if

$$\begin{aligned} z_1 &= 2z_2+\frac{3}{2}z_3 \\ z_4 &= -2z_2-\frac{3}{2}z_3 \\ -6z_2-\frac{9}{2}z_3+5z_2+4z_3 &= 0, \end{aligned}$$

if and only if

$$\begin{aligned} z_1 &= 2z_2+\frac{3}{2}z_3 \\ z_4 &= -2z_2-\frac{3}{2}z_3 \\ -\frac{1}{2}z_3-z_2 &= 0, \end{aligned}$$

if and only if

$$\begin{aligned} z_1 &= 2z_2 + \frac{3}{2}z_3 \\ z_4 &= -2z_2 - \frac{3}{2}z_3 \\ z_3 &= -2z_2, \end{aligned}$$

if and only if

$$\begin{aligned} z_1 &= -z_2 \\ z_3 &= -2z_2 \\ z_4 &= z_2. \end{aligned}$$

Therefore

$$G(2,T) = \{(-z_2, z_2, -2z_2, z_2) : z_2 \in \mathbb{C}\}.$$

Take

$$\begin{aligned} v_1 &= (-2,0,0,1), \\ v_2 &= (-1,1,-2,1). \end{aligned}$$

Then

$$\begin{aligned} (T-I)v_1 &= (0,-1,0,0), \\ (T-I)^2 v_1 &= (T-I)(T-I)v_1 \\ &= (T-I)((T-i)v_1) \\ &= (T-I)(0,-1,0,0) \\ &= (0,1,-1,0), \end{aligned}$$

$$\begin{aligned}(T-I)^3 v_1 &= (T-I)(T-I)^2 v_1\\ &= (T-I)((T-I)^2 v_1)\\ &= (T-I)(0,1,-1,0)\\ &= (0,0,0,0),\end{aligned}$$

and

$$\begin{aligned}(T-2I)v_2 &= (0,-2+2,1+1-2,-1+1)\\ &= (0,0,0,0).\end{aligned}$$

Set

$$\begin{aligned}u_1 &= (-2,0,0,1),\\ u_2 &= (0,-1,0,0),\\ u_3 &= (0,1,-1,0),\\ u_4 &= (-1,1,-2,1).\end{aligned}$$

Then

$$\mathcal{M}(I,(u_1,u_2,u_3,u_4),(e_1,e_2,e_3,e_4)) = \begin{pmatrix} -2 & 0 & 0 & 1\\ 0 & -1 & 0 & 0\\ 0 & 1 & -1 & 0\\ -1 & 1 & -2 & 1 \end{pmatrix}.$$

Let now, $a_{ij} \in \mathbb{C}$, $i,j \in \{1,2,3,4\}$, be such that

$$\begin{aligned}e_j &= a_{j1}u_1 + a_{j2}u_2 + a_{j3}u_3 + a_{j4}u_4\\ &= a_{j1}(-2,0,0,1) + a_{j2}(0,-1,0,0) + a_{j3}(0,1,-1,0)\\ &\quad + a_{j4}(-1,1,-2,1)\end{aligned}$$

$$\begin{aligned} &= (-2a_{j1},0,0,a_{j1})+(0,-a_{j2},0,0)+(0,a_{j3},-a_{j3},0) \\ &\quad +(-a_{j4},a_{j4},-2a_{j4},a_{j4}) \\ &= (-2a_{j1}-a_{j4},-a_{j2}+a_{j3}+a_{j4},-a_{j3}-2a_{j4},a_{j1}+a_{j4}), \end{aligned}$$

$j \in \{1,2,3,4\}$. *Then we get the systems*

$$\begin{aligned} -2a_{11}-a_{14} &= 1 \\ -a_{12}+a_{13}+a_{14} &= 0 \\ -a_{13}-2a_{14} &= 0 \\ a_{11}+a_{14} &= 0, \end{aligned}$$

and

$$\begin{aligned} -2a_{21}-a_{24} &= 0 \\ -a_{22}+a_{23}+a_{24} &= 1 \\ -a_{23}-2a_{24} &= 0 \\ a_{21}+a_{24} &= 0, \end{aligned}$$

and

$$\begin{aligned} -2a_{31}-a_{34} &= 0 \\ -a_{32}+a_{33}+a_{34} &= 0 \\ -a_{33}-2a_{34} &= 1 \\ a_{31}+a_{34} &= 0, \end{aligned}$$

and

$$
\begin{aligned}
-2a_{41} - a_{44} &= 0 \\
-a_{42} + a_{43} + a_{44} &= 0 \\
-a_{43} - 2a_{44} &= 0 \\
a_{41} + a_{44} &= 1.
\end{aligned}
$$

Therefore

$$
\begin{aligned}
a_{11} &= -1 \\
a_{12} &= -1 \\
a_{13} &= -2 \\
a_{14} &= 1,
\end{aligned}
$$

and

$$
\begin{aligned}
a_{21} &= 0 \\
a_{22} &= -1 \\
a_{23} &= 0 \\
a_{24} &= 0,
\end{aligned}
$$

and

$$
\begin{aligned}
a_{31} &= 0 \\
a_{32} &= -1 \\
a_{33} &= -1
\end{aligned}
$$

$$a_{34} = 0,$$

and

$$\begin{aligned} a_{41} &= -1 \\ a_{42} &= -2 \\ a_{43} &= -4 \\ a_{44} &= 2, \end{aligned}$$

and

$$\begin{aligned} e_1 &= -u_1 - u_2 - 2u_3 + u_4 \\ e_2 &= -u_2 \\ e_3 &= -u_2 - u_3 \\ e_4 &= -u_1 - 2u_2 - 4u_3 + 2u_4. \end{aligned}$$

Thus,

$$\mathscr{M}(I,(e_1,e_2,e_3,e_4),(u_1,u_2,u_3,u_4)) = \begin{pmatrix} -1 & -1 & -2 & 1 \\ 0 & -1 & 0 & 0 \\ 0 & -1 & -1 & 0 \\ -1 & -2 & -4 & 2 \end{pmatrix}.$$

Consequently the Jordan canonical form of the matrix of the operator T is

$$\mathscr{M}(I,(u_1,u_2,u_3,u_4),(e_1,e_2,e_3,e_4))\mathscr{M}(T,(e_1,e_2,e_3,e_4),(e_1,e_2,e_3,e_4))$$
$$\times \mathscr{M}(I,(e_1,e_2,e_3,e_4),(u_1,u_2,u_3,u_4))$$

$$= \begin{pmatrix} -2 & 0 & 0 & 1 \\ 0 & -1 & 0 & 0 \\ 0 & 1 & -1 & 0 \\ -1 & 1 & -2 & 1 \end{pmatrix} \begin{pmatrix} 0 & 1 & -1 & 1 \\ 0 & 0 & 1 & 0 \\ 0 & -1 & 2 & 0 \\ -2 & 1 & -2 & 3 \end{pmatrix}$$

$$\times \begin{pmatrix} -1 & -1 & -2 & 1 \\ 0 & -1 & 0 & 0 \\ 0 & -1 & -1 & 0 \\ -1 & -2 & -4 & 2 \end{pmatrix}$$

$$= \begin{pmatrix} -2 & 0 & 0 & 1 \\ 0 & -1 & 0 & 0 \\ 0 & 1 & -1 & 0 \\ -1 & 1 & -2 & 1 \end{pmatrix} \begin{pmatrix} -1 & -2 & -3 & 2 \\ 0 & -1 & -1 & 0 \\ 0 & -1 & -2 & 0 \\ -1 & -3 & -6 & 4 \end{pmatrix}$$

$$= \begin{pmatrix} 1 & 1 & 0 & 0 \\ 0 & 1 & 1 & 0 \\ 0 & 0 & 1 & 0 \\ 0 & 0 & 0 & 2 \end{pmatrix}.$$

Exercise 6.86. *Find the Jordan canonical form of the matrix of the operator $T \in \mathscr{L}(\mathbb{C}^n)$, where*

1. $n = 2$ *and*

$$\mathscr{M}(T) = \begin{pmatrix} 75 & 56 \\ -90 & 67 \end{pmatrix}.$$

2. $n = 3$ *and*

$$\mathscr{M}(T) = \begin{pmatrix} 5 & 8 & 16 \\ 4 & 1 & 8 \\ -4 & -4 & -11 \end{pmatrix}.$$

3. $n = 4$ *and*

$$\mathscr{M}(T) = \begin{pmatrix} -1 & -12 & 2 & 16 \\ -1 & 1 & 1 & 2 \\ -6 & -11 & 7 & 15 \\ -1 & -6 & 1 & 9 \end{pmatrix}.$$

4. $n = 5$ *and*

$$\mathscr{M}(T) = \begin{pmatrix} -5 & -2 & 0 & -4 & 8 \\ 2 & 1 & -2 & 2 & 0 \\ -3 & -1 & -2 & -3 & 8 \\ -1 & -1 & 1 & -2 & 0 \\ -4 & -1 & 0 & -2 & 7 \end{pmatrix}.$$

5. $n = 6$ *and*

$$\mathcal{M}(T) = \begin{pmatrix} 5 & 7 & 5 & 2 & -2 & -8 \\ 2 & 10 & 7 & 2 & -3 & -7 \\ -2 & -7 & -4 & -2 & 3 & 7 \\ 0 & 0 & -1 & 3 & 1 & 0 \\ -2 & -7 & -7 & -2 & 6 & 7 \\ 2 & 6 & 5 & 2 & -2 & -4 \end{pmatrix}.$$

Answer 6.87. *1.*

$$\begin{pmatrix} 3 & 0 \\ 0 & 5 \end{pmatrix}.$$

2.

$$\begin{pmatrix} -3 & 0 & 0 \\ 0 & -3 & 0 \\ 0 & 0 & 1 \end{pmatrix}.$$

3.

$$\begin{pmatrix} 1 & 0 & 0 & 0 \\ 0 & 3 & 0 & 0 \\ 0 & 0 & 5 & 0 \\ 0 & 0 & 0 & 7 \end{pmatrix}.$$

4.

$$\begin{pmatrix} -1 & 1 & 0 & 0 & 0 \\ 0 & -1 & 0 & 0 & 0 \\ 0 & 0 & -1 & 1 & 0 \\ 0 & 0 & 0 & -1 & 0 \\ 0 & 0 & 0 & 0 & 3 \end{pmatrix}.$$

5.

$$\begin{pmatrix} 3 & 1 & 0 & 0 & 0 & 0 \\ 0 & 3 & 1 & 0 & 0 & 0 \\ 0 & 0 & 3 & 0 & 0 & 0 \\ 0 & 0 & 0 & 3 & 1 & 0 \\ 0 & 0 & 0 & 0 & 3 & 0 \\ 0 & 0 & 0 & 0 & 0 & 1 \end{pmatrix}.$$

6.11 Advanced Practical Problems

Problem 6.88. *Let $T \in \mathscr{L}(\mathbb{C}^3)$ be defined by*

$$T(x,y,z) = (x+2y, x+y+2z, -y+z), \quad (x,y,z) \in \mathbb{C}^3.$$

Find the eigenvalues of T, the corresponding eigenspaces and generalized eigenspaces.

Answer 6.89.

$$\begin{aligned} \lambda &= 1, \\ E(1,T) &= \{(-2x,0,x) : x \in \mathbb{C}\}, \\ G(1,T) &= \mathbb{C}^3. \end{aligned}$$

Problem 6.90. *Let $T \in \mathscr{L}(\mathbb{C}^3)$ be defined by*

$$T(x,y,z) = (x-3y-4z, -x+3y+4z, x-3y-4z), \quad (x,y,z) \in \mathbb{C}^3.$$

Prove that T is nilpotent.

Hint 6.91. *Prove that $T^2 = 0$.*

Problem 6.92. *Let $T \in \mathscr{L}(V)$ be invertible. Prove that*

$$G(\lambda,T) = G\left(\frac{1}{\lambda}, T^{-1}\right) \tag{6.22}$$

for any $\lambda \in \mathbb{F}$, $\lambda \neq 0$.

Solution 6.93. *Let $v \in G(\lambda,T)$ be arbitrarily chosen. Then there is a $k \in \mathbb{N}$ such that*

$$(T-\lambda I)^k v = 0.$$

Hence,

$$(-\lambda)^k T^k \left(T^{-1} - \frac{1}{\lambda} I\right)^k v = 0$$

and

$$0 = T^{-k} 0$$

$$\begin{aligned} &= T^{-k}\left((-\lambda)^k T^k \left(T^{-1}-\frac{1}{\lambda}I\right)^k v\right) \\ &= (-\lambda)^k T^{-k} T^k \left(T^{-1}-\frac{1}{\lambda}I\right)^k v \\ &= (-\lambda)^k \left(T^{-1}-\frac{1}{\lambda}I\right)^k v, \end{aligned}$$

whereupon

$$\left(T^{-1}-\frac{1}{\lambda}I\right)^k v = 0.$$

Therefore $v \in G\left(\frac{1}{\lambda}, T^{-1}\right)$. *Since* $v \in G(\lambda, T)$ *was arbitrarily chosen and we get that it is an element of* $G\left(\frac{1}{\lambda}, T^{-1}\right)$, *we obtain the inclusion*

$$G(\lambda, T) \subseteq G\left(\frac{1}{\lambda}, T^{-1}\right). \tag{6.23}$$

Let now, $u \in G\left(\frac{1}{\lambda}, T^{-1}\right)$ *be arbitrarily chosen. Then there is an* $l \in \mathbb{N}$ *such that*

$$\left(T^{-1}-\frac{1}{\lambda}I\right)^l u = 0,$$

whereupon

$$\left(-\frac{1}{\lambda}\right)^l T^{-l}(T-\lambda I)^l u = 0.$$

Then

$$\begin{aligned} 0 &= T^l 0 \\ &= T^l\left(\left(-\frac{1}{\lambda}\right)^l T^{-l}(T-\lambda I)^l u\right) \\ &= \left(-\frac{1}{\lambda}\right)^l T^l T^{-l}(T-\lambda I)^l u \end{aligned}$$

$$= \left(-\frac{1}{\lambda}\right)^l (T-\lambda I)^l u.$$

Therefore

$$(T-\lambda I)^l u = 0.$$

Thus, $u \in G(\lambda, T)$. *Because* $u \in G\left(\frac{1}{\lambda}, T^{-1}\right)$ *was arbitrarily chosen and we get that it is an element of* $G(\lambda, T)$, *we arrive at the inclusion*

$$G\left(\frac{1}{\lambda}, T^{-1}\right) \subseteq G(\lambda, T).$$

By the last inclusion and the inclusion (6.23), *we get the equality* (6.22).

Problem 6.94. *Let* $T \in \mathscr{L}(V)$ *and* λ_l, $l \in \{1, \ldots, m\}$, *be distinct generalized eigenvalues of* T. *Prove that*

$$G(\lambda_l, T) \bigcap G(\lambda_k, T) = \{0\}$$

for any $l, k \in \{1, \ldots, m\}$, $l \neq k$.

Solution 6.95. *Assume that there are* $i, j \in \{1, \ldots, m\}$, $i \neq j$, *so that*

$$G(\lambda_i, T) \bigcap G(\lambda_j, T) \neq \{0\}. \tag{6.24}$$

Then there is a $v \in G(\lambda_i, T) \bigcap G(\lambda_j, T)$ *such that* $v \neq 0$. *Hence,* v *is a generalized eigenvector corresponding to* λ_i *and* λ_j. *Set*

$$w_i = v,$$

$$w_j = v.$$

Without loss of generality, suppose that $i < j$. *Let*

$$w_1, \ldots, w_{i-1}, w_{i+1}, \ldots, w_{j-1}, w_{j+1}, \ldots, w_m$$

be generalized eigenvectors corresponding to the generalized eigenvalues

$$\lambda_1, \ldots, \lambda_{i-1}, \lambda_{i+1}, \ldots, \lambda_{j-1}, \lambda_{j+1}, \ldots, \lambda_m,$$

respectively. Then, by Theorem 6.11, it follows that the list $w_1, \ldots, w_m$ *is linearly independent. This is a contradiction because* $w_i = w_j$. *Consequently* (6.24) *holds.*

Problem 6.96. *Let $T \in \mathscr{L}(V)$, m be a positive integer, and $v \in V$ be such that*

$$T^{m-1}v \neq 0, \quad T^m v = 0.$$

Prove that the list

$$v, Tv, \ldots, T^{m-1}v \tag{6.25}$$

is linearly independent.

Solution 6.97. *Let $a_0, \ldots, a_{m-1} \in \mathbb{F}$ be such that*

$$a_0 v + a_1 T v + a_2 T^2 v + \cdots + a_{m-1} T^{m-1} v = 0. \tag{6.26}$$

Since

$$T^m v = 0,$$

we have that

$$T^{m+k} v = 0, \quad k \in \mathbb{N}.$$

Hence, using (6.26), we find

$$\begin{aligned}
0 &= T^{m-1} 0 \\
&= T^{m-1}\left(a_0 v + a_1 T v + a_2 T^2 v + \cdots + a_{m-1} T^{m-1} v\right) \\
&= T^{m-1}(a_0 v) + T^{m-1}(a_1 T v) + T^{m-1}(a_2 T^2 v) + \cdots + T^{m-1}(a_{m-1} T^{m-1} v) \\
&= a_0 T^{m-1} v + a_1 T^m v + a_2 T^{m+1} v + \cdots + a_{m-1} T^{2m-1} v \\
&= a_0 T^{m-1} v,
\end{aligned}$$

whereupon

$$a_0 = 0.$$

Then, (6.26) takes the form

$$a_1 T v + a_2 T^2 v + \cdots + a_{m-1} T^{m-1} v = 0.$$

By the last equation, we obtain

$$0 = T^{m-2} 0$$

$$\begin{aligned}
&= T^{m-2}\left(a_1 Tv + a_2 T^2 v + \cdots + a_{m-1} T^{m-1} v\right) \\
&= T^{m-2}(a_1 Tv) + T^{m-2}(a_2 T^2 v) + \cdots + T^{m-2}(a_{m-1} T^{m-1} v) \\
&= a_1 T^{m-1} v + a_2 T^m v + \cdots + a_{m-1} T^{2m-3} v \\
&= a_1 T^{m-1} v,
\end{aligned}$$

whereupon

$$a_1 = 0.$$

Continue in this fashion, we get

$$a_j = 0, \quad j \in \{0, \ldots, m-1\}.$$

Consequently (6.25) *is linearly independent.*

Problem 6.98. *Let $N \in \mathscr{L}(V)$ be nilpotent. Prove that* 0 *is the only eigenvalue of V.*

Solution 6.99. *Let $m \in \mathbb{N}$ be such that $N^m = 0$. Let also, $\lambda \in \mathbb{F}$ be an eigenvalue of N. Then there is a $v \in V$, $v \neq 0$, such that*

$$Nv = \lambda v.$$

Assume that

$$N^l v = \lambda^l v \tag{6.27}$$

for some $l \in \mathbb{N}$. Then

$$\begin{aligned}
N^{l+1} v &= N N^l v \\
&= N(N^l v) \\
&= N(\lambda^l v) \\
&= \lambda^l N v \\
&= \lambda^{l+1} v.
\end{aligned}$$

Therefore, using the principal of the mathematical induction, the equality (6.27) *holds for any* $l \in \mathbb{N}$. *Hence, using that*

$$N^m v = 0,$$

we get

$$\begin{aligned} 0 &= N^m v \\ &= \lambda^m v. \end{aligned}$$

Because $v \neq 0$, *we find* $\lambda = 0$.

Problem 6.100. *Suppose that* $T \in \mathscr{L}(\mathbb{C}^3)$ *is defined by*

$$T(z_1, z_2, z_3) = (z_2, z_3, 0), \quad (z_1, z_2, z_3) \in \mathbb{C}^3.$$

Prove that there does not exist $S \in \mathscr{L}(\mathbb{C}^3)$ *such that*

$$S^2 = T.$$

Solution 6.101. *Assume that there exists* $S \in \mathscr{L}(\mathbb{C}^3)$ *such that*

$$S^2 = T. \tag{6.28}$$

By the definition of the operator T, *we get*

$$\begin{aligned} T^2(z_1, z_2, z_3) &= TT(z_1, z_2, z_3) \\ &= T(T(z_1, z_2, z_3)) \\ &= T(z_2, z_3, 0) \\ &= (z_3, 0, 0), \quad (z_1, z_2, z_3) \in \mathbb{C}^3, \end{aligned}$$

and

$$\begin{aligned} T^3(z_1, z_2, z_3) &= TT^2(z_1, z_2, z_3) \\ &= T(T^2(z_1, z_2, z_3)) \end{aligned}$$

$$= T(z_3,0,0)$$

$$= (0,0,0), \quad (z_1,z_2,z_3) \in \mathbb{C}^3.$$

Thus, T is nilpotent. By (6.28), *it follows that*

$$S^6 = T^3$$

$$= 0.$$

From here, we conclude that S is nilpotent. By Theorem 6.18, we get that

$$S^3 = 0.$$

Now, applying S to both sides of (6.28), *we find*

$$0 = S^3$$

$$= SS^2$$

$$= ST,$$

whereupon $S = 0$ and then $T = 0$. This is a contradiction.

Problem 6.102. *Prove or give a counterexample: the set of all nilpotent operators on V is a subspace of $\mathscr{L}(V)$.*

Solution 6.103. *Consider $\mathscr{L}(\mathbb{C}^3)$ and suppose that the operators $T_1, T_2 \in \mathscr{L}(\mathbb{C}^3)$ are defined by*

$$T_1(z_1,z_2,z_3) = (z_2,z_3,0),$$

$$T_2(z_1,z_2,z_3) = (0,0,z_1), \quad (z_1,z_2,z_3) \in \mathbb{C}^3.$$

By Problem 6.100, it follows that T_1 is a nilpotent operator. For T_2, we have

$$T_2^2(z_1,z_2,z_3) = T_2T_2(z_1,z_2,z_3)$$

$$= T_2(T_2(z_1,z_2,z_3))$$

$$= T_2(0,0,z_1)$$

$$= (0,0,0), \quad (z_1,z_2,z_3) \in \mathbb{C}^3,$$

i.e., T_2 is nilpotent. On the other hand,

$$\begin{aligned}(T_1+T_2)(z_1,z_2,z_3) &= T_1(z_1,z_2,z_3)+T_2(z_1,z_2,z_3)\\ &= (z_2,z_3,0)+(0,0,z_1)\\ &= (z_2,z_3,z_1), \quad (z_1,z_2,z_3) \in \mathbb{C}^3.\end{aligned}$$

Note that T_1+T_2 is not nilpotent because

$$(T_1+T_2)^l(1,1,1) = (1,1,1)$$

for any $l \in \mathbb{N}$. Therefore, in the general case the set of all nilpotent operators on V is not a subspace of $\mathscr{L}(V)$.

Problem 6.104. *Suppose $S,T \in \mathscr{L}(V)$ and ST is nilpotent. Prove that TS is nilpotent.*

Solution 6.105. *Let $m \in \mathbb{N}$ be such that*

$$(ST)^m = 0.$$

Then

$$(ST)^{m+1} = 0$$

and

$$\begin{aligned}(TS)^{m+2} &= TS(TS)^mTS\\ &= TS\underbrace{TSTS\dots TS}_{m \quad times}TS\\ &= T\underbrace{STST\dots ST}_{m+1 \quad times}S\\ &= 0.\end{aligned}$$

Therefore TS is nilpotent.

Problem 6.106. *Let $T \in \mathscr{L}(\mathbb{C}^2)$ be defined as in Problem 5.65. Find the multiplicities of the eigenvalues of T.*

Answer 6.107. 1.

Problem 6.108. *Let $T \in \mathscr{L}(\mathbb{C}^2)$ be defined as in Problem 5.67. Find the multiplicities of the eigenvalues of T.*

1. 3.

2. The multiplicity of $\lambda = 3$ equals 1. The multiplicity of $\lambda = -1$ equals 2.

3. 1.

Problem 6.109. *Let $T \in \mathscr{L}(\mathbb{C}^2)$ be defined as in Problem 5.65. Find the geometric multiplicities of the eigenvalues of T.*

Answer 6.110. 1.

Problem 6.111. *Let $T \in \mathscr{L}(\mathbb{C}^3)$ be defined as in Problem 5.67. Find the geometric multiplicities of the eigenvalues of T.*

Answer 6.112. 1.

Problem 6.113. *Let $T \in \mathscr{L}(\mathbb{C}^3)$ be defined as in Problem 6.90. Find*

$$\sqrt{I+T}(z_1, z_2, z_3), \quad (z_1, z_2, z_3) \in \mathbb{C}^3.$$

Answer 6.114.

$$\left(\frac{3}{2}z_1 - \frac{3}{2}z_2 - 2z_3, -\frac{1}{2}z_1 + \frac{5}{2}z_2 + 2z_3, \frac{1}{2}z_1 - \frac{3}{2}z_2 - z_3\right), \quad (z_1, z_2, z_3) \in \mathbb{C}^3.$$

Problem 6.115. *Let $T \in \mathscr{L}(\mathbb{C}^3)$ be defined as in Problem 5.65. Find its characteristic polynomials.*

Answer 6.116. *1.* $(z-i)(z+i)$.

2. $(z-2i)(z+2i)$.

Problem 6.117. *Let $T \in \mathscr{L}(\mathbb{C}^3)$ be defined as in Problem 5.67. Find its characteristic polynomials.*

Answer 6.118. *1.* $(z-1)^3$.

2. $(z-1)(z+1)^2$.

3. $(z-1)(z-3)(z-5)$.

Problem 6.119. *Determine which polynomials are monic.*

1. z^3+3z^2+z+1.

2. $-z^4+z^2+z+1$.

3. z^5-z^3.

4. $z^6-z^3+z^2$.

5. $-2z^9+z^8$.

Answer 6.120. 1. *Yes.*

2. *No.*

3. *Yes.*

4. *Yes.*

5. *No.*

Problem 6.121. *Find the minimal polynomial of the operator T, where*

1.
$$\mathcal{M}(T)=\begin{pmatrix} 1 & 2 \\ -1 & 0 \end{pmatrix}.$$

2.
$$\mathcal{M}(T)=\begin{pmatrix} 1 & -3 & 3 \\ -2 & -6 & 13 \\ -1 & -4 & 8 \end{pmatrix}.$$

3.
$$\mathcal{M}(T)=\begin{pmatrix} 1 & -3 & 4 \\ 4 & -7 & 8 \\ 6 & -7 & 7 \end{pmatrix}.$$

4.
$$\mathcal{M}(T)=\begin{pmatrix} 4 & -2 & -1 \\ -1 & 3 & -1 \\ 1 & -2 & 2 \end{pmatrix}.$$

5.

$$\mathscr{M}(T) = \begin{pmatrix} -1 & 0 & 0 \\ 0 & -1 & 0 \\ 0 & 0 & 2 \end{pmatrix}.$$

Answer 6.122. *1.* $p(z) = z^2 + z - 2$, $z \in \mathbb{F}$.

2. $p(z) = z^3 - 3z^2 + 3z - 1$, $z \in \mathbb{F}$.

3. $p(z) = z^3 - z^2 - 5z - 3$, $z \in \mathbb{F}$.

4. $p(z) = z^3 - 9z^2 + 23z - 15$, $z \in \mathbb{F}$.

5. $p(z) = z^2 - z + 2$, $z \in \mathbb{F}$.

Problem 6.123. *Suppose that $T \in \mathscr{L}(\mathbb{C}^4)$ is such that* 3, 5 *and* 8 *are eigenvalues of T. Prove that*

$$(T - 3I)^2(T - 5I)^2(T - 8I)^2 = 0. \tag{6.29}$$

Solution 6.124. *Let d_1, d_2, d_3 be the multiplicities of* 3, 5 *and* 8*, respectively. We have that*

$$d_1 + d_2 + d_3 = 4.$$

Hence, $d_1 \leq 2$, $d_2 \leq 2$, $d_3 \leq 2$ and (6.29) *holds.*

Problem 6.125. *Suppose that V is a complex vector space and $T \in \mathscr{L}(V)$ is such that* 5 *and* 6 *are eigenvalues of T and that T has no other eigenvalues. Prove that*

$$(T - 5I)^{n-1}(T - 6I)^{n-1} = 0. \tag{6.30}$$

Here $n = \dim V$.

Solution 6.126. *Let d_1 and d_2 be the multiplicities of* 5 *and* 8*, respectively. Then*

$$d_1 + d_2 = n.$$

Hence, $d_1 \leq n - 1$ and $d_2 \leq n - 1$ and (6.30) *holds.*

Problem 6.127. *Find the Jordan canonical form of the matrix of the operator $T \in \mathscr{L}(\mathbb{C}^n)$, where*

1. $n = 2$ *and*

$$\mathscr{M}(T) = \begin{pmatrix} 22 & -84 \\ 6 & -23 \end{pmatrix}.$$

2. $n = 2$ *and*

$$\mathcal{M}(T) = \begin{pmatrix} 7 & 2 \\ -8 & -1 \end{pmatrix}.$$

3. $n = 2$ *and*

$$\mathcal{M}(T) = \begin{pmatrix} -18 & 9 \\ -49 & 24 \end{pmatrix}.$$

4. $n = 2$ *and*

$$\mathcal{M}(T) = \begin{pmatrix} 2 & 1 \\ -25 & 12 \end{pmatrix}.$$

5. $n = 3$ *and*

$$\mathcal{M}(T) = \begin{pmatrix} 0 & 1 & 0 \\ -4 & 4 & 0 \\ -2 & -1 & 2 \end{pmatrix}.$$

6. $n = 3$ *and*

$$\mathcal{M}(T) = \begin{pmatrix} 2 & 6 & -15 \\ 1 & 1 & -5 \\ 1 & 2 & -6 \end{pmatrix}.$$

7. $n = 3$ *and*

$$\mathcal{M}(T) = \begin{pmatrix} 9 & -6 & -2 \\ 18 & -12 & -3 \\ 18 & -9 & -6 \end{pmatrix}.$$

8. $n = 3$ *and*

$$\mathcal{M}(T) = \begin{pmatrix} 4 & 6 & -15 \\ 1 & 3 & -5 \\ 1 & 2 & 4 \end{pmatrix}.$$

9. $n = 3$ *and*

$$\mathcal{M}(T) = \begin{pmatrix} 0 & -4 & 0 \\ 1 & -4 & 0 \\ 1 & -2 & 2 \end{pmatrix}.$$

10. $n = 3$ *and*

$$\mathcal{M}(T) = \begin{pmatrix} 12 & -6 & -2 \\ 18 & -9 & -3 \\ 18 & -9 & -3 \end{pmatrix}.$$

11. $n = 3$ *and*

$$\mathcal{M}(T) = \begin{pmatrix} 4 & -5 & 2 \\ 5 & -7 & 3 \\ 6 & -9 & 4 \end{pmatrix}.$$

12. $n = 3$ *and*

$$\mathcal{M}(T) = \begin{pmatrix} 5 & -3 & 2 \\ 6 & -4 & 4 \\ 4 & -4 & 5 \end{pmatrix}.$$

13. $n = 3$ *and*

$$\mathcal{M}(T) = \begin{pmatrix} 1 & -3 & 3 \\ -2 & -6 & 13 \\ -1 & -4 & 8 \end{pmatrix}.$$

14. $n = 3$ *and*

$$\mathcal{M}(T) = \begin{pmatrix} 7 & -12 & 6 \\ 10 & -19 & 10 \\ 12 & -24 & 13 \end{pmatrix}.$$

15. $n = 3$ *and*

$$\mathcal{M}(T) = \begin{pmatrix} 1 & -3 & 4 \\ 4 & -7 & 8 \\ 6 & -7 & 7 \end{pmatrix}.$$

16. $n = 3$ *and*

$$\mathcal{M}(T) = \begin{pmatrix} \alpha & 0 & 0 \\ 0 & \alpha & 0 \\ \alpha & 0 & \alpha \end{pmatrix}, \quad \alpha \neq 0.$$

17. $n = 3$ *and*

$$\mathcal{M}(T) = \begin{pmatrix} 3 & -1 & 0 \\ 6 & -3 & 2 \\ 8 & -6 & 5 \end{pmatrix}.$$

18. $n = 3$ *and*

$$\mathcal{M}(T) = \begin{pmatrix} 4 & 6 & -15 \\ 3 & 4 & -12 \\ 2 & 3 & -8 \end{pmatrix}.$$

19. $n = 3$ *and*

$$\mathcal{M}(T) = \begin{pmatrix} 4 & -5 & 7 \\ 1 & -4 & 9 \\ -4 & 0 & 5 \end{pmatrix}.$$

20. $n = 4$ *and*

$$\mathcal{M}(T) = \begin{pmatrix} 1 & -3 & 0 & 3 \\ -2 & -6 & 0 & 13 \\ 0 & -3 & 1 & 3 \\ -1 & -4 & 0 & 8 \end{pmatrix}.$$

21. $n=4$ *and*

$$\mathcal{M}(T)=\begin{pmatrix} 3 & -1 & 0 & 0 \\ 1 & 1 & 0 & 0 \\ 3 & 0 & 5 & -3 \\ 4 & -1 & 3 & -1 \end{pmatrix}.$$

22. $n=4$ *and*

$$\mathcal{M}(T)=\begin{pmatrix} 3 & -4 & 0 & 2 \\ 4 & -5 & -2 & 4 \\ 0 & 0 & 3 & -2 \\ 0 & 0 & 2 & -1 \end{pmatrix}.$$

23. $n=4$ *and*

$$\mathcal{M}(T)=\begin{pmatrix} 3 & -1 & 1 & -7 \\ 9 & -3 & -7 & -1 \\ 0 & 0 & 4 & -8 \\ 0 & 0 & 2 & -4 \end{pmatrix}.$$

24. $n=5$ *and*

$$\mathcal{M}(T)=\begin{pmatrix} 5 & -2 & 0 & -4 & 8 \\ 2 & 1 & -2 & 2 & 0 \\ -3 & -1 & -2 & -3 & 8 \\ -1 & -1 & 1 & -2 & 0 \\ -4 & -1 & 0 & -2 & 7 \end{pmatrix}.$$

25. $n=5$ *and*

$$\mathcal{M}(T)=\begin{pmatrix} 5 & 1 & 3 & -1 & -3 \\ 3 & 1 & 4 & 1 & -3 \\ 0 & -4 & 4 & 4 & 0 \\ 3 & 1 & 3 & 1 & -3 \\ 6 & -3 & 8 & 3 & -4 \end{pmatrix}.$$

26. $n=5$ *and*

$$\mathcal{M}(T)=\begin{pmatrix} -2 & -9 & -10 & -7 & 10 \\ 8 & 21 & 21 & 15 & -20 \\ -2 & -1 & 1 & 0 & 0 \\ -2 & -18 & -20 & -12 & 20 \\ -2 & -5 & -5 & -4 & 7 \end{pmatrix}.$$

27. $n = 6$ *and*

$$\mathcal{M}(T) = \begin{pmatrix} 5 & 7 & 5 & 2 & -2 & -2 \\ 2 & 10 & 7 & 2 & -3 & -7 \\ -2 & -7 & -4 & -2 & 3 & 7 \\ 0 & 0 & -1 & 3 & 1 & 0 \\ -2 & -7 & -7 & -2 & 6 & 7 \\ 2 & 6 & 5 & 2 & -2 & -4 \end{pmatrix}.$$

28. $n = 6$ *and*

$$\mathcal{M}(T) = \begin{pmatrix} -2 & -7 & -10 & 5 & 8 & 5 \\ -1 & 1 & -1 & 1 & 2 & 1 \\ -1 & -3 & 1 & 1 & 3 & 1 \\ -4 & -7 & -8 & 7 & 7 & 4 \\ -2 & -1 & -3 & 2 & 4 & 2 \\ -2 & -6 & -4 & 2 & 7 & 5 \end{pmatrix}.$$

29. $n = 6$ *and*

$$\mathcal{M}(T) = \begin{pmatrix} 13 & -3 & 1 & -3 & 9 & 3 \\ 3 & 2 & 1 & -1 & -3 & 1 \\ 9 & -4 & 5 & -3 & -9 & 3 \\ 15 & -7 & 2 & -1 & -15 & 5 \\ 18 & -5 & 0 & -6 & -14 & 6 \\ 42 & -14 & 0 & -14 & -42 & 18 \end{pmatrix}.$$

30. $n = 6$ *and*

$$\mathcal{M}(T) = \begin{pmatrix} 1 & 1 & 2 & -1 & 1 & 1 \\ -6 & 3 & 7 & -3 & 0 & 3 \\ -2 & 0 & 5 & -1 & 0 & 1 \\ -10 & 0 & 10 & -2 & 1 & 5 \\ 2 & 0 & -2 & 1 & 3 & -1 \\ -6 & 2 & 6 & -3 & 3 & 6 \end{pmatrix}.$$

Answer 6.128. *1.*

$$\begin{pmatrix} -2 & 0 \\ 0 & 1 \end{pmatrix}.$$

2.

$$\begin{pmatrix} 3 & 1 \\ 0 & 3 \end{pmatrix}.$$

3.

$$\begin{pmatrix} 3 & 1 \\ 0 & 3 \end{pmatrix}.$$

4.

$$\begin{pmatrix} 7 & 1 \\ 0 & 7 \end{pmatrix}.$$

5.

$$\begin{pmatrix} 2 & 0 & 0 \\ 0 & 2 & 1 \\ 0 & 0 & 2 \end{pmatrix}.$$

6.

$$\begin{pmatrix} -1 & 0 & 0 \\ 0 & -1 & 1 \\ 0 & 0 & -1 \end{pmatrix}.$$

7.

$$\begin{pmatrix} -3 & 0 & 0 \\ 0 & -3 & 1 \\ 0 & 0 & -3 \end{pmatrix}.$$

8.

$$\begin{pmatrix} 1 & 0 & 0 \\ 0 & 1 & 1 \\ 0 & 0 & 1 \end{pmatrix}.$$

9.

$$\begin{pmatrix} -2 & 0 & 0 \\ 0 & -2 & 1 \\ 0 & 0 & -2 \end{pmatrix}.$$

10.

$$\begin{pmatrix} 0 & 0 & 0 \\ 0 & 0 & 1 \\ 0 & 0 & 0 \end{pmatrix}.$$

11.

$$\begin{pmatrix} 1 & 0 & 0 \\ 0 & 0 & 1 \\ 0 & 0 & 0 \end{pmatrix}.$$

12.

$$\begin{pmatrix} 1 & 0 & 0 \\ 0 & 2 & 0 \\ 0 & 0 & 3 \end{pmatrix}.$$

13.

$$\begin{pmatrix} 1 & 1 & 0 \\ 0 & 1 & 1 \\ 0 & 0 & 1 \end{pmatrix}.$$

14.

$$\begin{pmatrix} 1 & 0 & 0 \\ 0 & 1 & 0 \\ 0 & 0 & -1 \end{pmatrix}.$$

15.

$$\begin{pmatrix} 3 & 0 & 0 \\ 0 & -1 & 1 \\ 0 & 0 & -1 \end{pmatrix}.$$

16.

$$\begin{pmatrix} \alpha & 0 & 0 \\ 0 & \alpha & 1 \\ 0 & 0 & \alpha \end{pmatrix}.$$

17.

$$\begin{pmatrix} 1 & 0 & 0 \\ 0 & 2+i & 0 \\ 0 & 0 & 2-i \end{pmatrix}.$$

18.

$$\begin{pmatrix} 1 & 0 & 0 \\ 0 & a & 0 \\ 0 & 0 & a^2 \end{pmatrix}, \quad a = -\frac{1}{2} \pm \frac{\sqrt{3}}{2} i.$$

19.

$$\begin{pmatrix} 1 & 0 & 0 \\ 0 & 2+3i & 0 \\ 0 & 0 & 2-3i \end{pmatrix}.$$

20.

$$\begin{pmatrix} 1 & 0 & 0 & 0 \\ 0 & 1 & 1 & 0 \\ 0 & 0 & 1 & 1 \\ 0 & 0 & 0 & 1 \end{pmatrix}.$$

21.

$$\begin{pmatrix} 2 & 1 & 0 & 0 \\ 0 & 2 & 0 & 0 \\ 0 & 0 & 2 & 1 \\ 0 & 0 & 0 & 2 \end{pmatrix}.$$

22.

$$\begin{pmatrix} 1 & 1 & 0 & 0 \\ 0 & 1 & 0 & 0 \\ 0 & 0 & -1 & 1 \\ 0 & 0 & 0 & -1 \end{pmatrix}.$$

23.

$$\begin{pmatrix} 0 & 1 & 0 & 0 \\ 0 & 0 & 0 & 0 \\ 0 & 0 & 0 & 1 \\ 0 & 0 & 0 & 0 \end{pmatrix}.$$

24.

$$\begin{pmatrix} -1 & 1 & 0 & 0 & 0 \\ 0 & -1 & 0 & 0 & 0 \\ 0 & 0 & -1 & 1 & 0 \\ 0 & 0 & 0 & -1 & 0 \\ 0 & 0 & 0 & 0 & 3 \end{pmatrix}.$$

25.

$$\begin{pmatrix} 2 & 1 & 0 & 0 & 0 \\ 0 & 2 & 1 & 0 & 0 \\ 0 & 0 & 2 & 0 & 0 \\ 0 & 0 & 0 & 2 & 0 \\ 0 & 0 & 0 & 0 & -1 \end{pmatrix}.$$

26.

$$\begin{pmatrix} 2 & 1 & 0 & 0 & 0 \\ 0 & 2 & 1 & 0 & 0 \\ 0 & 0 & 2 & 1 & 0 \\ 0 & 0 & 0 & 2 & 0 \\ 0 & 0 & 0 & 0 & 7 \end{pmatrix}.$$

27.

$$\begin{pmatrix} 3 & 1 & 0 & 0 & 0 & 0 \\ 0 & 3 & 1 & 0 & 0 & 0 \\ 0 & 0 & 3 & 0 & 0 & 0 \\ 0 & 0 & 0 & 3 & 1 & 0 \\ 0 & 0 & 0 & 0 & 3 & 0 \\ 0 & 0 & 0 & 0 & 0 & 1 \end{pmatrix}.$$

28.

$$\begin{pmatrix} 3 & 1 & 0 & 0 & 0 & 0 \\ 0 & 3 & 1 & 0 & 0 & 0 \\ 0 & 0 & 3 & 0 & 0 & 0 \\ 0 & 0 & 0 & 3 & 0 & 0 \\ 0 & 0 & 0 & 0 & 2 & 1 \\ 0 & 0 & 0 & 0 & 0 & 2 \end{pmatrix}.$$

29.

$$\begin{pmatrix} 4 & 1 & 0 & 0 & 0 & 0 \\ 0 & 4 & 1 & 0 & 0 & 0 \\ 0 & 0 & 4 & 0 & 0 & 0 \\ 0 & 0 & 0 & 4 & 0 & 0 \\ 0 & 0 & 0 & 0 & 4 & 0 \\ 0 & 0 & 0 & 0 & 0 & 3 \end{pmatrix}.$$

30.

$$\begin{pmatrix} 3 & 1 & 0 & 0 & 0 & 0 \\ 0 & 3 & 1 & 0 & 0 & 0 \\ 0 & 0 & 3 & 0 & 0 & 0 \\ 0 & 0 & 0 & 3 & 1 & 0 \\ 0 & 0 & 0 & 0 & 3 & 0 \\ 0 & 0 & 0 & 0 & 0 & 1 \end{pmatrix}.$$

Chapter 7

Solving Systems of Linear Dynamic Equations

Let $\mathbb{T}$ be a time scale with forward jump operator and delta differentiation operator σ and Δ, respectively.

7.1 Homogeneous Systems with Constant Coefficients

We will see how to use the Jordan canonical form to solve the systems

$$X^{\Delta} = \mathscr{M}(T)X \tag{7.1}$$

and

$$X^{\Delta} = -\mathscr{M}(T)X^{\sigma}, \tag{7.2}$$

where $T \in \mathscr{L}(\mathbb{F}^n)$ and $\mathscr{M}(T) = \mathscr{M}(T, (e_1, \ldots, e_n), (e_1, \ldots, e_n))$ is its matrix. Let $u_1, \ldots, u_n$ be the Jordan basis for $\mathscr{M}(T)$. With J we will denote the Jordan canonical form of the matrix $\mathscr{M}(T)$.

1. Write $\mathscr{M}(T)$ in the form

   $$\mathscr{M}(T) = \mathscr{M}(I, (e_1, \ldots, e_n), (u_1, \ldots, u_n))J\mathscr{M}(I, (u_1, \ldots, u_n), (e_1, \ldots, e_n)).$$

   Then (7.1) and (7.2) take the form

   $$X^{\Delta} = \mathscr{M}(I, (e_1, \ldots, e_n), (u_1, \ldots, u_n))J\mathscr{M}(I, (u_1, \ldots, u_n), (e_1, \ldots, e_n))X$$

and

$$X^{\Delta}=-\mathscr{M}(I,(e_1,\ldots,e_n),(u_1,\ldots,u_n))J\mathscr{M}(I,(u_1,\ldots,u_n),(e_1,\ldots,e_n))X^{\sigma},$$

respectively, whereupon

$$\mathscr{M}(I,(u_1,\ldots,u_n),(e_1,\ldots,e_n))X^{\Delta}=J\mathscr{M}(I,(u_1,\ldots,u_n),(e_1,\ldots,e_n))X$$

and

$$\mathscr{M}(I,(u_1,\ldots,u_n),(e_1,\ldots,e_n))X^{\Delta}=-J\mathscr{M}(I,(u_1,\ldots,u_n),(e_1,\ldots,e_n))X^{\sigma},$$

or

$$(\mathscr{M}(I,(u_1,\ldots,u_n),(e_1,\ldots,e_n))X)^{\Delta}=J(\mathscr{M}(I,(u_1,\ldots,u_n),(e_1,\ldots,e_n))X)$$

and

$$(\mathscr{M}(I,(u_1,\ldots,u_n),(e_1,\ldots,e_n))X)^{\Delta}=-J(\mathscr{M}(I,(u_1,\ldots,u_n),(e_1,\ldots,e_n))X)^{\sigma}.$$

2. Set
$$Y=\mathscr{M}(I,(u_1,\ldots,u_n),(e_1,\ldots,e_n))X.$$
Then we get the systems
$$Y^{\Delta}=JY$$
and
$$Y^{\Delta}=-JY^{\sigma}.$$

3. The solutions of the systems (7.1) and (7.2) are given by
$$X=\mathscr{M}(I,(e_1,\ldots,e_n),(u_1,\ldots,u_n))Y.$$

Example 7.1. *Let* $\mathbb{T}=2^{\mathbb{N}_0}$. *Consider the dynamic system*

$$\begin{aligned}
x_1^{\Delta} &= 6x_1\\
x_2^{\Delta} &= 3x_1+6x_2\\
x_3^{\Delta} &= 4x_1+2x_2+7x_3.
\end{aligned}$$

Here

$$\sigma(t)=2t,\quad t\in\mathbb{T},$$

and

$$\mathscr{M}(T) = \begin{pmatrix} 6 & 0 & 0 \\ 3 & 6 & 0 \\ 4 & 2 & 7 \end{pmatrix}.$$

By the computations in Example 6.83, we have that

$$\begin{aligned} u_1 &= (1,1,0), \\ u_2 &= (1,0,0), \\ u_3 &= (10,2,1) \end{aligned}$$

is a Jordan basis for $\mathscr{M}(T)$*, and*

$$\begin{aligned} \mathscr{M}(I,(u_1,u_2,u_3),(e_1,e_2,e_3)) &= \begin{pmatrix} 1 & 1 & 0 \\ 3 & 0 & 0 \\ 10 & 2 & 1 \end{pmatrix}, \\ \mathscr{M}(I,(e_1,e_2,e_3),(u_1,u_2,u_3)) &= \begin{pmatrix} 0 & \frac{1}{3} & 0 \\ 1 & -\frac{1}{3} & 0 \\ -2 & -\frac{8}{3} & 1 \end{pmatrix}, \\ J &= \begin{pmatrix} 6 & 1 & 0 \\ 0 & 6 & 0 \\ 0 & 0 & 7 \end{pmatrix}. \end{aligned}$$

Then, we consider the system

$$\begin{pmatrix} y_1^\Delta \\ y_2^\Delta \\ y_3^\Delta \end{pmatrix} = \begin{pmatrix} 6 & 1 & 0 \\ 0 & 6 & 0 \\ 0 & 0 & 7 \end{pmatrix} \begin{pmatrix} y_1 \\ y_2 \\ y_3 \end{pmatrix}$$

or

$$\begin{aligned} y_1^\Delta &= 6y_1 + y_2 \\ y_2^\Delta &= 6y_2 \end{aligned}$$

$$y_3^\Delta = 7y_3.$$

Hence,

$$\begin{aligned} y_3(t) &= c_3 e_7(t,1) \\ &= c_3 \prod_{s\in[1,t)} (1+7s), \\ y_2(t) &= c_2 e_6(t,1) \\ &= c_2 \prod_{s\in[1,t)} (1+6s), \quad t \in \mathbb{T}, \end{aligned}$$

where $c_2, c_3 \in \mathbb{C}$, *and*

$$\begin{aligned} y_1^\Delta(t) &= 6y_1(t) + c_2 e_6(t,1) \\ &= 6y_1(t) + c_2 \prod_{s\in[1,t)} (1+6s), \quad t \in \mathbb{T}. \end{aligned}$$

Therefore

$$\begin{aligned} y_1(t) &= c_1 e_6(t,1) + c_2 \int_1^t e_6(t,\sigma(\tau)) e_6(\tau,1)\Delta\tau \\ &= c_1 e_6(t,1) + c_2 \int_1^t \frac{1}{1+6\tau} e_6(t,\tau) e_6(\tau,1)\Delta\tau \\ &= c_1 e_6(t,1) + c_2 e_6(t,1) \int_1^t \frac{\Delta\tau}{1+6\tau} \\ &= c_1 \prod_{s\in[1,t)} (1+6s) + c_2 \left(\prod_{s\in[1,t)} (1+6s) \right) \left(\sum_{\tau\in[1,t)} \frac{\mu(\tau)}{1+6\tau} \Delta\tau \right) \\ &= c_1 \prod_{s\in[1,t)} (1+6s) + c_2 \left(\prod_{s\in[1,t)} (1+6s) \right) \left(\sum_{\tau\in[1,t)} \frac{\tau}{1+6\tau} \Delta\tau \right) \\ &= \left(c_1 + c_2 \sum_{\tau\in[1,t)} \frac{\tau}{1+6\tau} \right) \prod_{s\in[1,t)} (1+6s), \quad t \in \mathbb{T}, \end{aligned}$$

where $c_1 \in \mathbb{C}$ is a constant. Consequently

$$\begin{aligned}
y_1(t) &= \left(c_1 + c_2 \sum_{\tau\in[1,t)} \frac{\tau}{1+6\tau}\right) \prod_{s\in[1,t)} (1+6s) \\
y_2(t) &= c_2 \prod_{s\in[1,t)} (1+6s) \\
y_3(t) &= c_3 \prod_{s\in[1,t)} (1+7s), \quad t \in \mathbb{T}.
\end{aligned}$$

Consequently the solutions of the given system are

$$\begin{aligned}
\begin{pmatrix} x_1(t) \\ x_2(t) \\ x_3(t) \end{pmatrix} &= \mathscr{M}(I,(e_1,e_2,e_3),(u_1,u_2,u_3)) \begin{pmatrix} y_1(t) \\ y_2(t) \\ y_3(t) \end{pmatrix} \\
&= \begin{pmatrix} 0 & \frac{1}{3} & 0 \\ 1 & -\frac{1}{3} & 0 \\ -2 & -\frac{2}{3} & 1 \end{pmatrix} \begin{pmatrix} y_1(t) \\ y_2(t) \\ y_3(t) \end{pmatrix} \\
&= \begin{pmatrix} \frac{1}{3} y_2(t) \\ y_1(t)(-\frac{1}{3} y_2(t) \\ -2y_1(t) - \frac{2}{3} y_2(t) + y_3(t) \end{pmatrix}
\end{aligned}$$

$$= \begin{pmatrix} \frac{1}{3} c_2 \prod_{s\in[1,t)} (1+6s) \\ \left(c_1 + c_2 \sum_{\tau\in[1,t)} \frac{\tau}{1+6\tau}\right) \prod_{s\in[1,t)} (1+6s) - \frac{1}{3} c_2 \prod_{s\in[1,t)} (1+6s) \\ -2\left(c_1 + c_2 \sum_{\tau\in[1,t)} \frac{\tau}{1+6\tau} + \frac{1}{3} c_2\right) \prod_{s\in[1,t)} (1+6s) + c_3 \prod_{s\in[1,t)} (1+7s) \end{pmatrix}$$

$$= \begin{pmatrix} \frac{1}{3}c_2 \prod_{s\in[1,t)} (1+6s) \\ \left(c_1 + c_2\left(\sum_{\tau\in[1,t)} \frac{\tau}{1+6\tau} - \frac{1}{3}\right)\right) \prod_{s\in[1,t)} (1+6s) \\ -2\left(c_1 + c_2\left(\sum_{\tau\in[1,t)} \frac{\tau}{1+6\tau} + \frac{1}{3}\right)\right) \prod_{s\in[1,t)} (1+6s) + c_3 \prod_{s\in[1,t)} (1+7s) \end{pmatrix},$$

$t \in \mathbb{T}$.

Example 7.2. *Let $\mathbb{T} = 3\mathbb{Z}$. Consider the system*

$$\begin{aligned} x_1^\Delta &= 4x_2^\sigma + 2x_3^\sigma \\ x_2^\Delta &= -x_1^\sigma - 4x_2^\sigma - x_3^\sigma \\ x_3^\Delta &= -2x_3^\sigma. \end{aligned}$$

Here

$$\begin{aligned} \sigma(t) &= t+3, \\ \mu(t) &= 3, \quad t \in \mathbb{T}, \end{aligned}$$

and

$$\mathscr{M}(T) = \begin{pmatrix} 0 & -4 & -2 \\ 1 & 4 & 1 \\ 0 & 0 & 2 \end{pmatrix}.$$

By the computations in Example 6.84, we have that

$$\begin{aligned} u_1 &= (1,0,0), \\ u_2 &= (-2,-4,-2), \\ u_3 &= (0,0,1) \end{aligned}$$

is a Jordan basis for $\mathscr{M}(T)$ and

$$\mathscr{M}(I,(u_1,u_2,u_3),(e_1,e_2,e_3)) = \begin{pmatrix} 1 & 0 & 0 \\ -2 & -4 & -2 \\ 0 & 0 & 1 \end{pmatrix},$$

$$\mathscr{M}(I,(e_1,e_2,e_3),(u_1,u_2,u_3)) = \begin{pmatrix} 1 & 0 & 0 \\ -\frac{1}{2} & -\frac{1}{4} & -\frac{1}{2} \\ 0 & 0 & 1 \end{pmatrix},$$

$$J = \begin{pmatrix} 2 & 1 & 0 \\ 0 & 2 & 0 \\ 0 & 0 & 2 \end{pmatrix}.$$

Thus, we get the system

$$\begin{pmatrix} y_1^\Delta \\ y_2^\Delta \\ y_3^\Delta \end{pmatrix} = -\begin{pmatrix} 2 & 1 & 0 \\ 0 & 2 & 0 \\ 0 & 0 & 2 \end{pmatrix}\begin{pmatrix} y_1^\sigma \\ y_2^\sigma \\ y_3^\sigma \end{pmatrix},$$

or

$$y_1^\Delta = -2y_1^\sigma - y_2^\sigma$$

$$y_2^\Delta = -2y_2^\sigma$$

$$y_3^\Delta = -2y_3^\sigma.$$

Note that

$$\begin{aligned} \ominus 2(t) &= -\frac{2}{1+2\mu(t)} \\ &= -\frac{2}{1+6} \\ &= -\frac{2}{7}, \quad t \in \mathbb{T}. \end{aligned}$$

Therefore

$$y_2(t) = c_2 e_{-\frac{2}{7}}(t,0),$$

$$y_3(t) = c_3 e_{-\frac{2}{7}}(t,0), \quad t \in \mathbb{T},$$

where $c_2, c_3 \in \mathbb{C}$. *Observe that*

$$y_2^\sigma(t) = c_2 e_{-\frac{2}{7}}(\sigma(t),0)$$

$$\begin{aligned} &= c_2 e_{-\frac{2}{7}}(t+3,0) \\ &= c_2\left(1-\frac{6}{7}\right)e_{-\frac{2}{7}}(t,0) \\ &= \frac{1}{7}c_2 e_{-\frac{2}{7}}(t,0), \quad t\in\mathbb{T}. \end{aligned}$$

Now, we consider the equation

$$\begin{aligned} y_1^{\Delta}(t) &= -2y_1^{\sigma}(t)-y_2^{\sigma}(t) \\ &= -2y_1^{\sigma}(t)-c_2 e_{-\frac{2}{7}}(t+3,0), \quad t\in\mathbb{T}. \end{aligned}$$

For its solutions we have

$$\begin{aligned} y_1(t) &= c_1 e_{-\frac{2}{7}}(t,0)-c_2\int_0^t e_{-\frac{2}{7}}(t,\tau)e_{-\frac{2}{7}}(\tau+3,0)\Delta\tau \\ &= c_1 e_{-\frac{2}{7}}(t,0)-\frac{1}{7}c_2\int_0^t e_{-\frac{2}{7}}(t,\tau)e_{-\frac{2}{7}}(\tau,0)\Delta\tau \\ &= c_1 e_{-\frac{2}{7}}(t,0)-\frac{1}{7}c_2\int_0^t e_{-\frac{2}{7}}(t,0)\Delta\tau \\ &= c_1 e_{-\frac{2}{7}}(t,0)-\frac{1}{7}c_2 t e_{-\frac{2}{7}}(t,0) \\ &= \left(c_1-\frac{1}{7}c_2 t\right)e_{-\frac{2}{7}}(t,0), \quad t\in\mathbb{T}. \end{aligned}$$

Here $c_1\in\mathbb{C}$. Since

$$\begin{aligned} e_{-\frac{2}{7}}(t,0) &= \left(1-\frac{6}{7}\right)e^{\frac{t}{3}} \\ &= \frac{1}{7}e^{\frac{t}{3}}, \quad t\in\mathbb{T}, \end{aligned}$$

we get

$$y_1(t) = \frac{1}{7}\left(c_1-\frac{1}{7}c_2 t\right)e^{\frac{t}{3}}$$

$$\begin{aligned} y_2(t) &= \frac{1}{7}c_2e^{\frac{t}{3}} \\ y_3(t) &= \frac{1}{7}c_3e^{\frac{t}{3}}, \quad t \in \mathbb{T}. \end{aligned}$$

Therefore the solutions of the given system are as follows

$$\begin{aligned} \begin{pmatrix} x_1(t) \\ x_2(t) \\ x_3(t) \end{pmatrix} &= \begin{pmatrix} 1 & 0 & 0 \\ -\frac{1}{2} & -\frac{1}{4} & -\frac{1}{2} \\ 0 & 0 & 1 \end{pmatrix} \begin{pmatrix} y_1(t) \\ y_2(t) \\ y_3(t) \end{pmatrix} \\ &= \begin{pmatrix} y_1(t) \\ -\frac{1}{2}y_1(t) - \frac{1}{4}y_2(t) - \frac{1}{2}y_3(t) \\ y_3(t) \end{pmatrix} \\ &= \begin{pmatrix} \frac{1}{7}\left(c_1 - \frac{3}{7}c_2t\right)e^{\frac{t}{3}} \\ -\frac{1}{14}\left(c_1 - \frac{1}{7}c_2t\right)e^{\frac{t}{3}} - \frac{1}{28}c_2e^{\frac{t}{3}} - \frac{1}{14}c_3e^{\frac{t}{3}} \\ \frac{1}{7}c_3e^{\frac{t}{3}} \end{pmatrix} \\ &= \begin{pmatrix} \frac{1}{7}\left(c_1 - \frac{3}{7}c_2t\right)e^{\frac{t}{3}} \\ -\frac{1}{28}\left(2c_1 + c_2 + 2c_3 - \frac{3}{7}c_2t\right)e^{\frac{t}{3}} \\ \frac{1}{7}c_3e^{\frac{t}{3}} \end{pmatrix}, \quad t \in \mathbb{T}. \end{aligned}$$

Example 7.3. *Let* $\mathbb{T} = \mathbb{Z}$*. Consider the system*

$$\begin{aligned} x_1^\Delta &= x_2 - x_3 + x_4 \\ x_2^\Delta &= x_3 \\ x_3^\Delta &= -x_2 + 2x_3 \\ x_4^\Delta &= -2x_1 + x_2 - 2x_3 + 3x_4. \end{aligned}$$

Here

$$\begin{aligned}\sigma(t) &= t+1,\\ \mu(t) &= 1, \quad t \in \mathbb{T},\end{aligned}$$

and

$$\mathscr{M}(T) = \begin{pmatrix} 0 & 1 & -1 & 1 \\ 0 & 0 & 1 & 0 \\ 0 & -1 & 2 & 0 \\ -2 & 1 & -2 & 3 \end{pmatrix}.$$

By the computations in Example 6.85, we have that

$$\begin{aligned}u_1 &= (-2,0,0,1),\\ u_2 &= (0,-1,0,0),\\ u_3 &= (0,1,-1,0),\\ u_4 &= (-1,1,-2,1)\end{aligned}$$

is a Jordan basis for $\mathscr{M}(T)$ *and*

$$\begin{aligned}\mathscr{M}(I,(u_1,u_2,u_3,u_4),(e_1,e_2,e_3,e_4)) &= \begin{pmatrix} -2 & 0 & 0 & 1 \\ 0 & -1 & 0 & 0 \\ 0 & 1 & -1 & 0 \\ -1 & 1 & -2 & 1 \end{pmatrix},\\ \mathscr{M}(I,(e_1,e_2,e_3,e_4),(u_1,u_2,u_3,u_4)) &= \begin{pmatrix} -1 & -1 & -2 & 1 \\ 0 & -1 & 0 & 0 \\ 0 & -1 & -1 & 0 \\ -1 & -2 & -4 & 2 \end{pmatrix},\\ J &= \begin{pmatrix} 1 & 1 & 0 & 0 \\ 0 & 1 & 1 & 0 \\ 0 & 0 & 1 & 0 \\ 0 & 0 & 0 & 2 \end{pmatrix}.\end{aligned}$$

Then we get the system

$$\begin{pmatrix} y_1^\Delta \\ y_2^\Delta \\ y_3^\Delta \\ y_4^\Delta \end{pmatrix} = \begin{pmatrix} 1 & 1 & 0 & 0 \\ 0 & 1 & 1 & 0 \\ 0 & 0 & 1 & 0 \\ 0 & 0 & 0 & 2 \end{pmatrix} \begin{pmatrix} y_1 \\ y_2 \\ y_3 \\ y_4 \end{pmatrix} \quad on \quad \mathbb{T},$$

or

$$\begin{aligned} y_1^\Delta &= y_1 + y_2 \\ y_2^\Delta &= y_2 + y_3 \\ y_3^\Delta &= y_3 \\ y_4^\Delta &= 2y_4, \quad on \quad \mathbb{T}. \end{aligned}$$

We have

$$\begin{aligned} y_3(t) &= c_3 e_1(t,0) \\ y_4(t) &= c_4 e_2(t,0), \quad t \in \mathbb{T}, \end{aligned}$$

where $c_3, c_4 \in \mathbb{C}$. *Hence,*

$$\begin{aligned} y_2^\Delta(t) &= y_2(t) + y_3(t) \\ &= y_2(t) + c_3 e_1(t,0), \quad t \in \mathbb{T}, \end{aligned}$$

and then

$$\begin{aligned} y_2(t) &= c_2 e_1(t,0) + c_3 \int_0^t e_1(t,\sigma(\tau)) e_1(\tau,0) \Delta\tau \\ &= c_2 e_1(t,0) + \frac{1}{1} c_3 \int_0^t e_1(t,\tau) e_1(\tau,0) \Delta\tau \\ &= c_2 e_1(t,0) + \frac{1}{2} c_3 e_1(t,0) \int_0^t \Delta\tau \\ &= c_2 e_1(t,0) + \frac{1}{2} c_3 t e_1(t,0) \end{aligned}$$

$$= \left(c_2 + \frac{1}{2}c_3 t\right) e_1(t,0), \quad t \in \mathbb{T},$$

where $c_2 \in \mathbb{C}$, *and*

$$\begin{aligned} y_1^\Delta(t) &= y_1(t) + y_2(t) \\ &= y_1(t) + \left(c_2 + \frac{1}{2}c_3 t\right) e_1(t,0), \quad t \in \mathbb{Z}. \end{aligned}$$

Let

$$h(t) = \frac{1}{2}t^2 - \frac{1}{2}t, \quad t \in \mathbb{T}.$$

Then

$$\begin{aligned} h^\Delta(t) &= \frac{1}{2}(\sigma(t) + t) - \frac{1}{2} \\ &= \frac{1}{2}(t + 1 + t) - \frac{1}{2} \\ &= t + \frac{1}{2} - \frac{1}{2} \\ &= t, \quad t \in \mathbb{T}. \end{aligned}$$

Therefore

$$\begin{aligned} y_1(t) &= c_1 e_1(t,0) + \int_0^t \left(c_2 + \frac{1}{2}c_3\tau\right) e_1(t, \tau+1) e_1(\tau,0)\Delta\tau \\ &= c_1 e_1(t,0) + \frac{1}{2}\int_0^t \left(c_2 + \frac{1}{2}c_3\tau\right) e_1(t,\tau) e_1(\tau,0)\Delta\tau \\ &= \left(c_1 + \frac{1}{2}\int_0^t \left(c_2 + \frac{1}{2}c_3\tau\right)\Delta\tau\right) e_1(t,0) \\ &= \left(c_1 + \frac{1}{2}c_2 t + \frac{1}{4}c_3 \int_0^t \tau\Delta\tau\right) e_1(t,0) \\ &= \left(c_1 + \frac{1}{2}c_2 t + \frac{1}{4}c_3 \int_0^t h^\Delta(\tau)\Delta\tau\right) e_1(t,0) \end{aligned}$$

$$
\begin{aligned}
&= \left(c_1 + \frac{1}{2}c_2 t + \frac{1}{4}c_3 h(\tau)\Big|_{\tau=0}^{\tau=t}\right) e_1(t,0) \\
&= \left(c_1 + \frac{1}{2}c_2 t + \frac{1}{4}c_3\left(\frac{1}{2}t^2 - \frac{1}{2}t\right)\right) e_1(t,0) \\
&= \left(c_1 + \frac{1}{2}c_2 t - \frac{1}{8}c_3 t + \frac{1}{8}c_3 t^2\right) e_1(t,0) \\
&= \left(c_1 + \frac{1}{8}(4c_2 - c_3)t + \frac{1}{8}c_3 t^2\right) e_1(t,0), \quad t \in \mathbb{Z},
\end{aligned}
$$

where $c_1 \in \mathbb{C}$. Consequently

$$
\begin{aligned}
y_1(t) &= \left(c_1 + \frac{1}{8}(4c_2 - c_3)t + \frac{1}{8}c_3 t^2\right) e_1(t,0) \\
y_2(t) &= \left(c_1 + \frac{1}{2}c_3 t\right) e_1(t,0) \\
y_3(t) &= c_3 e_1(t,0) \\
y_4(t) &= c_4 e_2(t,0), \quad t \in \mathbb{T}.
\end{aligned}
$$

The solutions of the given system are as follows

$$
\begin{pmatrix} x_1(t) \\ x_2(t) \\ x_3(t) \\ x_4(t) \end{pmatrix} = \mathscr{M}(I,(e_1,e_2,e_3,e_4),(u_1,u_2,u_3,u_4)) \begin{pmatrix} y_1(t) \\ y_2(t) \\ y_3(t) \\ y_4(t) \end{pmatrix}
$$

$$
= \begin{pmatrix} -1 & -1 & -2 & 1 \\ 0 & -1 & 0 & 0 \\ 0 & -1 & -1 & 0 \\ -1 & -2 & -4 & 2 \end{pmatrix} \begin{pmatrix} \left(c_1 + \frac{1}{8}(4c_2 - c_3)t + \frac{1}{8}c_3 t^2\right) e_1(t,0) \\ \left(c_2 + \frac{1}{2}c_3 t\right) e_1(t,0) \\ c_3 e_1(t,0) \\ c_4 e_2(t,0) \end{pmatrix}
$$

$$= \begin{pmatrix} \left(-c_1 - \frac{1}{8}(4c_2 - c_3)t - \frac{1}{8}c_3t^2 - c_2 - \frac{1}{2}c_3t - 2c_3 + c_4\right) e_1(t,0) \\ -\left(c_2 + \frac{1}{2}c_3t\right) e_1(t,0) \\ -\left(c_2 + c_3 + \frac{1}{2}c_3t\right) e_1(t,0) \\ \left(-c_1 - \frac{1}{8}(4c_2 - c_3)t - \frac{1}{8}c_3t^2 - 2c_2 - c_3t - 4c_3\right) e_1(t,0) + 2c_4e_2(t,0) \end{pmatrix}$$

$$= \begin{pmatrix} \left(-c_1 - c_2 - 2c_3 + c_4 - \frac{1}{8}(4c_2 + 3c_3)t - \frac{1}{8}c_3t^2\right) e_1(t,0) \\ -\left(c_2 + \frac{1}{2}c_3t\right) e_1(t,0) \\ -\left(c_2 + c_3 + \frac{1}{2}c_3t\right) e_1(t,0) \\ \left(-c_1 - 2c_2 - 4c_3 - \frac{1}{8}(4c_2 + 3c_3)t - \frac{1}{8}c_3t^2\right) e_1(t,0) + 2c_4e_2(t,0) \end{pmatrix}$$

$$= \begin{pmatrix} 2\left(-c_1 - c_2 - 2c_3 + c_4 - \frac{1}{8}(4c_2 + 3c_3)t - \frac{1}{8}c_3t^2\right) e^t \\ -2\left(c_2 + \frac{1}{2}c_3t\right) e^t \\ -2\left(c_2 + c_3 + \frac{1}{2}c_3t\right) e^t \\ 2\left(-c_1 - 2c_2 - 4c_3 + 3c_4 - \frac{1}{8}(4c_2 + 3c_3)t - \frac{1}{8}c_3t^2\right) e^t \end{pmatrix}, \quad t \in \mathbb{T}.$$

Exercise 7.4. *Let $\mathbb{T}$ be arbitrarily chosen time scale. Using the Jordan canonical form, solve the following systems*

1.

$$\begin{aligned} x_1^\Delta &= 75x_1 + 56x_2 \\ x_2^\Delta &= -90x_1 + 67x_2. \end{aligned}$$

2.

$$\begin{aligned} x_1^\Delta &= 5x_1 + 8x_2 + 16x_3 \\ x_2^\Delta &= 4x_1 + x_2 + 8x_3 \end{aligned}$$

$$x_3^\Delta = -4x_1 - 4x_2 - 11x_3.$$

3.

$$\begin{aligned} x_1^\Delta &= -x_1 - 12x_2 + 2x_3 + 16x_4 \\ x_2^\Delta &= -x_1 + x_2 + x_3 + 2x_4 \\ x_3^\Delta &= -6x_1 - 11x_2 + 7x_3 + 15x_4 \\ x_4^\Delta &= -x_1 - 6x_2 + x_3 + 9x_4. \end{aligned}$$

4.

$$\begin{aligned} x_1^\Delta &= -5x_1 - 2x_2 - 4x_4 + 8x_5 \\ x_2^\Delta &= 2x_1 + x_2 - 2x_3 + 2x_4 \\ x_3^\Delta &= -3x_1 - x_2 - 2x_3 - 3x_4 + 8x_5 \\ x_4^\Delta &= -x_1 - x_2 + x_3 - 2x_4 \\ x_5^\Delta &= -4x_1 - x_2 - 2x_4 + 7x_5. \end{aligned}$$

5.

$$\begin{aligned} x_1^\Delta &= 5x_1 + 7x_2 + 5x_3 + 2x_4 - 2x_5 - 8x_6 \\ x_2^\Delta &= 2x_1 + 10x_2 + 7x_3 + 2x_4 - 3x_5 - 7x_6 \\ x_3^\Delta &= -2x_1 - 7x_2 - 4x_3 - 2x_4 + 3x_5 + 7x_6 \\ x_4^\Delta &= -x_3 + 3x_4 + x_5 \end{aligned}$$

$$x_5^\Delta = -2x_1 - 7x_2 - 7x_3 - 2x_4 + 6x_5 + 7x_6$$

$$x_6^\Delta = 2x_1 + 6x_2 + 5x_3 + 2x_4 - 2x_5 - 4x_6.$$

Answer 7.5. *1.*

$$x_1(t) = -7c_1e_3(t,t_0) + 4c_2e_5(t,t_0)$$

$$x_2(t) = 9c_1e_3(t,t_0) - 5c_2e_5(t,t_0), \quad t,t_0 \in \mathbb{T}, \quad c_1,c_2 \in \mathbb{C}.$$

2.

$$x_1(t) = -(c_1+2c_2)e_{-3}(t,t_0) - 2c_3e_1(t,t_0)$$

$$x_2(t) = c_1e_{-3}(t,t_0) - c_3e_1(t,t_0)$$

$$x_3(t) = c_2e_{-3}(t,t_0) + c_3e_1(t,t_0), \quad t,t_0 \in \mathbb{T}, \quad c_1,c_2,c_3 \in \mathbb{C}.$$

3.

$$x_1(t) = c_1e_1(t,t_0) + 2c_2e_3(t,t_0) + 2c_3e_5(t,t_0) + 22c_4e_7(t,t_0)$$

$$x_2(t) = c_2e_3(t,t_0) + c_3e_5(t,t_0) + 3c_4e_7(t,t_0)$$

$$x_3(t) = c_1e_1(t,t_0) + 2c_2e_3(t,t_0) + 4c_3e_5(t,t_0) + 18c_4e_7(t,t_0)$$

$$x_4(t) = c_2e_3(t,t_0) + c_3e_5(t,t_0) + 11c_4e_7(t,t_0), \quad t,t_0 \in \mathbb{T},$$

$c_1,c_2,c_3,c_4 \in \mathbb{C}$.

4.

$$x_1(t) = \left(2c_1 + 2c_2\int_{t_0}^{t} \frac{\Delta\tau}{1-\mu(\tau)} + c_5\right) e_{-1}(t,t_0)$$

$$x_2(t) = \left(c_2 + 2c_3 + 2c_4\int_{t_0}^{t} \frac{\Delta\tau}{1-\mu(\tau)} + 2c_4\right) e_{-1}(t,t_0)$$

$$\begin{aligned}
x_3(t) &= \left(2c_1+(2c_2+c_4)\int_{t_0}^{t}\frac{\Delta\tau}{1-\mu(\tau)}+c_3+c_5\right)e_{-1}(t,t_0)\\
x_4(t) &= \left(-c_2-c_3-c_4\int_{t_0}^{t}\frac{\Delta\tau}{1-\mu(\tau)}-c_4\right)e_{-1}(t,t_0)\\
x_5(t) &= \left(c_1+c_5+c_2\int_{t_0}^{t}\frac{\Delta\tau}{1-\mu(\tau)}\right)e_{-1}(t,t_0),
\end{aligned}$$

$$t,t_0\in\mathbb{T},\quad c_1,c_2,c_3,c_4,c_5\in\mathbb{C}.$$

5.

$$\begin{aligned}
x_1(t) &= \left(2\left(c_1+\int_{t_0}^{t}\frac{1}{1+3\mu(s)}\left(c_2+c_3\int_{t_0}^{s}\frac{\Delta\tau}{1+3\mu(\tau)}\right)\Delta s\right)\right.\\
&\quad\left.+c_2+(c_3+c_5)\int_{t_0}^{t}\frac{\Delta\tau}{1+3\mu(\tau)}+c_3+c_4+c_5+c_6\right)e_3(t,t_0)\\
x_2(t) &= \left(c_1+\int_{t_0}^{t}\frac{1}{1+3\mu(s)}\left(c_2+c_3\int_{t_0}^{s}\frac{\Delta\tau}{1+3\mu(\tau)}\right)\Delta s\right.\\
&\quad\left.+2c_2+(2c_3+c_5)\int_{t_0}^{t}\frac{\Delta\tau}{1+3\mu(\tau)}+c_3+c_4+c_5+c_6\right)e_3(t,t_0)\\
x_3(t) &= \left(-c_1-\int_{t_0}^{t}\frac{1}{1+3\mu(s)}\left(c_2+c_3\int_{t_0}^{s}\frac{\Delta\tau}{1+3\mu(\tau)}\right)\Delta s\right.\\
&\quad\left.-2c_2-(2c_3+c_5)\int_{t_0}^{t}\frac{\Delta\tau}{1+3\mu(\tau)}-c_4-c_6\right)e_3(t,t_0)\\
x_4(t) &= \left(c_4+c_5+c_5\int_{t_0}^{t}\frac{\Delta\tau}{1+3\mu(\tau)}\right)e_3(t,t_0)\\
x_5(t) &= \left(-c_1-\int_{t_0}^{t}\frac{1}{1+3\mu(s)}\left(c_2+c_3\int_{t_0}^{s}\frac{\Delta\tau}{1+3\mu(\tau)}\right)\Delta s\right.\\
&\quad\left.-2c_2-(2c_3+c_5)\int_{t_0}^{t}\frac{\Delta\tau}{1+3\mu(\tau)}-c_4+c_5-c_6\right)e_3(t,t_0)\\
x_6(t) &= \left(c_1+\int_{t_0}^{t}\frac{1}{1+3\mu(s)}\left(c_2+c_3\int_{t_0}^{s}\frac{\Delta\tau}{1+3\mu(\tau)}\right)\Delta s\right.
\end{aligned}$$

$$+c_2+(c_3+c_5)\int_{t_0}^{t}\frac{\Delta\tau}{1+3\mu(\tau)}+c_3+c_4+c_5+c_6\Big)e_3(t,t_0)$$

$$t,t_0\in\mathbb{T},\quad c_1,c_2,c_3,c_4,c_5,c_6\in\mathbb{C}.$$

7.2 Inhomogeneous Systems with Constant Coefficients

Consider the following inhomogeneous systems with constant coefficients

$$X^{\Delta}=\mathscr{M}(T)X+F(t),\quad t\in\mathbb{T}, \tag{7.3}$$

and

$$X^{\Delta}=-\mathscr{M}(T)X^{\sigma}+F(t),\quad t\in\mathbb{T}, \tag{7.4}$$

where $T\in\mathscr{L}(\mathbb{F}^n)$ and $\mathscr{M}(T)=\mathscr{M}(T,(e_1,\ldots,e_n),(e_1,\ldots,e_n))$ is its matrix. Let $u_1,\ldots,u_n$ be the Jordan basis for $\mathscr{M}(T)$. With J we will denote the Jordan canonical form of the matrix $\mathscr{M}(T)$.

1. Write $\mathscr{M}(T)$ in the form

   $$\mathscr{M}(T)=\mathscr{M}(I,(e_1,\ldots,e_n),(u_1,\ldots,u_n))J\mathscr{M}(I,(u_1,\ldots,u_n),(e_1,\ldots,e_n)).$$

   Then (7.3) and (7.4) take the form

   $$X^{\Delta}=\mathscr{M}(I,(e_1,\ldots,e_n),(u_1,\ldots,u_n))J\mathscr{M}(I,(u_1,\ldots,u_n),(e_1,\ldots,e_n))X$$
   $$+F(t)$$

   and

   $$X^{\Delta}=-\mathscr{M}(I,(e_1,\ldots,e_n),(u_1,\ldots,u_n))J\mathscr{M}(I,(u_1,\ldots,u_n),(e_1,\ldots,e_n))X^{\sigma}$$
   $$+F(t),$$

   respectively, whereupon

   $$\mathscr{M}(I,(u_1,\ldots,u_n),(e_1,\ldots,e_n))X^{\Delta}=J\mathscr{M}(I,(u_1,\ldots,u_n),(e_1,\ldots,e_n))X$$
   $$+\mathscr{M}(I,(u_1,\ldots,u_n),(e_1,\ldots,e_n))F(t),\quad t\in\mathbb{T},$$

and

$$\mathscr{M}(I,(u_1,\ldots,u_n),(e_1,\ldots,e_n))X^{\Delta} = -J\mathscr{M}(I,(u_1,\ldots,u_n),(e_1,\ldots,e_n))X^{\sigma}$$

$$+\mathscr{M}(I,(u_1,\ldots,u_n),(e_1,\ldots,e_n))F(t), \quad t \in \mathbb{T},$$

or

$$(\mathscr{M}(I,(u_1,\ldots,u_n),(e_1,\ldots,e_n))X)^{\Delta} = J(\mathscr{M}(I,(u_1,\ldots,u_n),(e_1,\ldots,e_n))X)$$

$$+\mathscr{M}(I,(u_1,\ldots,u_n),(e_1,\ldots,e_n))F(t), \quad t \in \mathbb{T},$$

and

$$(\mathscr{M}(I,(u_1,\ldots,u_n),(e_1,\ldots,e_n))X)^{\Delta}$$

$$= -J(\mathscr{M}(I,(u_1,\ldots,u_n),(e_1,\ldots,e_n))X)^{\sigma}$$

$$+\mathscr{M}(I,(u_1,\ldots,u_n),(e_1,\ldots,e_n))F(t), \quad t \in \mathbb{T}.$$

2. Set

$$\begin{aligned} Y &= \mathscr{M}(I,(u_1,\ldots,u_n),(e_1,\ldots,e_n))X, \\ Z(t) &= \mathscr{M}(I,(u_1,\ldots,u_n),(e_1,\ldots,e_n))F(t), \quad t \in \mathbb{T} \end{aligned}$$

Then we get the systems

$$Y^{\Delta} = JY + Z(t), \quad t \in \mathbb{T}.$$

and

$$Y^{\Delta} = -JY^{\sigma} + Z(t), \quad t \in \mathbb{T}.$$

3. The solutions of the systems (7.3) and (7.4) are given by

$$X = \mathscr{M}(I,(e_1,\ldots,e_n),(u_1,\ldots,u_n))Y.$$

Example 7.6. *Let* $\mathbb{T} = 2^{\mathbb{N}_0}$. *Consider the system*

$$\begin{aligned} x_1^\Delta &= 6x_1 + 1 \\ x_2^\Delta &= 3x_1 + 6x_2 - 1 \\ x_3^\Delta &= 4x_1 + 2x_2 + 7x_3 + 1, \quad t \in \mathbb{T}. \end{aligned}$$

Here

$$F(t) = \begin{pmatrix} 1 \\ -1 \\ 1 \end{pmatrix}, \quad t \in \mathbb{T}.$$

By the computations in Example 7.1, we have

$$\begin{aligned} Z(t) &= \mathscr{M}(I, (u_1, u_2, u_3), (e_1, e_2, e_3))F(t) \\ &= \begin{pmatrix} 1 & 1 & 0 \\ 3 & 0 & 0 \\ 10 & 2 & 1 \end{pmatrix} \begin{pmatrix} 1 \\ -1 \\ 1 \end{pmatrix} \\ &= \begin{pmatrix} 0 \\ 3 \\ 9 \end{pmatrix}, \quad t \in \mathbb{T}. \end{aligned}$$

Thus, we get the system

$$\begin{aligned} Y^\Delta &= \begin{pmatrix} y_1^\Delta \\ y_2^\Delta \\ y_3^\Delta \end{pmatrix} \\ &= JY + Z(t) \\ &= \begin{pmatrix} 6 & 1 & 0 \\ 0 & 6 & 0 \\ 0 & 0 & 7 \end{pmatrix} \begin{pmatrix} y_1 \\ y_2 \\ y_3 \end{pmatrix} + \begin{pmatrix} 0 \\ 3 \\ 9 \end{pmatrix} \\ &= \begin{pmatrix} 6y_1 + y_2 \\ 6y_2 \\ 7y_3 \end{pmatrix} + \begin{pmatrix} 0 \\ 3 \\ 9 \end{pmatrix} \end{aligned}$$

$$= \begin{pmatrix} 6y_1 + y_2 \\ 6y_2 + 3 \\ 7y_3 + 9 \end{pmatrix}, \quad t \in \mathbb{T},$$

i.e.,

$$\begin{aligned} y_1^\Delta &= 6y_1 + y_2 \\ y_2^\Delta &= 6y_2 + 3 \\ y_3^\Delta &= 7y_3 + 9, \quad t \in \mathbb{T}. \end{aligned} \tag{7.5}$$

By the last equation of the system (7.5), *we find*

$$\begin{aligned} y_3(t) &= c_1 e_7(t,1) + 9\int_1^t e_7(t,\sigma(\tau))\Delta\tau \\ &= c_1 e_7(t,1) + 9\int_1^t \frac{1}{1+7\tau} e_7(t,\tau)\Delta\tau \\ &= c_1 e_7(t,1) + 9\sum_{\tau\in[1,t)} \left(\frac{\tau}{1+7\tau} e_7(t,\tau)\right), \quad t \in \mathbb{T}. \end{aligned}$$

By the second equation of the system (7.5), *we get*

$$\begin{aligned} y_2(t) &= c_2 e_6(t,1) + 3\int_1^t e_6(t,\sigma(\tau))\Delta\tau \\ &= c_2 e_6(t,1) + 3\int_1^t \frac{1}{1+6\tau} e_6(t,\tau)\Delta\tau \\ &= c_2 e_6(t,1) + 3\sum_{\tau\in[1,t)} \left(\frac{\tau}{1+6\tau} e_6(t,\tau)\right), \quad t \in \mathbb{T}. \end{aligned}$$

By the first equation of the system (7.5), *we obtain*

$$\begin{aligned} y_1(t) &= c_3 e_6(t,1) + \int_1^t e_6(t,\sigma(\tau)) y_2(\tau)\Delta\tau \\ &= c_3 e_6(t,1) + \int_1^t \frac{1}{1+6\tau} e_6(t,\tau) y_2(\tau)\Delta\tau \\ &= c_3 e_6(t,1) + \sum_{\tau\in[1,t)} \left(\frac{\tau}{1+6\tau} e_6(t,\tau) y_2(\tau)\right) \end{aligned}$$

$$
\begin{aligned}
&= c_3 e_6(t,\tau) + \sum_{\tau\in[1,t)} \left(\frac{\tau}{1+6\tau} e_6(t,\tau) \right. \\
&\quad \left. \times \left(c_2 e_6(\tau,1) + 3 \sum_{s\in[1,\tau)} \left(\frac{s}{1+6s} e_6(\tau,s) \right) \right) \right),
\end{aligned}
$$

$t \in \mathbb{T}$. *Note that*

$$
\begin{aligned}
X(t) &= \begin{pmatrix} x_1(t) \\ x_2(t) \\ x_3(t) \end{pmatrix} \\
&= \mathscr{M}(I,(e_1,e_2,e_3),(u_1,u_2,u_3))Y \\
&= \begin{pmatrix} 0 & \frac{1}{3} & 0 \\ 1 & -\frac{1}{3} & 0 \\ -2 & -\frac{8}{3} & 1 \end{pmatrix} \begin{pmatrix} y_1(t) \\ y_2(t) \\ y_3(t) \end{pmatrix} \\
&= \begin{pmatrix} \frac{1}{3} y_2(t) \\ y_1(t) - \frac{1}{3} y_2(t) \\ -2y_1(t) - \frac{8}{3} y_2(t) + y_3(t) \end{pmatrix}, \quad t \in \mathbb{T},
\end{aligned}
$$

i.e.,

$$
\begin{aligned}
x_1(t) &= \frac{1}{3} y_2(t) \\
&= \frac{c_2}{3} e_6(t,1) + \sum_{\tau\in[1,t)} \left(\frac{\tau}{1+6\tau} e_6(t,\tau) \right), \\
x_2(t) &= y_1(t) - \frac{1}{3} y_2(t) \\
&= c_3 e_6(t,1) + \sum_{\tau\in[1,t)} \left(\frac{\tau}{1+6\tau} e_6(t,\tau) \right.
\end{aligned}
$$

$$
\begin{aligned}
& \times\left(c_2 e_6(\tau,1)+3\sum_{s\in[1,\tau)}\left(\frac{s}{1+6s}e_6(\tau,s)\right)\right)\Bigg)\\
& -\frac{c_2}{3}e_6(t,1)-\sum_{\tau\in[1,t)}\left(\frac{\tau}{1+6\tau}e_6(t,\tau)\right)\\
= & \left(c_3-\frac{c_2}{3}\right)e_6(t,1)+\sum_{\tau\in[1,t)}\left(\frac{\tau}{1+6\tau}e_6(t,\tau)\right.\\
& \left.\times\left(c_2 e_6(\tau,1)+3\sum_{s\in[1,\tau)}\left(\frac{s}{1+6s}e_6(\tau,s)\right)-1\right)\right),\\
x_3(t) = & -2y_1(t)-\frac{8}{3}y_2(t)+y_3(t)\\
= & -2c_3e_6(t,1)-2\sum_{\tau\in[1,t)}\left(\frac{\tau}{1+6\tau}e_6(t,\tau)\right.\\
& \left.\times\left(c_2 e_6(\tau,1)+3\sum_{s\in[1,\tau)}\left(\frac{s}{1+6s}e_6(\tau,s)\right)\right)\right)\\
& -\frac{8}{3}c_2e_6(t,1)-8\sum_{\tau\in[1,t)}\left(\frac{\tau}{1+6\tau}e_6(t,\tau)\right)\\
& +c_1e_7(t,1)+9\sum_{\tau\in[1,t)}\left(\frac{\tau}{1+7\tau}e_7(t,\tau)\right)\\
= & -2\left(c_3+\frac{4}{3}c_2\right)e_6(t,1)+c_1e_7(t,1)\\
& +9\sum_{\tau\in[1,t)}\left(\frac{\tau}{1+7\tau}e_7(t,\tau)\right)\\
& -2\sum_{\tau\in[1,t)}\left(\frac{\tau}{1+6\tau}e_6(t,\tau)\right.\\
& \left.\times\left(c_2 e_6(\tau,1)+3\sum_{s\in[1,\tau)}\left(\frac{s}{1+6s}e_6(\tau,s)\right)+4\right)\right),
\end{aligned}
$$

$t\in\mathbb{T}$, *is the general solution of the considered system. Here* $c_1,c_2,c_3\in\mathbb{C}$.

Exercise 7.7. *Solve the nonhomogeneous systems that correspond to the homogeneous systems in Exercise 7.4 with the following nonhomogeneous parts.*

1. $$F(t) = \begin{pmatrix} 1 \\ -1 \end{pmatrix}, \quad t \in \mathbb{T}.$$

2. $$F(t) = \begin{pmatrix} t \\ 1-t \\ t \end{pmatrix}, \quad t \in \mathbb{T}.$$

3. $$F(t) = \begin{pmatrix} e_1(t,t_0) \\ e_1(t,t_0)-1 \\ t \\ -1 \end{pmatrix}, \quad t, t_0 \in \mathbb{T}.$$

4. $$F(t) = \begin{pmatrix} -1+t \\ t \\ t^2 \\ \sin_1(t,t_0) \\ 1 \end{pmatrix}, \quad t, t_0 \in \mathbb{T}.$$

5. $$F(t) = \begin{pmatrix} -3 \\ -1 \\ 1+t \\ t \\ t^2 \\ 1-t \end{pmatrix}, \quad t \in \mathbb{T}.$$

7.3 IVPs for Inhomogeneous Systems with Constant Coefficients

Consider the following IVPs

$$\begin{aligned} X^{\Delta} &= \mathscr{M}(T)X + F(t), \quad t \in \mathbb{T}, \\ X(t_0) &= X_0, \end{aligned} \tag{7.6}$$

and

$$\begin{aligned} X^{\Delta} &= -\mathscr{M}(T)X^{\sigma}+F(t), \quad t\in\mathbb{T}, \\ X(t_0) &= X_0, \end{aligned} \tag{7.7}$$

where $T\in\mathscr{L}(\mathbb{F}^n)$ and $\mathscr{M}(T)=\mathscr{M}(T,(e_1,\ldots,e_n),(e_1,\ldots,e_n))$ is its matrix. Let $u_1,\ldots,u_n$ be the Jordan basis for $\mathscr{M}(T)$. With J we will denote the Jordan canonical form of the matrix $\mathscr{M}(T)$.

1. Write $\mathscr{M}(T)$ in the form

   $$\mathscr{M}(T)=\mathscr{M}(I,(e_1,\ldots,e_n),(u_1,\ldots,u_n))J\mathscr{M}(I,(u_1,\ldots,u_n),(e_1,\ldots,e_n)).$$

   Then, as in the previous sections, setting

   $$\begin{aligned} Y &= \mathscr{M}(I,(u_1,\ldots,u_n),(e_1,\ldots,e_n))X, \\ Z(t) &= \mathscr{M}(I,(u_1,\ldots,u_n),(e_1,\ldots,e_n))F(t), \quad t\in\mathbb{T}, \\ Y_0 &= \mathscr{M}(I,(u_1,\ldots,u_n),(e_1,\ldots,e_n))X_0, \end{aligned}$$

   we get the following IVPs

   $$\begin{aligned} Y^{\Delta} &= JY+Z(t), \quad t\in\mathbb{T}, \\ Y(t_0) &= Y_0. \end{aligned}$$

   and

   $$\begin{aligned} Y^{\Delta} &= -JY^{\sigma}+Z(t), \quad t\in\mathbb{T}, \\ Y(t_0) &= Y_0. \end{aligned}$$

2. The solutions of the IVPs (7.6) and (7.7) are given by

   $$X=\mathscr{M}(I,(e_1,\ldots,e_n),(u_1,\ldots,u_n))Y.$$

Example 7.8. *Consider the system in Example 7.6 subject to the initial condition*

$$X(1)=\begin{pmatrix} 1 \\ -1 \\ 1 \end{pmatrix}.$$

By the computations in Example 7.6, we find

$$\begin{aligned} x_1(1) &= \frac{c_2}{3} \\ &= 1, \\ x_2(1) &= c_3 - \frac{c_2}{3} \\ &= -1, \\ x_3(1) &= -2\left(c_3 + \frac{7}{3}c_2\right) + c_1 \\ &= 1, \end{aligned}$$

i.e., we get the system

$$\begin{aligned} \frac{c_2}{3} &= 1 \\ c_3 - \frac{c_2}{3} &= -1 \\ -2\left(c_3 + \frac{4}{3}c_2\right) + c_1 &= 1. \end{aligned}$$

Hence,

$$\begin{aligned} c_2 &= 3 \\ c_3 - 1 &= -1 \\ -2(c_3 - 4) + c_1 &= 1, \end{aligned}$$

or

$$\begin{aligned} c_2 &= 3 \\ c_3 &= 0 \end{aligned}$$

$$-8+c_1 = 1,$$

or

$$c_1 = 9$$

$$c_2 = 3$$

$$c_3 = 0.$$

Consequently the general solution of the considered system is given by

$$\begin{aligned}
x_1(t) &= e_6(t,1)+\sum_{\tau\in[1,t)}\left(\frac{\tau}{1+6\tau}e_6(t,\tau)\right)\\
x_2(t) &= -e_6(t,1)+\sum_{\tau\in[1,t)}\left(\frac{\tau}{1+6\tau}e_6(t,\tau)\right.\\
&\quad \left.\times\left(3e_6(\tau,1)+3\sum_{s\in[1,\tau)}\left(\frac{s}{1+6s}e_6(\tau,s)\right)-1\right)\right.\\
x_3(t) &= -8e_6(t,1)+9e_7(t,1)\\
&\quad +9\sum_{\tau\in[1,t)}\left(\frac{\tau}{1+7\tau}e_7(t,\tau)\right)\\
&\quad -2\sum_{\tau\in[1,t)}\left(\frac{\tau}{1+6\tau}e_6(t,\tau)\right.\\
&\quad \left.\times\left(3e_6(\tau,1)+3\sum_{s\in[1,\tau)}\left(\frac{s}{1+6s}e_6(\tau,s)\right)+4\right)\right),
\end{aligned}$$

$t\in\mathbb{T}$.

Exercise 7.9. *Solve the systems in Exercise 7.7 subject to the following initial conditions.*

1.
$$X(t_0)=\begin{pmatrix}1\\-1\end{pmatrix},\quad t_0\in\mathbb{T}.$$

2.

$$X(t_0) = \begin{pmatrix} 0 \\ 1 \\ 1 \end{pmatrix}, \quad t_0 \in \mathbb{T}.$$

3.

$$X(t_0) = \begin{pmatrix} 1 \\ -1 \\ 0 \\ -1 \end{pmatrix}, \quad t_0 \in \mathbb{T}.$$

4.

$$X(t_0) = \begin{pmatrix} -1 \\ 0 \\ 2 \\ 0 \\ 1 \end{pmatrix}, \quad t_0 \in \mathbb{T}.$$

5.

$$X(t_0) = \begin{pmatrix} -3 \\ -1 \\ 1 \\ 1 \\ 1 \\ 1 \end{pmatrix}, \quad t_0 \in \mathbb{T}.$$

7.4 Advanced Practical Problems

Problem 7.10. *Let $\mathbb{T}$ be arbitrary time scale. Solve the following homogeneous systems.*

1.

$$\begin{aligned} x_1^\Delta &= -50x_1 + 99x_2 \\ x_2^\Delta &= -20x_1 + 39x_2. \end{aligned}$$

2.

$$\begin{aligned} x_1^\Delta &= -18x_1 + 9x_2 \\ x_2^\Delta &= -49x_1 + 24x_2. \end{aligned}$$

3.

$$\begin{aligned} x_1^\Delta &= x_1 + x_2 \\ x_2^\Delta &= -16x_1 + 9x_2. \end{aligned}$$

4.

$$\begin{aligned} x_1^\Delta &= 2x_1^\sigma + x_2^\sigma \\ x_2^\Delta &= -25x_1^\sigma + 12x_2^\sigma. \end{aligned}$$

5.

$$\begin{aligned} x_1^\Delta &= -15x_1^\sigma + 9x_2^\sigma \\ x_2^\Delta &= -25x_1^\sigma + 15x_2^\sigma. \end{aligned}$$

6.

$$\begin{aligned} x_1^\Delta &= 3x_1 + 2x_3 \\ x_2^\Delta &= x_1 + 3x_2 + x_3 \\ x_3^\Delta &= x_2 + x_3. \end{aligned}$$

7.

$$\begin{aligned} x_1^\Delta &= 5x_1 + 8x_2 + 16x_3 \\ x_2^\Delta &= 4x_1 + x_2 + 3x_3 \\ x_3^\Delta &= -4x_1 - 4x_2 - 11x_3. \end{aligned}$$

8.

$$x_1^\Delta = 4x_1 + 2x_2 + 3x_3$$

$$x_2^\Delta = -x_1 + x_2 - 3x_3$$

$$x_3^\Delta = 2x_1 + 4x_2 + 9x_3.$$

9.

$$x_1^\Delta = 5x_1 + 2x_2 + x_3$$

$$x_2^\Delta = -x_1 + 2x_2 - x_3$$

$$x_3^\Delta = -x_1 - 2x_2 + 3x3.$$

10.

$$x_1^\Delta = 8x_1 - 3x_2 - 3x_3$$

$$x_2^\Delta = 4x_1 - 2x_3$$

$$x_3^\Delta = -2x_1 + x_2 + 3x_3.$$

11.

$$x_1^\Delta = -3x_1 + x_2 - x_3$$

$$x_2^\Delta = -7x_1 + 5x_2 - x_3$$

$$x_3^\Delta = -6x_1 + 6x_2 - 2x_3.$$

12.

$$x_1^\Delta = 3x_1$$

$$x_2^\Delta = 9x_1 - 5x_2 - 18x_3$$

$$x_3^\Delta = -4x_1 + 4x_2 + 12x_3.$$

13.

$$\begin{aligned} x_1^\Delta &= -6x_1 + 9x_2 \\ x_2^\Delta &= -6x_1 + 6x_2 - 2x_3 \\ x_3^\Delta &= 9x_1 - 9x_2 + 3x_3. \end{aligned}$$

14.

$$\begin{aligned} x_1^\Delta &= -18x_1 + 42x_2 + 168x_3 \\ x_2^\Delta &= x_1 - 7x_2 - 40x_3 \\ x_3^\Delta &= -2x_1 + 6x_2 + 27x_3. \end{aligned}$$

15.

$$\begin{aligned} x_1^\Delta &= -x_1 + x_2 - x_3 \\ x_2^\Delta &= -10x_1 + 6x_2 - 5x_3 \\ x_3^\Delta &= -6x_1 + 3x_2 - 2x_3. \end{aligned}$$

16.

$$\begin{aligned} x_1^\Delta &= -4x_2^\sigma + x_3^\sigma \\ x_2^\Delta &= 2x_1^\sigma - 6x_2^\sigma + x_3^\sigma \\ x_3^\Delta &= 4x_1^\sigma - 8x_2^\sigma. \end{aligned}$$

17.

$$\begin{aligned} x_1^\Delta &= -4x_1^\sigma + x_2^\sigma + 2x_3^\sigma \\ x_2^\Delta &= -5x_1^\sigma + x_2^\sigma + 3x_3^\sigma \end{aligned}$$

$$x_3^\Delta = -7x_1^\sigma + 2x_2^\sigma + 3x_3^\sigma.$$

18.

$$\begin{aligned} x_1^\Delta &= -4x_1^\sigma - 2x_2^\sigma + 5x_3^\sigma \\ x_2^\Delta &= -x_1^\sigma - x_2^\sigma + x_3^\sigma \\ x_3^\Delta &= -2x_1^\sigma - x_2^\sigma + 2x_3^\sigma. \end{aligned}$$

19.

$$\begin{aligned} x_1^\Delta &= 4x_1 - 3x_2 - 4x_3 + 3x_4 \\ x_2^\Delta &= 4x_1 - 3x_2 - 8x_3 + 7x_4 \\ x_3^\Delta &= -4x_1 + 4x_2 + 6x_3 - 4x_4 \\ x_4^\Delta &= -4x_1 + 4x_2 + 2x_3. \end{aligned}$$

20.

$$\begin{aligned} x_1^\Delta &= -8x_1 - 34x_2 + 30x_3 + 6x_4 \\ x_2^\Delta &= -9x_1 - 33x_2 + 30x_3 + 6x_4 \\ x_3^\Delta &= -7x_1 - 38x_2 + 33x_3 + 6x_4 \\ x_4^\Delta &= -36x_1 - 72x_2 + 72x_3 + 18x_4. \end{aligned}$$

21.

$$\begin{aligned} x_1^\Delta &= -5x_1 + 2x_2 - 14x_3 + 14x_4 \\ x_2^\Delta &= -4x_1 + x_2 - 14x_3 + 14x_4 \end{aligned}$$

$$\begin{aligned} x_3^\Delta &= -6x_1 + 2x_2 - 7x_3 + 8x_4 \\ x_4^\Delta &= -6x_1 + 2x_2 - 11x_3 + 12x_4. \end{aligned}$$

22.

$$\begin{aligned} x_1^\Delta &= 6x_1 + x_2 - x_3 + x_4 \\ x_2^\Delta &= x_1 + 6x_2 - x_3 + x_4 \\ x_3^\Delta &= -3x_1 + x_2 + 8x_3 - x_4 \\ x_4^\Delta &= -4x_1 + 4x_3 + 3x_4. \end{aligned}$$

23.

$$\begin{aligned} x_1^\Delta &= -4x_1 - 3x_2 + 9x_4 \\ x_2^\Delta &= 2x_1 + 2x_2 + x_3 - 2x_4 \\ x_3^\Delta &= 2x_1 + 3x_3 - 2x_4 \\ x_4^\Delta &= -6x_1 - 3x_2 + 11x_4. \end{aligned}$$

24.

$$\begin{aligned} x_1^\Delta &= -5x_1 - 2x_2 + x_3 + 2x_4 \\ x_2^\Delta &= -12x_1 - 7x_2 + 5x_3 + 6x_4 \\ x_3^\Delta &= -2x_1 - x_2 + x_4 \\ x_4^\Delta &= -26x_1 - 13x_2 + 7x_3 + 12x_4. \end{aligned}$$

25.

$$\begin{aligned}
x_1^{\Delta} &= -x_1+3x_2-11x_3+14x_4\\
x_2^{\Delta} &= 5x_2-12x_3+12x_4\\
x_3^{\Delta} &= -2x_1+3x_2-10x_3+14x_4\\
x_4^{\Delta} &= -2x_1+3x-2-11x_3+15x_4.
\end{aligned}$$

26.

$$\begin{aligned}
x_1^{\Delta} &= 8x_1-x_2-8x_3+12x_4\\
x_2^{\Delta} &= -6x_1+4x_2+12x_3-17x_4\\
x_3^{\Delta} &= -6x_1+2x_2+16x_3-19x_4\\
x_4^{\Delta} &= -6x_1+2x_2+12x_3-15x_4.
\end{aligned}$$

27.

$$\begin{aligned}
x_1^{\Delta} &= 4x_1+2x_2+x_3-6x_4\\
x_2^{\Delta} &= -16x_1+4x_2-16x_3\\
x_3^{\Delta} &= -6x_1-x_2-3x_3+4x_4\\
x_4^{\Delta} &= -6x_1+x_2-6x_3+2x_4.
\end{aligned}$$

28.

$$\begin{aligned}
x_1^{\Delta} &= -x_1-2x_2-x_3+3x_4\\
x_2^{\Delta} &= -6x_1-5x_2+x_3+6x_4
\end{aligned}$$

$$\begin{aligned} x_3^\Delta &= -6x_1 - 4x_2 + 6x_4 \\ x_4^\Delta &= -6x_1 - 7x_2 + x_3 + 8x_4. \end{aligned}$$

29.

$$\begin{aligned} x_1^\Delta &= -3x_1 - 4x_2 + 6x_4 \\ x_2^\Delta &= -x_1 - 02x_2 + x_3 + 2x_4 \\ x_3^\Delta &= -x_1 - 2x_2 + x_3 + 2x_4 \\ x_4^\Delta &= -3x_1 - 5x_2 + x_3 + 6x_4. \end{aligned}$$

30.

$$\begin{aligned} x_1^\Delta &= -x_1 + x_2 - 2x_3 \\ x_2^\Delta &= -10x_1 + x_2 + 2x_3 + 2x_4 \\ x_3^\Delta &= -5x_1 + x_2 + x_4 \\ x_4^\Delta &= -6x_1 + 2x_2 - 4x_3 + 2x_4. \end{aligned}$$

31.

$$\begin{aligned} x_1^\Delta &= -5x_1 + 3x_2 - 3x_3 + 6x_4 \\ x_2^\Delta &= -4x_1 + 3x_2 - 2x_3 + 4x_4 \\ x_3^\Delta &= -4x_1 + 2x_2 - x_3 + 4x_4 \\ x_4^\Delta &= -8x_1 + 3x_2 - 3x_3 + 9x_4. \end{aligned}$$

32.

$$x_1^\Delta = 8x_1 + 4x_2 - 3x_3 + x_4$$

$$x_2^\Delta = 12x_1 + 8x_2 - 6x_3 + 6x_4$$

$$x_3^\Delta = 24x_1 + 14x_2 - 10x_3 + 8x_4$$

$$x_4^\Delta = 6x_1 + 3x_2 - 3x_3 + 5x_4.$$

33.

$$x_1^\Delta = 7x_1 + 10x_2 - 6x_3 - 28x_4$$

$$x_2^\Delta = 2x_1 + 4x_2 - 2x_3 - 9x_4$$

$$x_3^\Delta = 2x_1 + 3x_2 - x_3 - 9x_4$$

$$x_4^\Delta = 2x_1 + 3x_2 - 2x_3 - 3x_4.$$

34.

$$x_1^\Delta = 4x_1 + 6x_2 - 5x_3 - x_4$$

$$x_2^\Delta = 5x_1 + 4x_2 - 5x_3$$

$$x_3^\Delta = 5x_1 + 6x_2 - 6x_3 - x_4$$

$$x_4^\Delta = 5x_1 + 5x_2 - 5x_3 - x_4.$$

35.

$$x_1^\Delta = x_2 - x_3 + x_4$$

$$x_2^\Delta = x_3$$

$$x_3^\Delta = -x_2 + 2x_3$$

$$x_4^\Delta = -2x_1 + x_2 - 2x_3 + 3x_4.$$

36.

$$
\begin{aligned}
x_1^{\Delta} &= -12x_1 + 15x_2 + 7x_3 - 11x_4 \\
x_2^{\Delta} &= -14x_1 + 16x_2 + 7x_3 - 10x_4 \\
x_3^{\Delta} &= -12x_1 + 14x_2 + 8x_3 - 11x_4 \\
x_4^{\Delta} &= -12x_1 + 12x_2 + 6x_3 - 7x_4.
\end{aligned}
$$

37.

$$
\begin{aligned}
x_1^{\Delta} &= 5x_1 + 2x_2 - x_3 - x_4 \\
x_2^{\Delta} &= 3x_1 + 10x_2 - 3x_3 - 3x_4 \\
x_3^{\Delta} &= 4x_1 + 8x_2 - 4x_4 \\
x_4^{\Delta} &= 3x_1 + 6x_2 - 3x_3 + x_4.
\end{aligned}
$$

38.

$$
\begin{aligned}
x_1^{\Delta} &= -5x_1 + 4x_2 - 3x_3 + 2x_4 \\
x_2^{\Delta} &= -8x_1 + 7x_2 - 4x_3 + 4x_4 \\
x_3^{\Delta} &= -10x_1 + 10x_2 - 6x_3 + 5x_4 \\
x_4^{\Delta} &= -2x_1 + 2x_2 - x_3.
\end{aligned}
$$

39.

$$
\begin{aligned}
x_1^{\Delta} &= 4x_1 + x_2 + 2x_3 - 2x_4 \\
x_2^{\Delta} &= -x_1 + 2x_2 + 2x_3
\end{aligned}
$$

$$
\begin{aligned}
x_3^\Delta &= 5x_3 - x_4 \\
x_4^\Delta &= x_1 + x_2 + 2x_3 + x_4.
\end{aligned}
$$

40.

$$
\begin{aligned}
x_1^\Delta &= 4x_1^\sigma + 4x_2^\sigma - 2x_3^\sigma - 2x_4^\sigma \\
x_2^\Delta &= -x_1^\sigma + 7x_2^\sigma - x_4^\sigma \\
x_3^\Delta &= -x_1^\sigma + x_2^\sigma + 6x_3^\sigma - x_4^\sigma \\
x_4^\Delta &= x_1^\sigma - 2x_2^\sigma + x_3^\sigma + 7x_4^\sigma.
\end{aligned}
$$

41.

$$
\begin{aligned}
x_1^\Delta &= 5x_1^\sigma - x_2^\sigma + 2x_3^\sigma + 4x_4^\sigma \\
x_2^\Delta &= 5x_1^\sigma - x_2^\sigma + 3x_3^\sigma + 4x_4^\sigma \\
x_3^\Delta &= 8x_1^\sigma - 5x_2^\sigma + 3x_3^\sigma + 4x_4^\sigma \\
x_4^\Delta &= -4x_1^\sigma + x_2^\sigma - x_3^\sigma - 3x_4^\sigma.
\end{aligned}
$$

42.

$$
\begin{aligned}
x_1^\Delta &= -2x_1^\sigma + 2x_2^\sigma + 3x_3^\sigma + 3x_4^\sigma \\
x_2^\Delta &= -4x_1^\sigma + 4x_2^\sigma + 4x_3^\sigma + 3x_4^\sigma \\
x_3^\Delta &= -x_1^\sigma - x_2^\sigma + 4x_3^\sigma + 3x_4^\sigma \\
x_4^\Delta &= -2x_1^\sigma + 2x_2^\sigma + x_3^\sigma + 2x_4^\sigma.
\end{aligned}
$$

43.

$$
x_1^\Delta = 10x_1 + 5x_2 - 2x_3 - 17x_4
$$

$$
\begin{aligned}
x_2^\Delta &= 8x_1 + 8x_2 - 4x_3 - 18x_4 \\
x_3^\Delta &= 6x_2 - 4x_3 - 6x_4 \\
x_4^\Delta &= 8x_1 + 3x_2 - x_3 - 13x_4.
\end{aligned}
$$

44.

$$
\begin{aligned}
x_1^\Delta &= 6x_1 - 6x_2 - 8x_3 + 3x_4 \\
x_2^\Delta &= -4x_1 + x_2 + 8x_3 + x_4 \\
x_3^\Delta &= 4x_1 - 4x_2 - 6x_3 + 2x_4 \\
x_4^\Delta &= -4x_1 - 2x_2 + 8x_3 + 4x_4.
\end{aligned}
$$

45.

$$
\begin{aligned}
x_1^\Delta &= 22x_1 - 5x_2 + x_3 - 5x_4 \\
x_2^\Delta &= 18x_1 - 2x_2 + 2x_3 - 5x_4 \\
x_3^\Delta &= 16x_1 - 4x_2 + 3x_3 - 4x_4 \\
x_4^\Delta &= 62x_1 - 16x_2 + 2x_3 - 13x_4.
\end{aligned}
$$

46.

$$
\begin{aligned}
x_1^\Delta &= 5x_1 + 2x_3 + x_4 \\
x_2^\Delta &= 2x_2 + x_3 + x_4 \\
x_3^\Delta &= -2x_1 - 2x_4 \\
x_4^\Delta &= 2x_1 - 2x_2 + 5x_3 + 7x_4.
\end{aligned}
$$

47.

$$\begin{aligned} x_1^\Delta &= -5x_1 + 2x_2 - 10x_3 + 10x_4 \\ x_2^\Delta &= -x_1 + 2x - 2 - x_3 - x_4 \\ x_3^\Delta &= -5x_1 + x_2 + 5x_4 \\ x_4^\Delta &= -10x_1 + 2x_2 - 10x_3 + 15x_4. \end{aligned}$$

48.

$$\begin{aligned} x_1^\Delta &= 2x_1^\sigma - 3x_2^\sigma - 2x_3^\sigma + 3x_4^\sigma \\ x_2^\Delta &= -4x_1^\sigma + 2x_2^\sigma + 4x_3^\sigma - 4x_4^\sigma \\ x_3^\Delta &= -4x_1^\sigma + 4x_3^\sigma - 3x_4^\sigma \\ x_4^\Delta &= -8x_1^\sigma + 4x_2^\sigma + 8x_3^\sigma - 8x_4^\sigma. \end{aligned}$$

49.

$$\begin{aligned} x_1^\Delta &= x_1^\sigma + 3x_2^\sigma - 6x_3^\sigma \\ x_2^\Delta &= -2x_1^\sigma + 3x_2^\sigma - 4x_3^\sigma + 6x_4^\sigma \\ x_3^\Delta &= -x_1^\sigma + x_2^\sigma - x_3^\sigma + 3x_4^\sigma \\ x_4^\Delta &= x_2^\sigma - 2x_3^\sigma + x_4^\sigma. \end{aligned}$$

50.

$$\begin{aligned} x_1^\Delta &= -x_1 + x_2 - 4x_3 + 2x_4 + 4x_5 \\ x_2^\Delta &= -x_1 - x_2 + 2x_3 + 2x_4 \end{aligned}$$

$$\begin{aligned}
x_3^\Delta &= -x_1 - x_2 + 2x_3 + 2x_4 \\
x_4^\Delta &= x_2 - 3x_3 + 2x_5 \\
x_5^\Delta &= -x_1 - x_2 + 2x_4 + 2x_5.
\end{aligned}$$

51.

$$\begin{aligned}
x_1^\Delta &= 5x_1 + x_2 + 3x_3 - x_4 - 3x_5 \\
x_2^\Delta &= 3x_1 + x_2 + 4x_3 + x_4 - 3x_5 \\
x_3^\Delta &= -4x_2 + 4x_3 + 4x_4 \\
x_4^\Delta &= 3x_1 + x_2 + 3x_3 + x_4 - 3x_5 \\
x_5^\Delta &= 6x_1 - 3x_2 + 8x_3 + 3x_4 - 4x_5.
\end{aligned}$$

52.

$$\begin{aligned}
x_1^\Delta &= 5x_1 + x_2 - 2x_5 \\
x_2^\Delta &= x_1 + 6x_2 + x_3 - 2x_4 - 2x_5 \\
x_3^\Delta &= x_1 + 2x_2 + 5x_3 - 2x_4 - 2x_5 \\
x_4^\Delta &= x_1 + 2x_2 + x_3 + 2x_4 - 2x_5 \\
x_5^\Delta &= 2x_1 + x_2 + x_5.
\end{aligned}$$

53.

$$\begin{aligned}
x_1^\Delta &= -2x_1^\sigma - 9x_2^\sigma - 10x_3^\sigma - 7x_4^\sigma + 10x_5^\sigma \\
x_2^\Delta &= 8x_1^\sigma + 21x_2^\sigma + 21x_3^\sigma + 15x_4^\sigma - 20x_5^\sigma
\end{aligned}$$

$$
\begin{aligned}
x_3^{\Delta} &= -2x_1^{\sigma} - x_2^{\sigma} + x_3^{\sigma} \\
x_4^{\Delta} &= -8x_1^{\sigma} - 18x_2^{\sigma} - 20x_3^{\sigma} - 12x_4^{\sigma} = 20x_5^{\sigma} \\
x_5^{\Delta} &= -2x_1^{\sigma} - 5x_2^{\sigma} - 5x_3^{\sigma} - 4x_4^{\sigma} + 7x_5^{\sigma}.
\end{aligned}
$$

54.

$$
\begin{aligned}
x_1^{\Delta} &= -x_1^{\sigma} + 2x_2^{\sigma} + 4x_3^{\sigma} + 2x_4^{\sigma} - x_5^{\sigma} \\
x_2^{\Delta} &= -3x_1^{\sigma} + 5x_2^{\sigma} + 4x_3^{\sigma} + x_4^{\sigma} - x_5^{\sigma} \\
x_3^{\Delta} &= -x_1^{\sigma} + 4x_3^{\sigma} + x_4^{\sigma} \\
x_4^{\Delta} &= -3x_1^{\sigma} + 2x_2^{\sigma} + 3x_3^{\sigma} + 4x_4^{\sigma} - x_5^{\sigma} \\
x_5^{\Delta} &= -2x_1^{\sigma} + x_2^{\sigma} + 3x_3^{\sigma} + x_4^{\sigma} + 2x_5^{\sigma}.
\end{aligned}
$$

55.

$$
\begin{aligned}
x_1^{\Delta} &= 9x_1 + 3x_2 + 3x_4 - 3x_5 - 12x_6 \\
x_2^{\Delta} &= 3x_2 + 3x_3 + x_4 - 2x_5 \\
x_3^{\Delta} &= -2x_2 + 8x_3 + x_4 - 2x_5 \\
x_4^{\Delta} &= -x_2 + 4x_4 + 2x_5 \\
x_5^{\Delta} &= -2x_2 + 2x_3 + 5x_5 \\
x_6^{\Delta} &= 2x_1 + x_2 + x_4 - x_5 - x_6.
\end{aligned}
$$

Problem 7.11. *Let $\mathbb{T}$ be an arbitrary time scale. Solve the corresponding nonhomogeneous systems to the homogeneous systems in Problem 7.10, where the corresponding nonhomogeneous parts are given as follows.*

1.

$$f(t)=\begin{pmatrix}\sin_2(t,t_0)\\ 1+t\end{pmatrix},\quad t,t_0\in\mathbb{T}.$$

2.

$$f(t)=\begin{pmatrix}e_1(t,t_0)\\ 0\end{pmatrix},\quad t,t_0\in\mathbb{T}.$$

3.

$$f(t)=\begin{pmatrix}-1+\cos_1(t,t_0)\\ 1+t+t^2\end{pmatrix},\quad t,t_0\in\mathbb{T}.$$

4.

$$f(t)=\begin{pmatrix}-1\\ \sin_3(t,t_0)\end{pmatrix},\quad t,t_0\in\mathbb{T}.$$

5.

$$f(t)=\begin{pmatrix}e_2(t,t_0)\\ e_1(t,t_0)\end{pmatrix},\quad t,t_0\in\mathbb{T}.$$

6.

$$f(t)=\begin{pmatrix}\sin_3(t,t_0)\\ \cos_1(t,t_0)\\ 1\end{pmatrix},\quad t,t_0\in\mathbb{T}.$$

7.

$$f(t)=\begin{pmatrix}-2\\ 1\\ 1+\sin_4(t,t_0)\end{pmatrix},\quad t,t_0\in\mathbb{T}.$$

8.

$$f(t)=\begin{pmatrix}2\\ 1\\ \sin_5(t,t_0)\end{pmatrix},\quad t,t_0\in\mathbb{T}.$$

9.

$$f(t)=\begin{pmatrix}-1\\ 1+\cos_1(t,t_0)+\sin_2(t,t_0)\\ 3\end{pmatrix},\quad t,t_0\in\mathbb{T}.$$

10.

$$f(t)=\begin{pmatrix}-1\\ 0\\ 2\cos_1(t,t_0)\end{pmatrix},\quad t,t_0\in\mathbb{T}.$$

11.

$$f(t) = \begin{pmatrix} 3 \\ 0 \\ -\cos_7(t,t_0) \end{pmatrix}, \quad t,t_0 \in \mathbb{T}.$$

12.

$$f(t) = \begin{pmatrix} 2+7\cos_1(t,t_0) \\ \sin_3(t,t_0) \\ 4\cos_2(t,t_0) \end{pmatrix}, \quad t,t_0 \in \mathbb{T}.$$

13.

$$f(t) = \begin{pmatrix} -1 \\ 2+\sin_1(t,t_0) \\ \cos_2(t,t_0) \end{pmatrix}, \quad t,t_0 \in \mathbb{T}.$$

14.

$$f(t) = \begin{pmatrix} -3 \\ -1+e_2(t,t_0) \\ e_2(t,t_0) \end{pmatrix}, \quad t,t_0 \in \mathbb{T}.$$

15.

$$f(t) = \begin{pmatrix} 0 \\ 1 \\ 3\cos_1(t,t_0) \end{pmatrix}, \quad t,t_0 \in \mathbb{T}.$$

16.

$$f(t) = \begin{pmatrix} -4 \\ -1 \\ \dfrac{1}{2+(\cos_2(t,t_0))^4} \end{pmatrix}, \quad t,t_0 \in \mathbb{T}.$$

17.

$$f(t) = \begin{pmatrix} 2 \\ -1 \\ 0 \end{pmatrix}, \quad t \in \mathbb{T}.$$

18.

$$f(t) = \begin{pmatrix} -3 \\ 1+t \\ t^3 \end{pmatrix}, \quad t \in \mathbb{T}.$$

19.

$$f(t) = \begin{pmatrix} 2 \\ 0 \\ 1+t \\ -t \end{pmatrix}, \quad t \in \mathbb{T}.$$

20.

$$f(t) = \begin{pmatrix} 1 \\ -1 \\ 2 - \cos_3(t, t_0) \\ \sin_1(t, t_0) \end{pmatrix}, \quad t, t_0 \in \mathbb{T}.$$

21.

$$f(t) = \begin{pmatrix} -1 \\ -3 \\ 4 \\ \sin_8(t, t_0) \end{pmatrix}, \quad t, t_0 \in \mathbb{T}.$$

22.

$$f(t) = \begin{pmatrix} -2 \\ -3 \\ 1 \\ 2e_2(t, t_0) \end{pmatrix}, \quad t, t_0 \in \mathbb{T}.$$

23.

$$f(t) = \begin{pmatrix} -3 \\ -1 \\ 1 \\ 2\sin_3(t, t_0) \end{pmatrix}, \quad t, t_0 \in \mathbb{T}.$$

24.

$$f(t) = \begin{pmatrix} -1 \\ 1 \\ \frac{1}{1+t^2} \\ -1 \end{pmatrix}, \quad t \in \mathbb{T}.$$

25.

$$f(t) = \begin{pmatrix} -3 \\ 0 \\ 1 \\ 2 \end{pmatrix}, \quad t \in \mathbb{T}.$$

26.

$$f(t) = \begin{pmatrix} -4 \\ -1 \\ 0 \\ 1 \end{pmatrix}, \quad t \in \mathbb{T}.$$

27.

$$f(t) = \begin{pmatrix} -2 \\ 1+t \\ 3t-1 \\ t^2 \end{pmatrix}, \quad t \in \mathbb{T}.$$

28.

$$f(t) = \begin{pmatrix} \sin_1(t,t_0) \\ 1-2\sin_1(t,t_0) \\ \cos_1(t,t_0) \\ t-t_0 \end{pmatrix}, \quad t,t_0 \in \mathbb{T}.$$

29.

$$f(t) = \begin{pmatrix} 2 \\ -1 \\ t_0 - e_1(t,t_0) \\ 0 \end{pmatrix}, \quad t,t_0 \in \mathbb{T}.$$

30.

$$f(t) = \begin{pmatrix} -1 \\ 2 \\ 1 \\ 3 \end{pmatrix}, \quad t \in \mathbb{T}.$$

31.

$$f(t) = \begin{pmatrix} -4 \\ 1 \\ 0 \\ 1 \end{pmatrix}, \quad t,t_0 \in \mathbb{T}.$$

32.

$$f(t) = \begin{pmatrix} -1 \\ -2 \\ -3 \\ 5 \end{pmatrix}, \quad t \in \mathbb{T}.$$

33.

$$f(t) = \begin{pmatrix} 0 \\ 1 \\ -2 \\ 1 \end{pmatrix}, \quad t \in \mathbb{T}.$$

34.

$$f(t) = \begin{pmatrix} \frac{1}{1+t_0^2+t^2} \\ t^2 \\ 1-t \\ t_0 - t^2 \end{pmatrix}, \quad t, t_0 \in \mathbb{T}.$$

35.

$$f(t) = \begin{pmatrix} 2 \\ -1 \\ 4 \\ t_0 \end{pmatrix}, \quad t, t_0 \in \mathbb{T}.$$

36.

$$f(t) = \begin{pmatrix} 1+t_0 \\ 3-2t_0 \\ e_4(t,t_0) \\ -1 \end{pmatrix}, \quad t, t_0 \in \mathbb{T}.$$

37.

$$f(t) = \begin{pmatrix} -3 \\ 1 \\ 0 \\ -4 \end{pmatrix}, \quad t \in \mathbb{T}.$$

38.

$$f(t) = \begin{pmatrix} -1 \\ 0 \\ t+1 \\ t^2 \end{pmatrix}, \quad t \in \mathbb{T}.$$

39.

$$f(t) = \begin{pmatrix} -3 \\ -1 \\ 0 \\ 1 \end{pmatrix}, \quad t, t_0 \in \mathbb{T}.$$

40.

$$f(t) = \begin{pmatrix} 2 \\ t \\ 1-t \\ 3 \end{pmatrix}, \quad t \in \mathbb{T}.$$

41.

$$f(t) = \begin{pmatrix} 4 \\ \frac{1}{t^2 + (\sin_1(t,t_0))^2} \\ -1 \\ 2 \end{pmatrix}, \quad t, t_0 \in \mathbb{T}.$$

42.

$$f(t) = \begin{pmatrix} 1 \\ 0 \\ 0 \\ -1 \end{pmatrix}, \quad t \in \mathbb{T}.$$

43.

$$f(t) = \begin{pmatrix} 4 \\ t \\ 1+t \\ 1-2t \end{pmatrix}, \quad t, t_0 \in \mathbb{T}.$$

44.

$$f(t) = \begin{pmatrix} 1 \\ -1 \\ 3 \\ -4 \end{pmatrix}, \quad t, t_0 \in \mathbb{T}.$$

45.

$$f(t) = \begin{pmatrix} -1 \\ 1 \\ 2 \\ 1 \end{pmatrix}, \quad t \in \mathbb{T}.$$

46.

$$f(t) = \begin{pmatrix} 2 \\ 1 \\ 0 \\ -2 \end{pmatrix}, \quad t, t_0 \in \mathbb{T}.$$

47.

$$f(t) = \begin{pmatrix} -1 \\ 0 \\ 0 \\ 5 \end{pmatrix}, \quad t \in \mathbb{T}.$$

48.

$$f(t) = \begin{pmatrix} -3 \\ -1 \\ 0 \\ 2 \end{pmatrix}, \quad t, t_0 \in \mathbb{T}.$$

49.

$$f(t) = \begin{pmatrix} 1 \\ -3 \\ 3+t^2 \\ 4 \end{pmatrix}, \quad t \in \mathbb{T}.$$

50.

$$f(t) = \begin{pmatrix} 4-t \\ 2 \\ 1 \\ 0 \\ 6 \end{pmatrix}, \quad t \in \mathbb{T}.$$

51.

$$f(t) = \begin{pmatrix} \sin_1(t, t_0) \\ 1 \\ 2 \\ 0 \\ 0 \end{pmatrix}, \quad t, t_0 \in \mathbb{T}.$$

52.

$$f(t) = \begin{pmatrix} e_1(t, t_0) \\ 3 \\ -1 \\ 4 \\ -2\sin_1(t, t_0) \end{pmatrix}, \quad t, t_0 \in \mathbb{T}.$$

53.

$$f(t) = \begin{pmatrix} -3 \\ -2 \\ -1 \\ 0 \\ 1 \end{pmatrix}, \quad t \in \mathbb{T}.$$

54.

$$f(t) = \begin{pmatrix} t \\ 1+t \\ 2-t^2 \\ 3 \\ 4-t \end{pmatrix}, \quad t \in \mathbb{T}.$$

55.

$$f(t) = \begin{pmatrix} 5 \\ \cos_3(t,t_0) \\ \sin_3(t,t_0)+\cos_3(t,t_0) \\ -1 \\ 0 \\ 1 \end{pmatrix}, \quad t, t_0 \in \mathbb{T}.$$

Problem 7.12. *Let $\mathbb{T}$ be an arbitrary time scale. Solve the IVPs for the homogeneous systems and the nonhomogeneous systems in Problem 7.10 and Problem 7.11, respectively, where the initial values are given as follows.*

1.

$$x(t_0) = \begin{pmatrix} -1 \\ 0 \end{pmatrix}, \quad t_0 \in \mathbb{T}.$$

2.

$$x(t_0) = \begin{pmatrix} 0 \\ 1 \end{pmatrix}, \quad t_0 \in \mathbb{T}.$$

3.

$$x(t_0) = \begin{pmatrix} 1 \\ 1 \end{pmatrix}, \quad t_0 \in \mathbb{T}.$$

4.

$$x(t_0) = \begin{pmatrix} -1 \\ 0 \end{pmatrix}, \quad t_0 \in \mathbb{T}.$$

5.

$$x(t_0) = \begin{pmatrix} 2 \\ -1 \end{pmatrix}, \quad t_0 \in \mathbb{T}.$$

6.

$$x(t_0) = \begin{pmatrix} 3 \\ -2 \\ 4 \end{pmatrix}, \quad t_0 \in \mathbb{T}.$$

7.

$$x(t_0) = \begin{pmatrix} -1 \\ 1 \\ 1 \end{pmatrix}, \quad t_0 \in \mathbb{T}.$$

8.

$$x(t_0) = \begin{pmatrix} -2 \\ 1 \\ 1 \end{pmatrix}, \quad t_0 \in \mathbb{T}.$$

9.

$$x(t_0) = \begin{pmatrix} -1 \\ -1 \\ 1 \end{pmatrix}, \quad t_0 \in \mathbb{T}.$$

10.

$$x(t_0) = \begin{pmatrix} 1 \\ 1 \\ 1 \end{pmatrix}, \quad t_0 \in \mathbb{T}.$$

11.

$$x(t_0) = \begin{pmatrix} -4 \\ 1 \\ 1 \end{pmatrix}, \quad t_0 \in \mathbb{T}.$$

12.

$$x(t_0) = \begin{pmatrix} 0 \\ -1 \\ -3 \end{pmatrix}, \quad t_0 \in \mathbb{T}.$$

13.

$$x(t_0) = \begin{pmatrix} -1 \\ 1 \\ 1 \end{pmatrix}, \quad t_0 \in \mathbb{T}.$$

14.

$$x(t_0) = \begin{pmatrix} -3 \\ 1 \\ 1 \end{pmatrix}, \quad t_0 \in \mathbb{T}.$$

15.

$$x(t_0) = \begin{pmatrix} 0 \\ 1 \\ 1 \end{pmatrix}, \quad t_0 \in \mathbb{T}.$$

16.

$$x(t_0) = \begin{pmatrix} -1 \\ -1 \\ 1 \end{pmatrix}, \quad t_0 \in \mathbb{T}.$$

17.

$$x(t_0) = \begin{pmatrix} -2 \\ 1 \\ 1 \end{pmatrix}, \quad t_0 \in \mathbb{T}.$$

18.

$$x(t_0) = \begin{pmatrix} -3 \\ 1 \\ 1 \end{pmatrix}, \quad t_0 \in \mathbb{T}.$$

19.

$$x(t_0) = \begin{pmatrix} 0 \\ 1 \\ 1 \\ 1 \end{pmatrix}, \quad t_0 \in \mathbb{T}.$$

20.

$$x(t_0) = \begin{pmatrix} 1 \\ 1 \\ 1 \\ 1 \end{pmatrix}, \quad t_0 \in \mathbb{T}.$$

21.

$$x(t_0) = \begin{pmatrix} -1 \\ 1 \\ 1 \\ 1 \end{pmatrix}, \quad t_0 \in \mathbb{T}.$$

22.

$$x(t_0) = \begin{pmatrix} -2 \\ -1 \\ 1 \\ 1 \end{pmatrix}, \quad t_0 \in \mathbb{T}.$$

23.

$$x(t_0) = \begin{pmatrix} -3 \\ 2 \\ -1 \\ 1 \end{pmatrix}, \quad t_0 \in \mathbb{T}.$$

24.

$$x(t_0) = \begin{pmatrix} 1-1 \\ -3 \\ 1 \\ 1 \end{pmatrix}, \quad t_0 \in \mathbb{T}.$$

25.

$$x(t_0) = \begin{pmatrix} 1 \\ 0 \\ 0 \\ 1 \end{pmatrix}, \quad t_0 \in \mathbb{T}.$$

26.

$$x(t_0) = \begin{pmatrix} -2 \\ 1 \\ 0 \\ 0 \end{pmatrix}, \quad t_0 \in \mathbb{T}.$$

27.

$$x(t_0) = \begin{pmatrix} 0 \\ 1 \\ 0 \\ 1 \end{pmatrix}, \quad t_0 \in \mathbb{T}.$$

28.

$$x(t_0) = \begin{pmatrix} 1 \\ 0 \\ 1 \\ 0 \end{pmatrix}, \quad t_0 \in \mathbb{T}.$$

29.

$$x(t_0) = \begin{pmatrix} 1 \\ 1 \\ 1 \\ 0 \end{pmatrix}, \quad t_0 \in \mathbb{T}.$$

30.

$$x(t_0) = \begin{pmatrix} 0 \\ 1 \\ 1 \\ 1 \end{pmatrix}, \quad t_0 \in \mathbb{T}.$$

31.

$$x(t_0) = \begin{pmatrix} 1 \\ -1 \\ -1 \\ 1 \end{pmatrix}, \quad t_0 \in \mathbb{T}.$$

32.

$$x(t_0) = \begin{pmatrix} -1 \\ -1 \\ 1 \\ 1 \end{pmatrix}, \quad t_0 \in \mathbb{T}.$$

33.

$$x(t_0) = \begin{pmatrix} -2 \\ 1 \\ 0 \\ 1 \end{pmatrix}, \quad t_0 \in \mathbb{T}.$$

34.

$$x(t_0) = \begin{pmatrix} -3 \\ 1 \\ 1 \\ 0 \end{pmatrix}, \quad t_0 \in \mathbb{T}.$$

35.

$$x(t_0) = \begin{pmatrix} 0 \\ 1 \\ 1 \\ -1 \end{pmatrix}, \quad t_0 \in \mathbb{T}.$$

36.

$$x(t_0) = \begin{pmatrix} -2 \\ 1 \\ 1 \\ 0 \end{pmatrix}, \quad t_0 \in \mathbb{T}.$$

37.

$$x(t_0) = \begin{pmatrix} -1 \\ 1 \\ 1 \\ 1 \end{pmatrix}, \quad t_0 \in \mathbb{T}.$$

38.

$$x(t_0) = \begin{pmatrix} 1 \\ 0 \\ 1 \\ -1 \end{pmatrix}, \quad t_0 \in \mathbb{T}.$$

39.

$$x(t_0) = \begin{pmatrix} 1 \\ 1 \\ 1 \\ 1 \end{pmatrix}, \quad t_0 \in \mathbb{T}.$$

40.

$$x(t_0) = \begin{pmatrix} -3 \\ -2 \\ -2 \\ 0 \end{pmatrix}, \quad t_0 \in \mathbb{T}.$$

41.

$$x(t_0) = \begin{pmatrix} -1 \\ 1 \\ 0 \\ 0 \end{pmatrix}, \quad t_0 \in \mathbb{T}.$$

42.

$$x(t_0) = \begin{pmatrix} 0 \\ 0 \\ 0 \\ 1 \end{pmatrix}, \quad t_0 \in \mathbb{T}.$$

43.

$$x(t_0) = \begin{pmatrix} -2 \\ 1 \\ 1 \\ 1 \end{pmatrix}, \quad t_0 \in \mathbb{T}.$$

44.

$$x(t_0) = \begin{pmatrix} -3 \\ -2 \\ -4 \\ 1 \end{pmatrix}, \quad t_0 \in \mathbb{T}.$$

45.

$$x(t_0) = \begin{pmatrix} 1 \\ 1 \\ 1 \\ -7 \end{pmatrix}, \quad t_0 \in \mathbb{T}.$$

46.

$$x(t_0) = \begin{pmatrix} 2 \\ 1 \\ -1 \\ 0 \end{pmatrix}, \quad t_0 \in \mathbb{T}.$$

47.

$$x(t_0) = \begin{pmatrix} -1 \\ -1 \\ 1 \\ 1 \end{pmatrix}, \quad t_0 \in \mathbb{T}.$$

48.

$$x(t_0) = \begin{pmatrix} -3 \\ -2 \\ 1 \\ 1 \end{pmatrix}, \quad t_0 \in \mathbb{T}.$$

49.

$$x(t_0) = \begin{pmatrix} 1 \\ 0 \\ 0 \\ 0 \end{pmatrix}, \quad t_0 \in \mathbb{T}.$$

50.

$$x(t_0) = \begin{pmatrix} -2 \\ 1 \\ 1 \\ 1 \\ 1 \end{pmatrix}, \quad t_0 \in \mathbb{T}.$$

51.

$$x(t_0) = \begin{pmatrix} -1 \\ 2 \\ -1 \\ -1 \\ -1 \end{pmatrix}, \quad t_0 \in \mathbb{T}.$$

52.

$$x(t_0) = \begin{pmatrix} 1 \\ 0 \\ 0 \\ 0 \\ -1 \end{pmatrix}, \quad t_0 \in \mathbb{T}.$$

53.

$$x(t_0) = \begin{pmatrix} -3 \\ -4 \\ -1 \\ 1 \\ 0 \end{pmatrix}, \quad t_0 \in \mathbb{T}.$$

54.

$$x(t_0) = \begin{pmatrix} -1 \\ -2 \\ -1 \\ 0 \\ 0 \end{pmatrix}, \quad t_0 \in \mathbb{T}.$$

55.

$$x(t_0) = \begin{pmatrix} 0 \\ 1 \\ 1 \\ -1 \\ 1 \\ 0 \end{pmatrix}, \quad t_0 \in \mathbb{T}.$$

Bibliography

[1] R. Agarwal and M. Bohner. Quadratic functionals for second order matrix equations on time scales, Nonlinear Analysis, 33(1998), pp. 675-692.

[2] Sh. Axler. Linear algebra done right, Springer, 2015.

[3] M. Bohner. Calculus of variation on time scales, Dynamic Systems and Applications, 13(2004), pp. 339-349.

[4] M. Bohner and G. Sh. Guseinov. Double integral calculus of variations on time scales, Computers and Mathematics with Applications, 54(2007), pp. 45-57.

[5] M. Bohner and A. Peterson. Dynamic equations on time scales: an introduction with applications, Birkhäuser, Boston, 2001.

[6] M. Bohner and S. Georgiev. Multivariable dynamic calculus on time scales, Springer, 2017.

[7] R. Ferreira, A. Malinowska and D. Torres. Optimality conditions for the calculus of variations with higher-order delta derivatives, Applied Mathematic Letters, 24(2011), pp. 87-92.

[8] R. Hilscher. Linear Hamiltonian systems on time scales: positivity of quadratic functionals, Mathematical and Computer Modelling, 32(2000), pp. 507-527.

[9] S. Weintraub. Jordan canonical form: theory and practice, Morgan & Claypool Publishers, 2009.

Index